PHARMACOLOGY OF MICRONUTRIENTS

CURRENT TOPICS IN NUTRITION AND DISEASE

Series Editors

Anthony A. Albanese
The Burke Rehabilitation Center
White Plains, New York

David Kritchevsky
The Wistar Institute
Philadelphia, Pennsylvania

VOLUME 1: BONE LOSS: CAUSES, DETECTION, AND THERAPY
Anthony A. Albanese

VOLUME 2: CHROMIUM IN NUTRITION AND DISEASE
Günay Saner

VOLUME 3: NUTRITION FOR THE ELDERLY
Anthony A. Albanese

VOLUME 4: NUTRITIONAL PHARMACOLOGY
Gene A. Spiller, *Editor*

VOLUME 5: NEW TRENDS IN NUTRITION, LIPID RESEARCH, AND CARDIOVASCULAR DISEASES
Nicolás G. Bazán, Rodolfo Paoletti, and
James M. Iacono, *Editors*

VOLUME 6: CLINICAL, BIOCHEMICAL, AND NUTRITIONAL ASPECTS OF TRACE ELEMENTS
Ananda S. Prasad, *Editor*

VOLUME 7: CLINICAL APPLICATIONS OF RECENT ADVANCES IN ZINC METABOLISM
Ananda S. Prasad, Ivor E. Dreosti, and Basil S. Hetzel, *Editors*

VOLUME 8: ANIMAL AND VEGETABLE PROTEINS IN LIPID METABOLISM
AND ATHEROSCLEROSIS
Michael J. Gibney and David Kritchevsky, *Editors*

VOLUME 9: DIET, NUTRITION, AND CANCER: FROM BASIC RESEARCH
TO POLICY IMPLICATIONS
Daphne A. Roe, *Editor*

VOLUME 10: MALNUTRITION: DETERMINANTS AND CONSEQUENCES
Philip L. White and Nancy Selvey, *Editors*

VOLUME 11: ENERGY INTAKE AND ACTIVITY
Ernesto Pollitt and Peggy Amante, *Editors*

VOLUME 12: ABSORPTION AND MALABSORPTION OF MINERAL NUTRIENTS
Noel W. Solomons and Irwin H. Rosenberg, *Editors*

VOLUME 13: VITAMIN B-6: ITS ROLE IN HEALTH AND DISEASE
Robert D. Reynolds and James E. Leklem, *Editors*

VOLUME 14: DIETARY FIBER AND OBESITY
Per Björntorp, George V. Vahouny, and David Kritchevsky, *Editors*

VOLUME 15: NUTRITIONAL DISEASES: RESEARCH DIRECTIONS IN COMPARATIVE PATHOBIOLOGY
Dante G. Scarpelli and George Migaki, *Editors*

VOLUME 16: BASIC AND CLINICAL ASPECTS OF NUTRITION AND BRAIN DEVELOPMENT
David K. Rassin, Bernard Haber, and Boris Drujan, *Editors*

VOLUME 17: FAT DISTRIBUTION DURING GROWTH AND LATER HEALTH OUTCOMES
Claude Bouchard and Francis E. Johnston, *Editors*

VOLUME 18: ESSENTIAL AND TOXIC TRACE ELEMENTS IN HUMAN HEALTH AND DISEASE
Ananda S. Prasad, *Editor*

VOLUME 19: CLINICAL AND PHYSIOLOGICAL APPLICATIONS OF VITAMIN B-6
James E. Leklem and Robert D. Reynolds, *Editors*

VOLUME 20: PHARMACOLOGY OF MICRONUTRIENTS
Nestor W. Flodin

PHARMACOLOGY OF MICRONUTRIENTS

Nestor W. Flodin
Department of Biochemistry
College of Medicine
University of South Alabama
Mobile, Alabama

Alan R. Liss, Inc., New York

Address all Inquiries to the Publisher
Alan R. Liss, Inc., 41 East 11th Street, New York, NY 10003

Copyright © 1988 Alan R. Liss, Inc.

All rights reserved. This book is protected by copyright. No part of it, except brief excerpts for review, may be reproduced, stored in a retrieval system, or transmitted in any form or by any means electronic, mechanical, photocopying, recording, or otherwise, without written permission from the publisher.

Printed in the United States of America.

While the authors, editors, and publisher believe that drug selection and dosage and the specifications and usage of equipment and devices, as set forth in this book, are in accord with current recommendations and practice at the time of publication, they accept no legal responsibility for any errors or omissions, and make no warranty, express or implied, with respect to material contained herein. In view of ongoing research, equipment modifications, changes in governmental regulations and the constant flow of information relating to drug therapy, drug reactions and the use of equipment and devices, the reader is urged to review and evaluate the information provided in the package insert or instructions for each drug, piece of equipment or device for, among other things, any changes in the instructions or indications of dosage or usage and for added warnings and precautions.

Library of Congress Cataloging-in-Publication Data
Flodin, Nestor W. (Nestor Winston), 1915–

Pharmacology of micronutrients.
  (Current topics in nutrition and disease ; v. 20)
  Includes index.
  1. Vitamin therapy. 2. Trace elements—Therapeutic use. 3. Pharmacology. I. Title. II. Series.
  [DNLM: 1. Trace Elements—pharmacology. 2. Vitamins—pharmacology. W1 CU82R v.20 / QU 130 F628h]
  RM259.F57 1988   615'.328   88-9487
  ISBN 0-8451-1619-3

Dedication

To my wife Betty, without whose help and support this book would not have been written.

Contents

Preface
Nestor W. Flodin ix
Abbreviations xi

SECTION I: PHARMACOLOGY OF THE VITAMINS

1. Vitamin A 3
2. Vitamin D 31
3. Vitamin E 59
4. Vitamin K 93
5. Thiamin (Vitamin B-1) 103
6. Riboflavin (Vitamin B-2) 117
7. Niacin and Niacinamide 129
8. Vitamin B-6 139
9. Folic Acid 161
10. Vitamin B-12 179
11. Pantothenic Acid 189
12. Biotin 193
13. Vitamin C 201

SECTION II: PHARMACOLOGY OF THE TRACE ELEMENTS

14. Chromium 247
15. Copper 255
16. Selenium 269
17. Zinc 285
Appendix 321
Index 323

Preface

Many important new discoveries have been made in recent years concerning the actual or potential medical value of treating specific diseases with vitamins and trace elements administered in larger than nutritional quantities. I have long felt that there was a need for a compilation of these findings which, somewhat in the manner of the familiar PDR® on drugs,* would provide information for each micronutrient with respect to its actions, medical conditions for which its usefulness has been investigated, dosage sizes and length of treatment, precautions and side effects, and interactions with drugs and other nutrients.

Retirement from active academic duties finally gave me the time to compile and summarize the available information on the pharmacology of micronutrients, based on a survey of data reported by qualified investigators in peer-reviewed biomedical journals. This volume may therefore be viewed as a state-of-the-art summary of what is known about the pharmacological properties of the vitamins and trace elements. It will prove of value, I hope, as an instructional tool in departments of pharmacology and as a reference source for physicians and dietitians with a special interest in the physiological actions of micronutrients. I hope it will also stimulate further research interest in this broad and important area of clinical medicine.

In its seventeen chapters, this book discusses studies of the clinical effects of the four fat-soluble vitamins, nine water-soluble vitamins, and four trace elements (chromium, copper, selenium, and zinc). Iron, fluorine, and iodine have not been included, as their clinical uses are well covered in existing textbooks and adequately taught in medical schools. The other trace and ultratrace elements—manganese, molybdenum, nickel, arsenic, and vanadium—are under active investigation, but to date no clear-cut clinical applications have been found.

Each chapter opens with an introductory section summarizing the chemical and physical properties, biochemical and physiological functions, signs and symptoms of deficiency, food sources, and recommended daily allowances (RDAs). This is followed by sections on absorption, metabolism, and excretion; available dosage forms; clinical studies (the largest portion of the book); toxicity and side effects; and interactions. As may be inferred from this de-

---

*PDR® is a registered trademark of Medical Economics Co., Inc.

scription, this is not a book on dietetics or nutritional biochemistry, and therefore the chapters should not be looked upon as comprehensive reviews.

However, the sections on clinical studies, toxicity and side effects, and interactions represent quite complete surveys of the peer-reviewed biomedical literature through 1987. Each chapter is well referenced.

There is no extensive discussion of the classical deficiency diseases such as scurvy, beriberi, and pellagra. This is for two principal reasons. First, this subject is already thoroughly taught in most medical schools, even though the average physician in western countries is unlikely ever to see a fullblown case in a lifetime of medical practice. Second, these diseases are readily treatable with dietary therapy or with small nutritional amounts of vitamins, whereas this text is concerned with the physiological effects and potential medical usefulness of micronutrients when administered in larger than nutritional amounts.

The reader who is interested in a particular clinical area or disease state should consult the index. Details are given in the text on the effects of micronutrient treatment, often with case histories, as well as on recommended dosage regimens based on published reports written by clinicians experienced in the study and practice of micronutrient therapy.

Physicians who wish to try micronutrient therapy should consult the original journal reports if further details of treatment are desired, and of course, before prescribing a micronutrient in pharmacologic doses, just as with a drug, they should familiarize themselves with its toxicity, side effects, and interactions as discussed in the chapter on that micronutrient. The physician should also maintain a degree of skepticism with regard to the micronutrient's effectiveness, since micronutrients, unlike drugs, are not patentable and therefore have not drawn the support of the pharmaceutical industry for clinical trials involving hundreds or thousands of patients. However, many double-blind, placebo-controlled studies involving small numbers of subjects have been carried out. Moreover, in comparison with most drugs, the micronutrients, unless grossly abused, have low toxicity and few side effects, so they may often safely be given a trial to determine whether a beneficial clinical response may be obtained.

I wish to express my gratitude to Hemmige Bhagavan, Ph.D., Helen Hittner, M.D., Frank Kretzer, Ph.D., and Derrick Lonsdale, M.D., for review of portions of the text and for many valuable comments and suggestions.

Self-prescribed dosing with larger than nutritional amounts of micronutrients is not recommended and could be hazardous to health. Pharmacologic dosages of micronutrients should not be taken without the supervision of a physician. Mention of a trade name or proprietary product or the description of a particular mode of treatment in this book does not constitute a recommendation or endorsement by the author or the publisher.

Nestor W. Flodin, Ph.D.

Abbreviations

AE	acrodermatitis enteropathica	GSHPx	glutathione peroxidase
ARDS	adult respiratory distress syndrome	G6PD	glucose-6-phosphate dehydrogenase
ATP	adenosine triphosphate	GTF	glucose tolerance factor
BCG	bacillus Guérin-Calmette vaccine	HDL	high-density lipoprotein
b.i.d.	twice daily	HDLC	high-density lipoprotein cholesterol
BPD	bronchopulmonary dysplasia	HDN	hemorrhagic disease of the newborn
Ca	calcium	HPN	home parenteral nutrition
CF	cystic fibrosis	h.s.	at bedtime
CH_3	methyl	Ig	immunoglobulin
CIS	carcinoma in situ	IHD	ischemic heart disease
CN	cyano	IM	intramuscular
CNS	central nervous system	INH	isonicotinyl hydrazide
CoA	coenzyme A	i.p.	intraperitoneal
Cr	chromium	IV	intravenous
CSF	cerebrospinal fluid	IVH	intraventricular hemorrhage
Cu	copper	kcal	kilocalorie
d	day	kg	kilogram
DAO	diamine oxidase	LDH	lactate dehydrogenase
dl	deciliter	LDL	low-density lipoprotein
DMBA	dimethylbenzanthracene	MAO	monoamine oxidase
DNA	deoxyribonucleic acid	mcg(μg)	microgram
EDTA	ethylenediamine tetraacetate	mg	milligram
EEG	electroencephalogram	MNL	mononuclear leukocytes
EGOT	erythrocyte glutamate oxaloacetate transaminase	Na	sodium
		NAD	nicotinamide adenine dinucleotide
EGPT	erythrocyte glutamate pyruvate transaminase	NADP	nicotinamide adenine dinucleotide phosphate
EGR	erythrocyte glutathione reductase	ng	nanogram
EKG	electrocardiogram	NK-cell	natural killer cell
EPP	erythropoietic protoporphyria	NPO	not taking nourishment by mouth
ETK	erythrocyte transketolase	OC	oral contraceptives
FAD	flavin adenine dinucleotide	OGTT	oral glucose tolerance test
FDPase	fructose-1,6-diphosphatase	OTC	over the counter (nonprescription)
Fe	iron	PBC	primary biliary cirrhosis
FMN	flavin mononucleotide	PCM	protein-calorie malnutrition
g	gram	PDR®	Physician's Desk Reference
GFR	glomerular filtration rate	pg	picogram
GI	gastrointestinal	PHA	phytohemagglutinin
GPT	glutamate pyruvate transaminase	PLP	pyridoxal-5'-phosphate
GSH	glutathione (reduced)	PMN	polymorphonuclear lymphocytes

PMS	premenstrual syndrome	**SGOT**	serum glutamate oxaloacetate transaminase
PTH	parathyroid hormone		
RBC	red blood cells	**SGPT**	serum glutamate pyruvate transaminase
RBP	retinol-binding protein		
RDA	recommended daily allowance	**SIDS**	sudden infant death syndrome
RDS	respiratory distress syndrome	**SLE**	systemic lupus erythematosus
RNA	ribonucleic acid	**SOD**	superoxide dismutase
ROP	retinopathy of prematurity	**SRBC**	sheep red blood cells
SCA	sickle cell anemia	**t.i.d.**	three times daily
Se	selenium	**TPN**	total parenteral nutrition
Se-Cys	selenocysteine	**TPP**	thiamin pyrophosphate
Se-Met	selenomethionine	**UV**	ultraviolet
		Zn	zinc

SECTION I: PHARMACOLOGY OF THE VITAMINS

1
Vitamin A

Introduction	3
Absorption, Metabolism, and Excretion	6
Dosage Forms	8
Clinical Studies	8
Infections and Stress	8
The Preterm Infant	9
Neonatal Cholestasis	9
Down's Syndrome	9
Alcoholism and Cirrhosis	10
Malabsorption	11
Crohn's Disease	12
Glaucoma	12
Dermatologic Applications	13
Acne vulgaris	13
Aging and photoaging	13
Psoriasis	14
Ichthyosiform dermatoses	14
Pityriasis rubra pilaris	15
Darier's disease	15
Oral lichen planus	15
Photosensitivity	16
Cancer	16
Retinol and carotene	16
Retinoic acid	19
Other Potential Uses	19
Toxicity and Side Effects	21
Interactions	22
References	24

INTRODUCTION

The term vitamin A is used to designate all-*trans*-retinol or vitamin A alcohol (see Fig. 1 for structure) and related compounds with similar physiologic

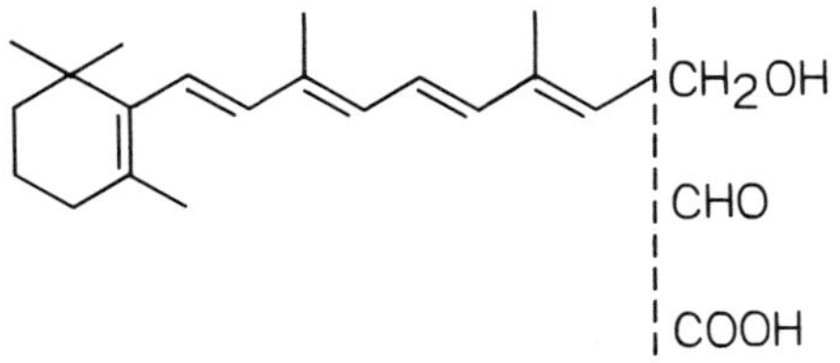

4 / Pharmacology of Micronutrients

Fig. 1. Structural formula of vitamin A.

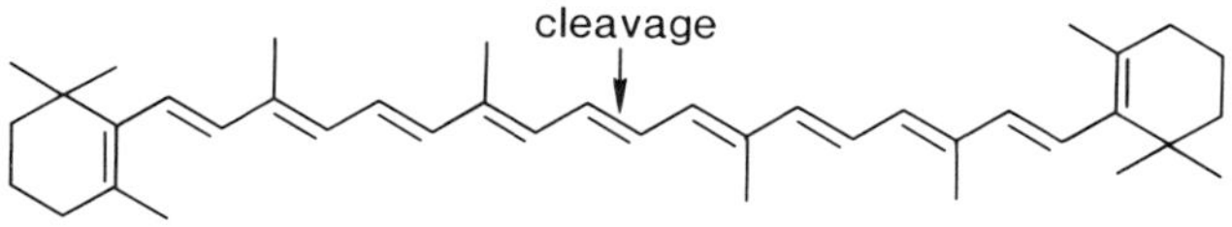

Fig. 2. Structural formula of beta-carotene (provitamin A).

activity, including retinyl esters, retinaldehyde (also termed retinene or retinal), and retinoic acid. Dehydroretinol, a form of vitamin A occurring in certain fresh-water fish, has an additional double bond in the ring. Retinaldehyde is the active form in the retina. Further oxidation in the body cells yields retinoic acid, which is active in most functions of vitamin A, but not in the retina, since there is no biochemical mechanism for its reduction to retinaldehyde. Retinoic acid is not stored and, together with its further oxidation products, is an excretion product of vitamin A metabolism [1–3].

In a dietary sense, the term vitamin A is extended to include certain carotenes of vegetables and fruit (provitamin A) that can be enzymatically processed in the body to yield vitamin A. beta-Carotene (see Fig 2) is an example. Carotenes and vitamin A are readily oxidized and inactivated when isolated and exposed to air, but are fairly stable in foods, where they are protected by natural antioxidants and by fats and oils. Pharmaceutical preparations generally require encapsulation or coatings for protection from oxidation [1–3].

Vitamin A is essential for growth and development, reproduction, visual function, and normal differentiation and function of mucosal epithelium [1–5]. The Wald cycle in the retina [6] involves isomerization of all-*trans*-retinol to 11-*cis*-retinol, followed by oxidation to 11-*cis*-retinaldehyde, which acts as a chromophoric group for light-sensitive rhodopsin proteins in the retinal rods and cones. Light energy activates isomerization of 11-*cis*- to all-*trans*-retinaldehyde and its release from these proteins, causing nerve activity that is transmitted to the brain's optic center. The freed all-*trans*-retinaldehyde may be isomerized back to 11-*cis*-retinaldehyde for recombination with opsin to regenerate rhodopsin, or it may be reduced enzymatically to all-*trans*-retinol [1,4,6,7].

The molecular mechanisms of other vitamin A functions are not as well understood as that of its function in the retina. However, there is evidence that one mode of activity is similar to that of steroid hormones—involving combination of the vitamin with intracellular binding proteins and interaction with nuclear DNA to stimulate production of messenger RNAs for specific proteins [8,9]. Another mechanism of action that appears especially important for normal cell membrane composition and function, as well as for mucus secretion, is the intracellular synthesis of mannosyl retinyl phosphate, which is used in the production of glycoproteins [4,8,10,11]. Normal mucous membrane function is of course essential for resistance to infection and is dependent on adequate vitamin A nutriture, but, in addition, supplemental vitamin A in larger than ordinary nutritional amounts has been shown to enhance both humoral and cell-mediated immune response in laboratory animals [12–14]. In Indonesian children, mild vitamin A deficiency is positively associated with increased mortality, infections being the probable major cause[15]. Laboratory animal studies have demonstrated enhancement of T- and B-lymphocyte responses to mitogens by both beta-carotene and canthaxanthin, a carotenoid not convertible to vitamin A [16]. A single report on a human study indicated an increase in T lymphocytes with the OKT 4 marker in adults taking 180 mg/d of beta-carotene for 2 weeks [17].

Normally 90% or more of total body vitamin A is stored in the liver, esterified with long-chain fatty acids, mainly palmitic [1,18]. The normal range of plasma vitamin A is 20–80 mcg/dl, falling below this range only when liver stores are depleted [2,19]. Actually, levels below 40 mcg/dl are suspect, as deficient dark adaptation or night blindness, a sign of early or mild vitamin A deficiency, was noted by Carney and Russell [20] in 6 of 18 patients with plasma vitamin A levels of 30–39 mcg/dl; incidence of night blindness increased markedly at lower plasma levels until finally no subjects with levels below 15 mcg/dl had normal dark adaptation. Signs of more severe vitamin A deficiency, such as xerophthalmia, are rarely seen in western countries. However, marginal vitamin A deficiency states are often seen in chronic alcoholism, and a recent survey among teenage girls in North Carolina noted a prevalence of about 20% of plasma retinol levels in the range of 20–39 mcg/dl [21].

The fatty portions of many animal protein foods are good dietary sources of vitamin A in the form of long-chain fatty acid esters, including dairy products, liver, kidney, heart, and many fish. Liver oils of marine fish are sources of therapeutic amounts [1,2]. Margarine is usually fortified with vitamin A. Green and yellow vegetables and yellow fruits are good sources of provitamin

A, the exact vitamin A value depending on the types of carotene present [22,23]. Dietary and pharmaceutical amounts of vitamin A activity are usually expressed as International Units (IU), although official nutrition boards now prefer to state such activity in terms of retinol equivalents, where 1.0 IU equals 0.3 mcg of all-*trans*-retinol, with other conversion factors for retinyl esters and carotenes [2]. The RDA for adults and pregnant women is 1,000 retinol equivalents, or about 3,300 IU, increasing to about 4,000 IU for lactation [24]. The average American diet more than meets the RDA [22].

The carotenoids are not considered to be essential dietary constituents per se except in the case of strict vegetarians, but there is increasing evidence that they have important physiologic roles other than as sources of vitamin A. For example, beta-carotene is the most effective quencher known for singlet oxygen (a highly reactive oxygen metabolite) [25]. In addition, in vitro and in vivo studies showed beta-carotene to have antimutagenic properties not exhibited by retinol [26,27]. Unlike retinol, which is not a good antioxidant, beta-carotene is an effective one, particularly at the low oxygen tensions found in body tissues other than the lung and in red cells [25]. This observation has led to much research, epidemiologic and experimental, on possible anticancer properties of beta-carotene, inasmuch as it has been shown that increased concentrations of reactive oxygen and peroxy organic compounds and radicals can promote initiated cells to neoplastic growth [28]. The carotenes, therefore, are an important part of the body's defenses against free radicals and toxic oxygen metabolites, complementing the protective effects of vitamin C and E, glutathione peroxidase, catalase, superoxide dismutase, ceruloplasmin, bilirubin, and uric acid [29].

ABSORPTION, METABOLISM, AND EXCRETION

Vitamin A occurring in foods as long-chain fatty acid esters is released as retinol by the action of digestive lipases and esterases in the small intestine. Bile and pancreatic secretions are important for release of the vitamin. After absorption into the mucosal cells, retinol is re-esterified, mainly with palmitate, stearate, and oleate, and incorporated into chylomicrons, which are transported in the lymphatic circulation. Absorption of dietary vitamin A approaches 90%. Retinol in excess of needs is stored in the liver as fatty acid esters, chiefly palmitate [1–3].

Provitamin A carotenes are also absorbed with the lipids of the diet. At normal dietary intakes, most of these carotenes are enzymatically degraded in the mucosal cell to retinol and other products, though some enters the circulation as such; the liver may participate in conversion of carotenes to retinol.

Small amounts of retinoic acid may also be produced in the enterocytes; these enter the portal circulation [1-3,22,23,30].

The level of vitamin A in the peripheral circulation is closely controlled by the liver, which in response to physiologic need hydrolyzes stored retinyl esters and releases retinol, not as such but in combination with a hepatically synthesized retinol-binding protein (RBP). In plasma, this complex combines further with prealbumin, which is believed to protect the retinol-RBP complex from filtration by the kidney. Retinol is delivered to the body tissues via cell surface receptors that recognize RBP. Following release of retinol to the cell, the RBP comes away in a slightly changed form that has less affinity for prealbumin and is filtered and degraded by the kidney. Within the body cells, retinol can be oxidized to retinaldehyde and then irreversibly to retinoic acid. Intracellular binding proteins, differing from RBP, have been identified for retinol and retinoic acid [1-3].

The liver, in addition to acting as a storage depot for 90% or more of the body's vitamin A, functions as a catabolic and excretory organ for the vitamin. Retinol and retinoic acid are converted by the liver into their beta-glucuronides at a low and fairly constant rate when liver vitamin A stores are at normal levels and excreted into bile; some is reabsorbed by enterohepatic circulation. If vitamin A intakes are high, the rates of hepatic catabolism of vitamin A and biliary excretion of metabolites rise markedly. Several, but not all, oxidation products of retinol and retinoic acid have been identified. Acidic metabolites with shortened side chains tend to be excreted in the urine. Generally about twice as great a quantity of vitamin A metabolites is excreted in the feces as in the urine [1-3,18].

Human volunteers given single pharmacological doses of retinol (200,000-500,000 IU) after overnight fasting showed sharp rises in plasma content of retinyl palmitate (with some stearate), peaking at about 4 hours after ingestion of the dose, while plasma retinol concentration showed little change. By the eighth hour after dosage, the concentrations of retinyl palmitate had fallen below that of retinol, and they approached initial fasting levels of the ester by the 16th. If high daily dosage was continued, fasting levels of plasma retinyl palmitate and stearate gradually rose [31]. Studies in laboratory animals indicate that plasma turnover of retinol is high in both vitamin A-adequate and vitamin A-deficient animals, apparently involving active recycling among liver, plasma, peripheral tissues, and possibly interstitial fluid [32].

Data on the pharmacokinetics of carotenes in man are scanty. Studies on laboratory rats consuming a diet containing 0.2% beta-carotene showed that plasma became saturated within 3 days, while liver, adrenal, and ovary had not reached saturation in 147 days. The half-life varied from less than 3 days for plasma to 9 days for liver and 18 days for muscle. Most beta-carotene was stored in the liver, with small amounts in the skin, and none detectable in body fat [33]. The human on high beta-carotene intake differs in requiring 2–4 weeks for plasma levels to plateau [34] and also in that he accumulates beta-carotene in subcutaneous fat [33]. In male adult volunteers, beta-carotene administered in purified form produced much larger plasma responses, measured at 32 hours after ingestion, than equivalent amounts given as carrots [35].

It is of interest that in a comparison of 125 habitual smokers with 125 age- and race-matched nonsmokers, plasma carotene and vitamin C levels were about 20% lower in the smokers than in the nonsmokers [36]. The finding has relevance because of the indications that carotene status is of importance in resistance to carcinogenesis (see Cancer section under Clinical Studies, pp. 16–19). Tests in men and women show that exposure of the skin to sunlight or UV-A light causes significant photodegradation of carotenoids and a lowering of their plasma levels [37].

DOSAGE FORMS

Vitamin A is available over the counter (OTC) as capsules containing 10,000, 25,000, and 50,000 IU per capsule. The OTC marketing of 25,000- and 50,000-IU dosage forms is a highly questionable practice in view of potential teratogenicity and chronic toxicity problems (see Toxicity and Side Effects section, pp. 21–22). Coated tablets are also marketed. A water-miscible preparation containing 50,000 IU/ml is available for administration by intramuscular injection to patients with malabsorption problems. Tretinoin (all-*trans*-retinoic acid) for topical use is marketed as a solution (0.05%), a cream (0.05% and 0.10%), and a gel (0.01% and 0.025%) [38].

beta-Carotene is supplied in 30-mg capsules for oral administration (Solatene®, Roche Laboratories). It is also available over the counter as capsules and coated tablets supplying 15 mg beta-carotene.

CLINICAL STUDIES
Infections and Stress

Catabolism and excretion of vitamin A may be accelerated by acute and chronic infections and by stress. Massive urinary excretion of vitamin A may occur in cancer, tuberculosis, pneumonia, urinary tract infections, chronic nephritis, appendicitis, and prostate diseases [2,3,39,40]. In general, plasma vitamin A levels are depressed during pyrexia, as in children with rheumatic fever and in patients with infectious hepatitis [39]. In view of the importance of vitamin A nutriture to immune response [1–3, 12–17], precautionary oral supplementation with 10,000 IU of vitamin A daily may be advisable under these conditions.

The Preterm Infant

It has been repeatedly shown that plasma levels of vitamin A below 20 mcg/dl are present in 60–80% of preterm infants [41–43]. In 42 preterm infants, averaging 32 weeks gestational age, Brandt et al. [41] found mean serum vitamin A levels one-third lower than mean levels in 51 full-term infants. Since vitamin A is necessary for proper differentiation and maintenance of mucus-secreting cells, these investigators consider low vitamin A status to be a possible predisposing factor for necrotizing enterocolitis in the premature infant. It appears that placental transfer of vitamin A is rather inefficient, as more than one-third of term neonates also have low vitamin A levels [42]. Immaturity of hepatic RBP production is probably a factor, but in addition the plasma vitamin A/RBP molar ratio is lower in the preterm than in the term infant [42].

Hustead et al. [43] found that preterm infants who developed bronchopulmonary dysplasia (BPD) had lower plasma retinol concentrations at birth and at day 21 than preterm infants who did not develop BPD; all infants studied were receiving recommended intakes of vitamin A. The recommended intake for the neonate is 1,500 IU/d orally [24], and this intake should be increased for the preterm infant with a low plasma retinol level. Woodruff et al. recommend an aqueous dispersion of vitamin A alcohol for oral administration to preterm infants and noted a better plasma response to 5,000 IU/d than to 3,000 IU/d [44]. If intravenous feeding is necessary, the recommended amount of retinol for the preterm infant is 230 IU/kg/d [45]. The plasma retinol level should be monitored. Gutcher et al. [46] have discussed precautions to avoid losses and ensure reliable delivery of vitamin A in total parenteral nutrition of the preterm infant. Preparations containing retinol as the palmitate ester, such as Roche Laboratories's Berocca PN®, were more reliable than those containing unesterified retinol or retinyl acetate.

Neonatal Cholestasis

Balistreri [47] has comprehensively reviewed the subject of neonatal cholestasis, including predisposing factors and causes of the syndrome, differential diagnosis of cholestasis, and management methods. Nutritional management is complex and involves a number of special nutrient supplements. Among these is 10,000–15,000 IU/d of vitamin A orally in water-dispersible form (Aquasol A®).

Down's Syndrome

Children with Down's syndrome are prone to respiratory and pyrexial infections. Palmer [48] studied the effect of vitamin A supplementation over a

6-month period in 12 Down's patients who received a monthly oral dose of 1,000 IU/kg of water-miscible vitamin A, compared with 11 patients who were not treated. All Down's patients at the start of the study, as well as those who did not receive treatment over the 6-month period, had a significantly higher incidence of respiratory and gastrointestinal infections in comparison with their normal siblings and age-matched normal controls. In addition, they had significantly lower plasma vitamin A levels, impaired absorption of vitamin A from an oily vehicle, increased serum IgG and secretory IgA, and decreased IgM and serum albumin. With vitamin A treatment, there was a decreased incidence of infections, improvement in absorptive ability, increase in plasma vitamin A, and reduction of serum IgG. Since the monthly vitamin supplement was not excessively large, it appears that a daily supplement of 5,000 IU of vitamin A would usually be adequate for the Down's syndrome patient, but a water-miscible preparation is recommended in view of the absorption problems of these children.

Alcoholism and Cirrhosis

Low plasma vitamin A values and night blindness are frequently seen in chronic alcoholics. Low vitamin A status often stems in part from poor dietary habits, but in addition there are adverse metabolic effects of alcohol on vitamin A metabolism, notably a reduction in hepatic storage of the vitamin relative to normal individuals on the same dietary intake. Hepatic vitamin A drops to very low levels in cirrhosis, together with RBP synthesis and plasma vitamin A levels, and alcoholic cirrhotics therefore frequently display night blindness [49].

Alcoholics with cirrhosis usually respond with attainment of normal dark adaptation within 1–4 weeks of supplementation with 10,000 IU/d of vitamin A orally [20,50,51]. However, if there is zinc deficiency or general undernutrition (protein-calorie malnutrition), there may be a lack of response. Zinc appears to be necessary for hepatic RBP synthesis, and retinol dehydrogenase of the retina is a Zn-dependent enzyme [49,52]. Treatment with 220 mg/d of zinc sulfate for 1–2 weeks, in addition to vitamin A, may return dark adaptation to normal [52,53]. McClain et al. [54] found that supplementation of chronic alcoholic cirrhotics with vitamin A and zinc corrected night blindness and their low levels of vitamin A, Zn, and RBP. In four patients complaining of hypogonadism and impotence, abstinence from alcohol together with treatment with vitamin A and zinc reversed the impotence and raised serum levels of testosterone, vitamin A, RBP, and Zn [54].

Protein-calorie malnutrition, as well as deficiencies of other micronutrients, e.g., riboflavin, vitamin C, or pantothenic acid, may adversely affect dark adaptation in alcoholics with normal serum vitamin A and Zn levels, as reported by Dutta et al. [55]. They describe normalization of dark adaptation function, normalization of low serum albumin levels, and mild elevation of RBP and prealbumin levels in two underweight patients with pancreatic insufficiency of alcoholic origin (who were abstinent from alcohol for at least 6 months), when they were treated with a high-protein (150 g/d), high-calorie (3,500 kcal/d) diet for 6 weeks; vitamin A and Zn supplements were not needed. The physician should therefore keep in mind that night blindness may occasionally have a cause other than vitamin A deficiency.

Brissot et al. [56], in a study of 34 cases of idiopathic hemochromatosis, 33 cases of alcoholic cirrhosis, 10 cases of nonalcoholic cirrhosis, and 35 normal controls, found that there were markedly reduced plasma vitamin A and RBP levels in both alcoholic and nonalcoholic cirrhosis, but low Zn values were seen only in the alcoholic cirrhosis group. In idiopathic hemochromatosis, serum vitamin A was significantly below normal, but RBP concentration was normal and Zn levels close to control levels. Clinical signs of vitamin A deficiency, such as ichthyosis and visual disturbances, may occur in these patients and may require treatment with vitamin A alone, vitamin A and Zn, or vitamin A, Zn, and protein-calorie enrichment of the diet.

Malabsorption

Inadequate absorption of vitamin A, along with other dietary lipids, may occur in patients with pancreatic insufficiency, small bowel bypass, liver disease, and regional enteritis [20]. Patients with ulcerative colitis may also show reduced absorption of the vitamin [57,58]. A daily supplement of 10,000 IU of vitamin A, preferably in water-dispersible formulation, may be sufficient to maintain adequate vitamin A nutriture in many or most of these patients, but some may require larger amounts, particularly after small bowel resection. For example, Wechsler [59] describes a case of vitamin A deficiency that developed in a teen-age girl after partial jejunoileal bypass for morbid obesity. Despite an ample food intake, she developed phrynoderma around the elbows and knees, together with night blindness. Oral therapy with 50,000 IU vitamin A t.i.d. reversed the skin lesions in 1 month and the night blindness in 2 months; she was subsequently maintained on 50,000 IU daily. For patients who are still more refractory, intramuscular injection of 100,000–200,000 IU at intervals of a few months may be necessary. *In female patients of child-bearing potential, it must be kept in mind that vitamin A in high dosage may be teratogenic (see Toxicity and Side Effects section, pp. 21–22).*

Cystic fibrosis patients display reduced plasma levels of vitamin A, sometimes associated with reduced plasma levels of Zn, RBP, and prealbumin [52]. Generous dosages of water-dispersible vitamin A (10,000–25,000 IU/d) are usually prescribed, but Ekvall and Mitchell [60] found that cystic fibrosis children could be maintained on a 5,000-IU daily dose of water-dispersible retinyl palmitate provided that 100 IU of water-dispersible d-alpha-tocopheryl acetate was given at the same time. Vitamin E is known to protect vitamin A from oxidation both in the intestinal lumen and in the tissues [2]. If plasma Zn and RBP levels are low in these patients, a zinc supplement should be added and attention paid to the diet to ensure an adequate intake of high-quality protein.

Crohn's Disease

Skogh et al. [61] give an interesting account of return of normal bowel function in a 31-year-old woman with Crohn's disease who had had an ileocecal resection with clinical benefit, but who still had moderate diarrhea about twice a day. Initial treatment with oral vitamin A (retinyl palmitate 50,000 IU t.i.d.) and vitamin E (d-alpha-tocopheryl acetate 100 IU t.i.d.) was given for psoriasis, which cleared in 2 weeks. The most striking effect was return of normal bowel function. Subsequent experiments with discontinuance of the vitamins separately showed that the favorable effect on bowel function was due to vitamin A alone. Normal bowel function continued when the patient received 100,000 IU oral vitamin A daily; 50,000 IU proved inadequate. The authors comment that, since vitamin A profoundly affects the metabolism and differentiation of epithelial mucosa, supplements may restore normal function of the intestinal barrier in Crohn's disease. Although only a single case, these findings would bear follow-up in other treatment centers.

Glaucoma

Evans [62] has described successful treatment of hundreds of cases of primary glaucoma in West African patients with a combination of 180,000 IU vitamin A, 200 IU vitamin E, and 3,000 mg vitamin C per day orally. Since many of these patients are protein-deficient, a 20-g/d protein supplement was also given. In most cases, this treatment brought ocular tension within normal limits within 1 week; earlier trials showed that supplements half as large brought ocular tension to normal in 2–4 weeks. Administration of a carbonic anhydrase inhibitor and/or a miotic with the nutrient therapy gave no additional benefit. No comment is made by Evans on maintenance after normal tension was achieved, but presumably much lower supplemental levels would suffice and, in the case of vitamin A, would be necessary to avoid eventual development of toxic symptoms. Although it is not known whether the glaucoma of West Africa and that observed in industrially developed nations is of similar etiology, it would seem worthwhile to explore more thoroughly the value of nutrient therapy of glaucoma in ophthalmologic centers, *again bearing in mind the possibility of teratogenic activity of vitamin A in the case of* women of child-bearing potential (see Toxicity and Side Effects section, pp. 21–22).

Dermatologic Applications

Acne vulgaris. All-*trans*-retinoic acid (tretinoin) is a well-established drug, approved by FDA, for topical treatment of acne. It is supplied as a 0.1% and 0.05% cream, a 0.025% and 0.01% gel, and a 0.05% solution (Retin-A®, Ortho Pharmaceutical). Efficacy of tretinoin for this disorder was first established by Kligman and coworkers [63]. Tretinoin appears to increase the rate of production of loose horny cells in the follicular canal, preventing formation of comedones and unseating existing ones [63]. Matsuoka [64] recommends starting treatment with daily applications of the 0.10% cream or a lower strength gel or solution. The 0.05% cream is preferable for patients with sensitive skin. Since tretinoin is a strong skin irritant, application should be avoided around the mouth, eyes, and angles of the nose. During the first few weeks of therapy, there may be erythema, exfoliation, and a pustular flare-up caused by disruption of pre-existing comedones, but thereafter side effects usually become rare, so that more frequent or stronger applications may be tolerated. If adverse reactions are too severe, frequency of application should be reduced to once every other day. The drug should be applied to thoroughly dry skin, at least 30 minutes after washing. Since tretinoin causes thinning of the epidermis, the patient should use an effective sun screen during exposure to sunlight in the summer months. Benzoyl peroxide and tretinoin therapy may be combined to advantage in many patients, e.g., by applying benzoyl peroxide in the morning and tretinoin at night [64].

Aging and photoaging. Great public interest has been aroused by the recent publications of Kligman et al. [65] and Weiss et al. [66] reporting that long-term vitamin A acid (tretinoin) treatment of photoaged skin in middle-aged and elderly Caucasians reversed much of the damage caused by sunlight exposure. The Kligman study, though not double-blind, demonstrated the ability of 0.05% tretinoin cream (Retin-A®) to reverse epidermal atrophy and dyplasia, disperse previously clumped melanin granules, promote angiogenesis and formation of new collagen in the papillary dermis, and cause exfoliation of retained follicular horn [65]. The Weiss study used double-blind procedures, comparing 0.1% tretinoin cream with vehicle, although the dermatitis and drying induced by tretinoin probably vitiated the double-blind aspect of the study except for the examination of histologic sections [66]. Of 30 patients who completed the 16-week study, each of whom applied tretinoin ointment to the dorsal aspect of one forearm and placebo vehicle to the other, all showed slight to great improvement in the tretinoin-treated arm with respect to diminution of fine wrinkling, and most showed improvement in coarse wrinkling and roughness. The majority of letigines and solar-induced freckles showed a mild to moderate reduction in color. Fifteen patients received tretinoin to the face, and 15 received vehicle cream. Fourteen of the 15 tretinoin group experienced slight-to-moderate improvement in fine wrinkling. Small improvements were also observed in coarse wrinkling and tactile roughness. Restoration of skin pinkness occurred in tretinoin-treated skin and was regarded favorably by the patients [66].

Biopsy specimens were not taken from irritated areas, but tretinoin-treated forearms showed an increase of 273% in mean epidermal thickness. However, there was no evidence of edema. Granular layer thickness was enhanced, and the number of mitotic figures increased over placebo-treated skin [66].

Initially, most patients experienced tretinoin-induced dermatitis, which lasted 2–12 weeks. As the dermatitis began to subside, improvements as noted above became evident and progressed to the conclusion of the study. Patients were supplied with emollient cream to combat the dermatitis, as well as a mild soap. A sunscreen preparation was provided for those who had to be under prolonged sunlight exposures. Eleven patients were given topical anti-inflammatory steroid ointments, none of which were used for more than 3 days, to mitigate tretinoin-induced inflammation [66].

These findings have received great acclamation, both public and medical, and it has been proposed that tretinoin ointment may be useful in combating so-called intrinsic aging due to normal senescence as well as the long-term changes induced by excessive sunlight exposure [66,67]. We will undoubtedly see much more clinical investigation of these potentialities in the near future, including the possibility of reversing neoplastic cellular changes.

Psoriasis. Fry et al. [68] obtained significant clinical improvement in 11 of 12 patients with psoriasis treated with 0.1% all-*trans*-retinoic acid in petrolatum ointment twice daily over a 2-week controlled trial period. In two patients who used the ointment with an occlusive polyethylene dressing, the psoriatic lesions healed completely. During a 4-week follow-up period without treatment, six of nine patients developed some relapse but not to the original severity of the lesions. Frost and Weinstein [69] compared effectiveness of 0.1, 0.2, and 0.3% all-*trans*-retinoic acid in petrolatum or cold cream in 26 psoriasis patients, obtaining fair-to-good clinical response in 24 of 26 patients in 1–2 weeks, with histologic sections showing a change from parakeratosis to normal-appearing stratum corneum as clinical improvement occurred. The 0.2% and 0.3% concentrations were more effective than the 0.1% ointment, but irritation of intertriginous areas occurred in 3 of 11 patients receiving the 0.3% ointment; this subsided when therapy was suspended for 2 days.

Ichthyosiform dermatoses. Frost and Weinstein [69] obtained good clinical responses to 0.2% and 0.3% petrolatum ointments of all-*trans*-retinoic acid in cases of lamellar ichthyosis and epidermolytic hyperkeratosis, but not in ichthyosis vulgaris; there was modest improvement in X-linked ichthyosis. Grice et al. [70] found that a 3-week treatment with 0.1% retinoic acid in yellow petrolatum was effective in treating keratosis pilaris occurring in six patients with the autosomal dominant type of ichthyosis.

Pityriasis rubra pilaris. Very high oral dosages of vitamin A in water-dispersible form (Aquasol A®, USV Pharmaceutical) have been used over strictly limited time periods, usually 2 weeks, in adult patients with pityriasis rubra pilaris. For example, Randle et al. [71] prescribed 1×10^6 IU/d for 5–14 days, obtaining good clinical response; six of seven patients followed up after treatment did not suffer relapses in observation periods ranging from 1 to 10 years. Toxic effects during high-dose vitamin A administration included somnolence, fatigue, headache, desquamation, dryness of the mucous membranes, and mild reversible changes in liver enzymes. *Such high doses of vitamin A are of course contraindicated in women of child-bearing potential because of the known teratogenic properties of excessive vitamin A (see Toxicity and Side Effects section, pp. 21–22).*

Darier's disease. Thomas et al. [72] treated three adult patients with this heritable disorder with 1×10^6 IU/d of water-dispersible vitamin A (Aquasol A®) orally for 14 days and obtained 50–80% improvement in the skin lesions. Desquamation was minimal; side effects consisted of drowsiness, mild frontal headache, dry lips, and dry nose. Two patients had minimal recurrences 6 weeks after treatment, while one maintained her improvement for at least a year without further therapy. *Again, note that high dosages of vitamin A are contraindicated for women of child-bearing potential.*

A topical treatment of Darier's disease with all-*trans*-retinoic acid is described by Fulton et al. [73]. They recommend application of 0.5% retinoic acid in a vehicle of 95% ethanol, ethylene glycol monomethyl ether, and propylene glycol (1:1:2) under occlusive wrap for 48 hours. Remissions in some treated skin areas last 3 months or longer; other areas may be held in remission by weekly applications without occlusion.

Oral lichen planus. Günther [74] has reported therapy of lichen planus of the oral mucous membrane with all-*trans*-retinoic acid given either orally or topically (six patients each). Oral dosage was 10 mg retinoic acid in hard gelatin capsules, taken 3 or 4 times daily an hour before meals. For topical treatment, 0.1% retinoic acid in orabase paste was applied to the affected areas daily or every other day. Clinical assessments were made after varying periods of treatment, range, 3–7 weeks on oral therapy, 2–5 weeks on topical therapy. Both treatments produced a striking improvement of the buccal mucosa, but less on the tongue. Side effects were minor. Local erythemas in topical application could be avoided by application every other day. On oral dosage, two patients had slight elevations of SGPT and one had a transient headache at start of therapy. Most patients maintained their improvement over the following 6 months without further treatment, but three had slight-to-moderate recurrences 3 weeks to 2 months after withdrawal of therapy. *The contraindication for use of retinol and retinyl esters in women of child-bearing potential applies equally to retinoic acid.*

Photosensitivity. Mathews-Roth [75,76] has reviewed her own studies and those of others on the use of beta-carotene to treat the photosensitivity of erythropoietic protoporphyria (EPP). More than 75% of patients have had a significant improvement in their tolerance to sunlight exposure. Yellowing of the skin occurs at the dosages used. Otherwise, no side effects have been noted except for occasional loosening of the stools in a few patients. Dosages are tailored to the individual and usually range from 180 to 300 mg/d for adult patients, taken orally as 30-mg capsules (Solatene®, Roche Laboratories) for a minimum of 3 months. Carotene blood levels should be maintained at or above 800 mcg/dl.

The drug displays some effectiveness against other photosensitivities, for example, polymorphous light eruption, hydroa vaccineforme, and porphyria variegata; Mathews-Roth suggests that beta-carotene should be given a trial in any kind of photosensitivity disease in which there has been no response to other forms of therapy [75,76]. beta-Carotene is thought to work by virtue of its ability to quench the singlet oxygen and free radicals generated in the skin by photosensitizing chemicals such as protoporphyrin.

beta-Carotene does not have sunscreen properties, but may increase the tanning action of sunlight, as opposed to the burning effect [75,76].

Despite the maintenance of high plasma levels of beta-carotene, plasma and liver retinol levels remain normal, as there appears to be a feedback inhibition of intestinal and hepatic enzymes that convert carotene to vitamin A once normal retinol levels are attained [75,76].

Cancer

Retinol and carotene. A number of properties of vitamin A—its role in ensuring normal differentiation of mucus-secreting epithelial tissues, its control of cell membrane composition and function, and its necessity for normal immune function [1–5,8,12–17,77,78]—led to the expectation that vitamin A nutriture might be important in prevention of cancer. Vitamin A deficiency diminishes glutathione levels in lung and liver of experimental animals [79], an important observation in view of glutathione's importance for reduction of lipid peroxides, which can promote initiated cells into neoplastic growth [28]. Basu et al. [78] have reviewed the large body of evidence indicating that vitamin A deficiency increases susceptibility to chemical carcinogenesis in laboratory animals. Vitamin A at doses above nutritional levels has enhanced protection against experimentally induced carcinoma of the trachea and bronchus, esophagus, stomach and intestine, lung, and breast [78]. Brevard et al. [80] studied the effects on colorectal histology of young male rats after 2 weeks of feeding diets containing 2, 5, or 10 times the recommended daily allowance of vitamin A as either retinol or beta-carotene. The increase from the lowest level to 5 times the recommended allowance significantly increased the numbers of goblet and mast cells and had effects on mucin composition and quantity that could improve immune defenses against cancer. Further increase in retinol equivalents had little additional effect. There was no difference in results whether the retinol equivalents were provided as retinol or beta-carotene. Temple and Basu [81] report that dietary beta-carotene exerts a significant protective effect against chemically induced colon cancers in mice.

The carotenes have displayed ability to prevent or delay the appearance of chemically, virally, or ultraviolet light-induced experimental cancer [82–86]. Although part of the antitumor activity of carotenes, especially beta-carotene, may depend on conversion in the body to vitamin A, such activity in laboratory animals is also shown by canthaxanthin, a carotene that is not a provitamin A [83,84,86]. Tomita et al. [87] found that oral beta-carotene caused a strong augmentation of tumor immunity (rejection) responses in mice injected with syngeneic Meth A fibrosarcoma cells. The cancer-preventive activity of carotenes per se is generally attributed to their ability to quench agents that may damage cell membranes and DNA, e.g., singlet oxygen and peroxy-radicals [25,86,88,89]. Burton and Ingold [25] have advanced evidence that beta-carotene is an unusual type of lipid antioxidant in that it exhibits good radical-trapping antioxidant behavior only at partial pressures of oxygen significantly below 150 torr, the partial pressure of oxygen in ambient air. Such low-oxygen partial pressures are found in most tissues under physiologic conditions. At higher oxygen pressures, beta-carotene loses its antioxidant activity. The chain-reaction-breaking action of beta-carotene complements that of vitamin E, which is effective at high oxygen concentrations, such as those prevailing in lung and red cell membranes [25]. Cigarette smoking is reported to depress the level of carotenes in the blood [36].

There is ample epidemiologic evidence of lowered cancer risk in individuals consuming diets rich in retinol equivalents (vitamin A plus provitamin A), as well as in individuals whose plasma retinol levels are at the upper end of the normal range [78]. In general, dietary vitamin A is associated with decreased risk of epithelial cancers [90]. Wylie-Rosett et al. [91], in a study of uterocervical cytology in 87 women with severe dysplasia or carcinoma in situ (CIS), compared with 82 normal controls, found that there was a substantially greater risk for severe dysplasia or CIS in women with low total vitamin A or low beta-carotene intakes. Basu et al. [92] reported observation of negative correlation between recurrence of colorectal cancer and plasma levels of retinol and RBP in a group of patients who had had colorectal cancers surgically removed; it must be noted, of course, that low plasma retinol levels in the presence of reduced levels of its carrier protein, RBP, may reflect primarily protein-calorie malnutrition.

The main difficulty with epidemiologic observations based on calculations of dietary vitamin A or measurements of plasma retinol is that they cannot reveal whether the apparent resistance to cancer associated with a good state of vitamin A nutriture should be attributed to retinol or carotene. Peto et al. [93] carefully reviewed this question a few years ago, and epidemiologic studies by Shekelle et al. [94] and Colditz et al. [95] tend to support the consumption of carotene-containing vegetables and fruit, rather than consumption of animal foods containing preformed retinol, as the dominant factor in conferring cancer resistance. Smith and Jick [96] found no evidence, in a study of 800 newly diagnosed cancer patients and 3,433 patients with nonmalignant conditions, that regular intake of pharmaceutical preparations containing retinyl esters had any protective effect against carcinogenesis. On the other hand, Stich et al. [97] found equal benefits in reducing DNA breakage in the buccal mucosal cells of Filipino betel nut and tobacco chewers whether they were treated with oral retinol (100,000 IU/week) or beta-carotene (300,000 IU/week) over a 3-month period.

Several epidemiologic studies of more recent date, based on micronutrient analyses of blood samples obtained before onset of cancer, have established significant negative correlations between plasma carotene levels and risk of lung cancer [98–101].

From the evidence obtained so far, it would appear judicious, as cancer-preventive measures, to maintain a good state of vitamin A nutriture and to consume an abundance of carotene-containing fruits and vegetables. It must be borne in mind that the latter foods contain not only carotenes but other substances that may be beneficial in cancer prevention, such as fiber, vitamin C, folic acid, and perhaps still-unidentified cancer-preventive agents [95]. The question of whether beta-carotene itself is cancer preventive may be resolved in the next few years by a prospective randomized trial now in progress in 20,000 U.S. physicians who are taking capsules containing beta-carotene or placebo [102].

There are a few reports of the use of toxic or near-toxic dosages of retinol or its esters as adjunctive therapy in treatment of established cancer. For example, Meyskens et al. [103] noted a favorable but statistically nonsignificant trend toward a reduced relapse rate when patients with surgically treated stages I and II malignant melanoma were treated over a 3-year period with BCG (bacillus Calmette Guérin vaccine) plus 100,000 IU oral retinol daily, compared with patients on BCG alone. Sakamoto et al. [104] used oral high-dose vitamins A, C, and E as an adjunct to conventional chemotherapy over a 12-month period in 20 patients with a variety of established cancers (lymphoma, leukemia,

Hodgkin's disease, multiple myeloma, ependymoma, uterine sarcoma, clear-cell carcinoma of the kidney, oat-cell and non-oat-cell carcinoma of the lung, pancreatic carcinoma, colon carcinoma, and breast carcinoma). Vitamin dosages were: vitamin A (Aquasol A®) two capsules of 50,000 IU q.i.d., vitamin E capsules of 400 IU q.i.d., and 2,000 mg vitamin C q.i.d., all doses taken after meals and at bedtime. After 12 months, 15 of the 20 patients had a complete disappearance or greater than 50% reduction in tumor size, whereas only 8 of the 20 had malignancies that would normally have responded well to chemotherapy. Only one patient showed any symptom of vitamin A toxicity, namely a breast cancer patient who developed arthritic-like knee pain after 5–6 months; she was kept free of this pain by reducing vitamin A dosage to 50,000 IU every other day. No hepatomegaly or blood analysis abnormalities were noted in these patients.

Retinoic acid. All-*trans*-retinoic acid has been tried against tumors of the skin and mucous membranes both topically and orally. Bollag and Ott [105] evaluated 0.1–0.3% retinoic acid topical preparations in 60 patients with actinic keratoses, obtaining complete regression in 24 and partial regression in 17. In 16 with basal cell epitheliomata, there were five complete and ten partial regressions. Bollag, in his 1983 review of studies by himself and coworkers [106], states that oral all-*trans*-retinoic acid has given positive clinical results in patients with actinic keratoses, basal cell carcinomas, leukoplakias of the mouth, tongue, and larynx (90% complete or partial regressions), and papillomas of the urinary bladder (73% complete or partial regressions). Other tumors did not respond. However, like retinol and retinyl esters, all-*trans*-retinoic acid taken orally in therapeutically active amounts, besides being teratogenic, eventually results in unacceptable toxicity, causing cancer researchers to turn to investigation of synthetic retinoids as drugs of potentially greater efficacy and lower toxicity [106].

Levine and Meyskens [107] obtained encouraging results by topical retinoic acid treatment of metastatic cutaneous melanomas in two patients. The treatment involved application of one drop of all-*trans*-retinoic acid, 0.05% solution (Retin A®), to each lesion once a day, followed by covering with occlusive tape (Blenderm®). The patients were seen every 4 weeks to assess progress, and after 12 weeks skin biopsy specimens were again taken from the treated areas. Complete regression of a total of 25 lesions was noted in one patient, a 54-year-old man, and a partial response in the other, a 59-year-old woman.

Other Potential Uses

Animal studies have shown that a marginal vitamin A status is deleterious to wound healing after surgery and that provision of supplemental vitamin A in either modest (5 times the recommended intake) [108] or "toxic" levels [109,110] improved the rate of wound healing. Vitamin A supplements reversed the retardation of skin wound healing in the rat induced by corticosteroid administration [111]. beta-Carotene fed at requirement level doubled wound strength at 5 days in rats compared with feeding the requirement as retinyl acetate [108]. These observations deserve follow-up in humans.

Seng et al. [112] showed that little active collagenase is released after mild burns of the cornea in vitamin A-adequate rats, but much more in the corneas of vitamin A-deficient rats, resulting in slower recovery and occasional ulceration. Brinckerhoff and coworkers [113] reported that all-*trans*-retinoic acid inhibited collagenase production in cultures of rheumatoid synovial cells obtained from patients with rheumatoid arthritis and that 13-*cis*-retinoic acid (isotretinoin) given to rats with adjuvant arthritis reduced inflammation and collagenase production [114]. These findings are being followed up clinically, using synthetic retinoids because of their lower toxicity [38].

Kinley and Krause [115] reported in 1959 that oral administration of 100,000 IU/d of retinyl acetate for 4-6 months significantly reduced (by an average of 30%) the elevated serum cholesterol levels of patients with known atherosclerosis (previous myocardial infarction), but had no effect on individuals with normal cholesterol levels. No untoward symptoms were reported, although blood levels of vitamin A at the end of the study were elevated slightly (mean, 107 mcg/dl versus normal range of 20-80 mcg/dl). There appears to have been no follow-up of this observation, undoubtedly because of the potential for toxicity on long-continued administration of high-dose vitamin A and because of the advent of several effective cholesterol-lowering drugs.

Patients with hemolytic and iron-storage disease tend to have low plasma vitamin A levels, probably because of iron-catalyzed oxidation of the vitamin [56,116]. Supplemental vitamin A, with vitamin E for antioxidant protection, may be needed. Attention should be paid to plasma Zn and RBP levels, with Zn and high-quality protein supplements as needed.

Borisov [117], studying the vitamin A status of 376 healthy young male Russian athletes, found indications of a heightened vitamin A requirement, depending in part on the degree of physical and emotional stress to which they were exposed. Marginal plasma retinol levels of 10-30 mcg/dl were present in 46% of the athletes, while 8% had values below 10 mcg/dl. Borisov estimated their requirements at 7,000-8,000 IU/d during winter-spring and 8,000-10,000 IU/d during autumn. Regular strenuous exercise may therefore call for a modest vitamin A supplement if the diet does not provide enough of the vitamin, but further clinical studies are needed to confirm these observations.

TOXICITY AND SIDE EFFECTS

The various forms of vitamin A—retinol, retinyl esters, and retinoic acid (but not beta-carotene)—produce similar toxic symptoms when given in excessive amounts. Administration of single oral doses of 200,000–300,000 IU of vitamin A to preschool children has produced such symptoms as headache (probably due to increased intracranial pressure), somnolence, irritability, vomiting, and desquamation. In neonates, hypervitaminosis A causes bulging of the fontanelles [2].

Adults are more tolerant of large single doses, but may manifest some of the same symptoms with oral doses of 1×10^6 IU or larger [2]. Some clinicians have continued daily dosage of this magnitude for as long as 14 days in patients with skin diseases; the side effects observed, i.e., somnolence, fatigue, headache, and dry mucous membranes, were considered acceptable by the investigators in view of the clinical benefits [71,72].

In chronic dosage of 25,000 IU/d or more in preschool children, or 100,000 IU/d or more in adults, the same symptoms may develop as those seen with very large single doses. In addition, there may be alopecia, anorexia, papilledema, lip fissuring, hypercalcemia, bone thickening (exostoses), bone and joint pain, and hepatomegaly [2]. These signs of toxicity may take many months or, in adults, even several years to manifest [2]. As hepatic capacities for storage of retinyl esters and for catabolism of retinol are gradually overwhelmed, increasing amounts of retinyl esters appear in plasma, not bound to RBP; these are powerfully membranolytic [1–3, 18, 118]. Occasional instances have been reported of chronic toxicity in adults taking 50,000 IU/d [119] or even as little as 25,000 IU/d of supplemental vitamin A [120] over long periods. Individual capacities to handle large loads of vitamin A evidently vary considerably.

The symptoms and physical changes of hypervitaminosis A are reversible upon withdrawal of vitamin A (1–3). *However, chronic vitamin A toxicity in children may produce permanent long bone deformities by premature closure of the epiphyses* [121].

Most reported cases of severe vitamin A toxicity have occurred in individuals who were self-medicating themselves with large daily doses without supervision by a physician [2]. The patient must be cautioned against taking retinol or retinyl esters in daily doses larger than 10,000 IU without medical supervision.

In advanced renal failure, daily supplements of vitamin A as low as 5,000 IU may induce elevated plasma retinol levels and symptoms of chronic toxicity, e.g., pruritus and hair loss, as the kidney no longer contributes adequately to metabolism of RBP and oxidation of retinol to retinoic acid [122–124]. *Supplemental vitamin A is therefore contraindicated in renal dialysis patients and others with advanced chronic kidney disease.* If multivitamin preparations are being taken, they should be free of vitamin A.

Administration of vitamin A in greater than RDA amounts is contraindicated in pregnant women or women contemplating pregnancy, as excessive maternal vitamin A intake has definite teratogenic potential in experimental animals and possibly in humans [125,126]. This should be kept in mind in prescribing oral vitamin A (or synthetic retinoids) in treating acne or other skin conditions in young women.

beta-Carotene in high doses has no known side effects other than yellowing of the skin and an occasional loosening of the stools in a few patients [75,76]. High doses of beta-carotene have not been noted to lead to hypervitaminosis A, as there appears to be feedback inhibition of the carotene conversion to retinol when plasma retinol levels rise above normal [75,76]. No teratogenic effects have been reported in human pregnancy when vitamin A or retinol equivalents were ingested at RDA levels. Vitamin A intake in pregnancy should not be restricted to levels much lower than the RDA, as abruptio placentae is associated with below-normal plasma levels of vitamin A and beta-carotene, as well as of vitamins C and E and folic acid. This complication appears to be a condition of multiple vitamin deficiency in which the criticality of vitamin A nutriture is unclear but may be of importance [127].

Supplementation with pharmaceutical retinol or retinyl esters appears unnecessary in the elderly if they are receiving RDA amounts of retinol equivalents from dietary retinol and carotenoids. Phelan et al. [128] report that elderly subjects (aged 60–94 years) showed rises of plasma retinyl esters (an index of potential toxic effects) when receiving oral vitamin A supplements, whereas young adults did not show this increase; vitamin A of dietary origin, including carotenes, did not elevate retinyl esters in the plasma of the elderly subjects.

INTERACTIONS

In laboratory animals receiving normal intakes of dietary vitamin A as retinyl ester or beta-carotene, chronic ethanol consumption lowers hepatic storage of the vitamin [129,130]. When vitamin A is supplied to the rat at an elevated level (about 30 times requirement), which is high enough to reduce food consumption by 10–20% but not high enough to harm the liver if no alcohol is given, chronic ethanol consumption causes hepatic inflammation, fibrosis, and necrosis, as well as other abnormalities [131]. Chronic alcohol consumption therefore is a contraindication for high-dose vitamin A (and perhaps synthetic retinoid) therapy. Ingestion of phenobarbital or butylated hydroxytoluene (BHT) potentiated the adverse effect of ethanol on hepatic vitamin A storage in laboratory rats [132].

Most commonly employed drugs are not known to interfere with the metabolism or requirement of vitamin A. However, neomycin sulfate reduces absorption of the vitamin, and phenobarbital and caffeine may lower vitamin A reserves somewhat [2]. In animal experiments [133], injection of corticosterone caused a rapid loss of vitamin A from the plasma, liver, adrenals, and thymus; this could be reversed by simultaneous injection of retinol in corn oil, suggesting that vitamin A supplementation may be desirable during chronic glucocorticoid therapy. However, if glucocorticoids are being used as immunosuppressives, this pharmacologic effect may be vitiated by supplemental vitamin A [134]. Only limited human studies have been carried out on drug interactions with vitamin A, but Leo et al. [135] recently reported on 11 patients with moderate drug-induced liver changes in whom hepatic vitamin A storage was less than 10% of normal, although serum vitamin A and RBP levels were only slightly depressed. Drugs taken by these patients included alpha-methyldopa, prednisone, benzodiazepine, hydrocortisone, heroin, methadone, phenelzine sulfate, phenytoin, phenothiazine, and "multiple drugs." The authors call for consideration of vitamin A supplementation in patients with such low hepatic vitamin A levels and suggest that the recommended daily allowance may have to be increased for population segments subjected to chronic drug use.

Oral contraceptives (estrogen-progestin combinations) raise plasma retinol and RBP levels while lowering hepatic vitamin A storage, apparently stimulating hepatic RBP synthesis and export of the vitamin A-RBP complex [2]. Whether this increases the vitamin A requirement is not known.

Zinc nutriture is closely tied to vitamin A metabolism in that Zn appears to play a role in hepatic RBP synthesis and is a cofactor for retinol dehydrogenase in the retina. Zinc deficiency alone or in association with vitamin A deficiency may cause night blindness [52–54].

There are important interrelationships between vitamin A and vitamin E. Vitamin E supplements prevent or alleviate the toxicity of massive doses of vitamin A in laboratory animals [136,137]. In normal human adults given 200 IU of alpha-tocopheryl acetate per day orally without a vitamin A supplement, plasma vitamin A and RBP levels fell significantly in 3 weeks [138]. In subjects taking 30,000 IU of retinyl acetate daily, but no vitamin E, plasma retinol concentration rose steadily, while if 200 IU of vitamin E was taken with the 30,000 IU of vitamin A, there was a small but significant decline in plasma retinol by the third week. The authors suggest that vitamin E may either enhance hepatic vitamin A uptake or increase vitamin A transport to and utilization by peripheral tissues [138]. An essential function of vitamin E, jointly with vitamin A, in the retina is indicated by the fact that degeneration of photoreceptor cells in the retina of young rats is far greater when they are fed diets deficient in both vitamins than when they are fed diets deficient in either one of the vitamins separately [139].

The toxic effects of excess vitamin A are reduced by feeding extra vitamin D in the rat [140], dog [141], and chick and turkey poult [142]. Vitamin A-induced injury to cultured human lymphoblastoid cells was counteracted in part by separate additions of taurine, zinc, and vitamin E, each of which had membrane-stabilizing effects, and counteracted completely when all three substances were present at the same time [143].

REFERENCES

1. Pitt GAJ: Vitamin A. In: *Fat-Soluble Vitamins*, AT Diplock (ed). Lancaster, PA: Technomic Publishing Co., 1985, pp 1–75.
2. Olson JA: Vitamin A. In: *Handbook of Vitamins*, LJ Machlin (ed). New York: Marcel Dekker, 1984, pp 1–43.
3. Lui NST, Roels OA: Vitamin A and carotene. In: *Modern Nutrition in Health and Disease*, 6th ed., RS Goodhart, ME Shils (eds). Philadelphia: Lea & Febiger, 1980, pp 142–159.
4. Weber F: Biochemical mechanisms of vitamin A action. Proc Nutr Soc 42:31–41, 1983.
5. Zile MH, Cullum ME: The function of vitamin A: Current concepts. Proc Soc Exp Biol Med 172:139–152, 1983.
6. Wald G: The visual function of vitamin A. Vitam Horm 18:417–430, 1960.
7. O'Brien DF: The chemistry of vision. Science 218: 961–966, 1982.
8. Wolf G. Levin LV, Bolmer SD: Multiple functions of vitamin A: Nuclear and extranuclear. In: *Modulation and Mediation of Cancer by Vitamins*, FL Meyskens, KN Prasad (eds). New York: Karger, 1983, pp 146–152.
9. Anon.: Intracellular retinol-binding protein and keratin messenger-RNA: Evidence for a nuclear function of vitamin A. Nutr Rev 40:154–157, 1982.
10. Wolf G: Retinol-linked sugars in glycoprotein synthesis. Nutr Rev 37:265–267, 1979.
11. Anon.: Retinol involvement in epidermal glycoprotein synthesis. Nutr Rev 37:265–267, 1979.
12. Athanassiades TJ: Adjuvant effect of vitamin A palmitate and analogs on cell-mediated immunity. J Natl Cancer Inst 67:1153–1156, 1981.
13. Davis CY, Sell JL: Effect of all-*trans*-retinol and retinoic acid nutriture on the immune system of chicks. J Nutr 113:1914–1919, 1983.
14. Malkovsky M, Edwards AJ, Hunt R, et al.: T-cell-mediated enhancement of host-versus-graft reactivity in mice fed a diet enriched in vitamin A acetate. Nature 302:338–340, 1983.
15. Sommer A, Tarwotjo I, Hussaini G, Susanto D: Increased mortality in children with mild vitamin A deficiency. Lancet 2:585–588, 1983.
16. Bendich A, Shapiro SS: Effect of beta-carotene and canthaxanthin on the immune response of the rat. J Nutr 116:2254–2262, 1986.
17. Alexander M, Newmark H, Miller R: Oral beta-carotene can increase the number of OKT 4^+ cells in human blood. Immunol Lett 9:221–224, 1985.
18. Olson JA: New approaches to methods for the assessment of nutritional status of the individual. Am J Clin Nutr 35:1166–1168, 1982.
19. Pitt GAJ: The assessment of vitamin A status. Proc Nutr Soc 40:173–178, 1981.
20. Carney EA, Russell RM: Correlation of dark adaptation test results with serum vitamin A levels in diseased adults. J Nutr 110:552–557, 1980.
21. Sumner SK, Liebman M, Wakefield LM: Vitamin A status of adolescent girls. Nutr Rep Int 35:423–431, 1987.

22. Bauernfeind JC: Carotenoid vitamin A precursors and analogs in foods and feeds. J Agr Food Chem 20:456–473, 1972.
23. Simpson KL: Relative value of carotenoids as precursors of vitamin A. Proc Nutr Soc 42:7–17, 1983.
24. Food and Nutrition Board: *Recommended Dietary Allowances*, 9th ed. Washington, DC: National Academy of Sciences, 1980.
25. Burton GW, Ingold KU: Beta-carotene: An unusual type of antioxidant. Science 224:569–573, 1984.
26. Belisario MA et al.: Inhibition of cyclophosphamide mutagenicity by beta-carotene. Biomed Pharmacother 39:445–448, 1985.
27. Renner HW: Anticlastogenic effect of beta-carotene in Chinese hamsters. Time and dose response studies with different mutagens. Mutat Res 144:251–256, 1985.
28. Cerutti PA: Prooxidant states and tumor promotion. Science 227:375–381, 1985.
29. Machlin LJ, Bendich A: Free radical tissue damage: Protective role of antioxidant nutrients. FASEB J 1:441–445, 1987.
30. Ott DB, Lachance PA: Retinoic acid—a review. Am J Clin Nutr 32:2522–2531, 1979.
31. Goodman GE, Alberts DS, Peng YM, et al.: Pharmacokinetics and phase I trial of retinol and 13-*cis*-retinoic acid. In: *Modulation and Mediation of Cancer by Vitamins*, FL Meyskens, KN Prasad (eds). New York: Karger, 1983, pp 311–316.
32. Lewis KC, Green MH, Underwood BA: Vitamin A turnover in rats as influenced by vitamin A status. J Nutr 111:1135–1144, 1981.
33. Shapiro SS, Mott DJ, Machlin LJ: Kinetic characteristics of beta-carotene uptake and depletion in rat tissue. J Nutr 114:1924–1933, 1984.
34. Mathews-Roth MM, Pathak MA, Parrish J, et al.: A clinical trial of the effects of oral beta-carotene on the responses of human skin to solar radiation. J Invest Dermatol 59:349–383, 1972.
35. Brown ED, Bieri JG, Micozzi M, et al.: Plasma response in normal men to various carotenoid sources. Fed Proc 46:1189A, 1987.
36. Chow CK, Thacker RR, Changchit C, et al.: Lower levels of vitamin C and carotenes in plasma of cigarette smokers. J Am Coll Nutr 5:305–312, 1986.
37. Roe DA: Photodegradation of carotenoids in human subjects. Fed Proc 46:1886–1889, 1987.
38. Mendel HG, Cohn VH: Fat-soluble vitamins. Vitamins A, K, and E. In: *Goodman and Gilman's The Pharmacologic Basis of Therapeutics*, 7th ed., LS Goodman, TW Rall, F Murad (eds). New York: Macmillan, 1985, pp 1573–1591.
39. Moore T: *Vitamin A*. New York: Elsevier Publishing Co., 1957, pp 357–364, 418–441.
40. Roels OA: Vitamin A physiology. JAMA 214:34, 1970.
41. Brandt RB, Mueller DG, Schroeder JR, et al.: Serum vitamin A in premature and term neonates. J Pediatr 92:101–104, 1978.
42. Shenai JP, Chytil F, Jhaveri A, Stahlman MT: Plasma vitamin A and retinol-binding protein in premature and term neonates. J Pediatr 99:302–305, 1981.
43. Hustead VA, Gutcher GR, Anderson SA, et al.: Relationship of vitamin A (retinol) status to lung disease in the preterm infant. J Pediatr 105:610–615, 1984.
44. Woodruff CW, Latham CB, Mactier H, Hewett JE: Vitamin A status of preterm infants: Correlation between plasma retinol concentration and retinol dose response. Am J Clin Nutr 46:985–988, 1987.
45. AMA Department of Food and Nutrition: Guidelines for multivitamin preparations for parenteral use. J Paren Ent Nutr 3:258–262, 1979.
46. Gutcher GR, Lax AA, Farrell PM: Vitamin A losses to plastic intravenous infusion devices and an improved method of delivery. Am J Clin Nutr 40:8–13, 1984.

47. Balistreri WF: Neonatal cholestasis. J Pediatr 106:171–184, 1985.
48. Palmer S: Influence of vitamin A nutriture on the immune response: Findings in children with Down's syndrome. Int J Vitam Nutr Res 48:188–216, 1978.
49. Bonjour JP: Vitamins and alcoholism. IX. Vitamin A. Int J Vitam Nutr Res 51:166–177, 1981.
50. Sandberg MA, Rosen JB, Berson EL: Cone and rod function in vitamin A deficiency with chronic alcoholism and in retinitis pigmentosa. Am J Ophthalmol 84:658–665, 1977.
51. Mobarhan S, Russell RM, Underwood BA, et al.: Evaluation of the relative dose response test for vitamin A nutriture in cirrhotics. Am J Clin Nutr 34:2664–2270, 1981.
52. Solomons NW, Russell RM: The interaction of vitamin A and zinc: Implications for human nutrition. Am J Clin Nutr 33:2031–2040, 1980.
53. Morrison SA, Russell RM, Carney EA, Oaks EV: Zinc deficiency: A cause of abnormal dark adaptation in cirrhotics. Am J Clin Nutr 31:276, 1978.
54. McClain CJ, Van Thiel DH, Parker S, et al.: Alterations in zinc, vitamin A, and retinol-binding protein in chronic alcoholics: A possible mechanism for night blindness and hypogonadism. Alcoholism Clin Exp Res 3:135–141, 1979.
55. Dutta SK, Russell RM, Lakanapal V: Abnormal dark adaptation in adult patients with protein-energy malnutrition: Correction by protein-energy repletion. Nutr Res 1:443–448, 1981.
56. Brissot P, LeTreut A, Dien G, et al.: Hypovitaminemia A in idiopathic hemochromatosis and hepatic cirrhosis. Digestion 17:469–478, 1978.
57. Bercovitz Z, Page RC: Metabolic and vitamin studies in chronic ulcerative colitis. Ann Intern Med Phys Physio-Biol 20:239, 1944.
58. Sharman IM, Dick AP, Farthing MJG, et al.: Carotenoid and retinol levels in the blood of ulcerative colitis patients. Proc Nutr Soc 38:54A, 1979.
59. Wechsler HL: Vitamin A deficiency following small-bowel bypass surgery for obesity. Arch Dermatol 115:73–75, 1979.
60. Ekvall S, Mitchell A: The effect of supplemental vitamin E on vitamin A serum levels in cystic fibrosis. Int J Vit Nutr Res 48:325–332, 1978.
61. Skogh M, Sundquist T, Tagesson C: Vitamin A in Crohn's disease. Lancet 1:766, 1980.
62. Evans SC: Ophthalmic nutrition and prevention of eye disorder and blindness. Nutr Metab 21(suppl 1):268–272, 1977.
63. Kligman AM, Fulton JE Jr, Plewig G: Topical vitamin A acid in acne vulgaris. Arch Dermatol 99:469–476, 1969.
64. Matsuoka LY: Acne. J Pediatr 103:849–854, 1983.
65. Kligman AM, Grove GL, Hirose R, et al.: Topical tretinoin for photoaged skin. J Am Acad Dermatol 15:836–859, 1986.
66. Weiss JS, Ellis CN, Headington JT, et al.: Topical tretinoin improves photoaged skin. A double-blind vehicle-controlled study. JAMA 259:527–532, 1988.
67. Gilchrest BA: Editorial: At last! A medical treatment for skin aging. JAMA 259:569–570, 1988.
68. Fry L, Macdonald A, McMinn RMH: Effect of retinoic acid in psoriasis. Br J Dermatol 83:391–396, 1970.
69. Frost P, Weinstein GD: Topical administration of vitamin A acid for ichthyosiform dermatoses and psoriasis. JAMA 207:1863–1868, 1969.
70. Grice K, Sattar H, Baker H: Urea and retinoic acid in ichthyosis and their effect on transepidermal water loss and water holding capacity of stratum corneum. Acta Derm Venereol (Stockh) 53:114–118, 1973.

71. Randle HW, Diaz-Perez JL, Winkelmann RK: Toxic doses of vitamin A for pityriasis rubra pilaris. Arch Dermatol 116:888–892, 1980.
72. Thomas R III, Cooke JP, Winkelmann RK: High-dose vitamin A therapy for Darier's disease. Arch Dermatol 118:891–895, 1982.
73. Fulton JE, Gross PR, Cornelius CE, Kligman AM: Darier's disease, treatment with topical vitamin A acid. Arch Dermatol 98:396–399, 1968.
74. Günther S: The therapeutic value of retinoic acid (vitamin A acid) in lichen planus of the oral mucous membrane. Dermatologica 147:130–136, 1973.
75. Mathews-Roth MM: Photosensitization by porphyrins and prevention of photosensitization by carotenoids. J Natl Cancer Inst 69:279–285, 1982.
76. Mathews-Roth MM: Photoprotection by carotenoids. Fed Proc 46:1890–1893, 1987.
77. Sporn MB, Roberts AB: Role of retinoids in differentiation and carcinogenesis. Cancer Res 43:3034–3040, 1983.
78. Basu TK, Chan U, Fields A: Vitamin A (retinol) and epithelial cancer in man. In: *Vitamins, Nutrition, and Cancer* KN Prasad (ed). Basel: Karger, 1984, pp 33–45.
79. Dogra SC, Khanduja KL, Sharma RR: Effect of vitamin A deficiency on the levels of glutathione and glutathione-S-transferase activity in rat lung and liver. Experientia 38:903–904, 1982.
80. Brevard PB, Anderton LG, Magee AC: In vivo effects of retinoids on the histological changes in colorectal tissue. Nutr Rep Int 31:635–648, 1985.
81. Temple NJ, Basu TK: Protective effect of beta-carotene against colon tumors in mice. J Natl Cancer Inst 78:1211–1214, 1987.
82. Epstein JH: Effects of beta-carotene on ultra-violet induced cancer formation in the hairless mouse. Photochem Photobiol 25:211–213, 1977.
83. Santamaria L, Bianchi A, Arnboldi A, Andreoni L: Prevention of the benzo(a)pyrene photocarcinogenic effect by beta-carotene and canthaxanthin. Med Biol Environ 9:113–120, 1981.
84. Mathews-Roth MM: Antitumor activity of beta-carotene, canthaxanthin and phytoene. Oncology 39:33–37, 1982.
85. Seifter E, Rettura G, Padawer J, Levenson SM: Maloney murine sarcoma virus tumors in CBA/J mice: Chemoprevention and chemotherapeutic actions of supplemental beta-carotene. J Natl Cancer Inst 68:835–840, 1982.
86. Santamaria L, Bianchi A, Arnboldi A et al.: Dietary carotenoids block photocarcinogenic enhancement by benzo(a)pyrene and inhibit its carcinogenesis in the dark. Experientia 39:1043–1045, 1983.
87. Tomita Y, Himeno K, Nomoto K, et al.: Augmentation of tumor immunity against syngeneic tumors in mice by beta-carotene. J Natl Cancer Inst 78:679–681, 1987.
88. Krinsky NI, Deneke SM: Interaction of oxygen and oxy-radicals with carotenoids. J Natl Cancer Inst 69:205–210, 1982.
89. Flavin DF, Kolbye AC Jr: Nutritional factors with the potential to inhibit critical pathways of tumor promotion. In: *Modulation and Mediation of Cancer by Vitamins*, FL Meyskens, KN Prasad (eds). New York: Karger, 1983, pp 24–38.
90. Middleton B, Byers T, Marshall J, Graham S: Dietary vitamin A and cancer—a multisite case-control study. Nutr Cancer 8:107–116, 1986.
91. Wylie-Rosett JA, Romney SL, Slagle NS, et al.: Influence of vitamin A on cervical dysplasia and carcinoma in situ. Nutr Cancer 6:49–57, 1984.
92. Basu TK, Chan UM, Fields ALA, McPherson TA: Retinol and postoperative colorectal cancer patients. Br J Cancer 51:61–65, 1985.
93. Peto R, Doll R, Buckley JD, Sporn MB: Can dietary beta-carotene materially reduce human cancer rates? Nature 290:201–208, 1981.

94. Shekelle RB, Lepper M, Liu S, et al.: Dietary vitamin A and risk of cancer in the Western Electric study. Lancet 2:1186–1190, 1981.
95. Colditz GA, Branch LG, Lipnick RJ, et al.: Increased green and yellow vegetable intake and lowered cancer deaths in an elderly population. Am J Clin Nutr 41:32–36, 1985.
96. Smith PG, Jick H: Cancers among users of preparations containing vitamin A. Cancer 42:808–811, 1978.
97. Stich HF, Rosin MP, Vallejera MO: Reduction with vitamin A and beta-carotene administration of proportion of micronucleated buccal mucosal cells in Asian betel nut and tobacco chewers. Lancet 1:1204–1206, 1984.
98. Stahelin HB, Rosel F, Buess E, Brubacher G: Cancer, vitamins and plasma lipids: Prospective Basel study. J Natl Cancer Inst 73:1463–1468, 1984.
99. Salonen JT, Salonen R, Lappetelainen R, et al.: Risk of cancer in relation to serum concentrations of selenium and vitamins A and E: Matched case-control analysis of prospective data. Br Med J 290:417–420, 1985.
100. Menkes MS, Comstock GW, Vuilleumier JP, et al.: Serum beta-carotene, vitamins A and E, selenium, and the risk of lung cancer. N Engl J Med 315:1250–1254, 1986.
101. Pastorino U, Pisani P, Berrino F et al.: Vitamin A and female lung cancer: A case-control study on plasma and diet. Nutr Cancer 10:171–179, 1987.
102. Editorial: Vitamin A and cancer. Lancet 2:325–326, 1984.
103. Meyskens FL Jr, Alberts DS, Aapro MS, Surwit EA: Clinical studies of vitamin A and its synthetic derivatives in cancer prevention and treatment at the University of Arizona. In: *Modulation and Mediation of Cancer by Vitamins*, FL Meyskens and KN Prasad (eds). New York: Karger, 1983, pp 306–310.
104. Sakamoto A, Chougoule PB, Prasad KN: Retrospective analysis of the effect of vitamins A, C, and E in human neoplasms. In: *Modulation and Mediation of Cancer by Vitamins*, FL Meyskens and KN Prasad (eds). New York: Karger, 1983, pp 330–333.
105. Bollag W, Ott F: Vitamin A-Säure in der Tumortherapie. Schweiz Med Wochenschr 101:17, 1971.
106. Bollag W: Vitamin A and retinoids: From nutrition to pharmacotherapy in dermatology and oncology. Lancet 1:860–863, 1983.
107. Levine N, Meyskens FL: Topical vitamin-A-acid therapy for cutaneous metastatic melanoma. Lancet 2:224–226, 1980.
108. Gerber LE, Erdman JW Jr: Effect of dietary retinyl acetate, beta-carotene and retinoic acid on wound healing in rats. J Nutr 112:1555–1564, 1982.
109. Seifter E, Crowley LV, Rettura G, et al.: Influence of vitamin A on wound healing in rats with femoral fracture. Ann Surg 181:863–841, 1975.
110. Lee IH, Tong TG: Mechanism of action of retinyl compounds in wound healing. I. Structural relationships of retinyl compounds and wound healing. J Pharm Sci 59:851–854, 1970.
111. Ehrlich HP, Tarver H: Effects of beta-carotene, vitamin A and glucocorticoids on collagen synthesis in wounds. Proc Soc Exp Biol Med 137:936–938, 1971.
112. Seng WL, Glogowski JA, Wolf G, et al.: The effect of thermal burns on the release of collagenase from corneas of vitamin A-deficient and control rats. Invest Ophthalmol Vis Sci 19:1461–1470, 1980.
113. Brinckerhoff CE, et al.: Inhibition by retinoic acid of collagenase production in rheumatoid synovial cells. N Engl J Med 303:432–436, 1980.
114. Brinckerhoff CE, Coffey JW, Sullivan AC: Inflammation and collagenase production in rats with adjuvant arthritis reduced with 13-*cis*-retinoic acid. Science 221:756–758, 1983.
115. Kinley LJ, Krause RF: Influence of vitamin A on cholesterol blood levels. Proc Soc Exp Biol Med 102:353–355, 1959.

116. Katerelos C, Constantopoulos A, Agathopoulos A, et al.: Serum levels of retinol, retinol-binding protein, carotenoids and triglycerides in children with beta-thalassemia major. Acta Haematol 62:100–105, 1979.
117. Borisov IM: Retinol requirements of students engaged in sports. Teor Prakt Fiz Kult 5:58–61, 1977.
118. Nutrition Note: Retinol binding protein in man and rat. Nutr Rev 35:253–254, 1977.
119. Clinical Nutrition Case: The pathophysiological basis of vitamin A toxicity. Nutr Rev 40:272–274, 1982.
120. Herbert V: Toxicity of 25,000 IU vitamin A supplements in "health" food users. Am J Clin Nutr 36:185–186, 1982.
121. Caffey J: *Pediatric X-Ray Diagnosis*, 7th ed. Chicago: Year-Book Medical Publishers, 1978, pp 1466–1470, 1625.
122. Yatzidis H, Digenis P, Fountas P: Hypervitaminosis A accompanying advanced chronic renal failure. Br Med J 3:352–353, 1975.
123. Shmunes E: Hypervitaminosis A in a patient with alopecia receiving renal dialysis. Arch Dermatol 115:882–883, 1979.
124. Cundy T, Earnshaw M, Heynen G, Kanis JA: Vitamin A and hyperparathyroid bone disease in uremia. Am J Clin Nutr 38:914–920, 1983.
125. Bates CJ: Vitamin A in pregnancy and lactation. Proc Nutr Soc 42:65–79, 1983.
126. Editorial: Vitamin A and teratogenesis. Lancet 1:319–320, 1985.
127. Sharma S, et al.: Comparison of blood levels of vitamin A, beta-carotene, and vitamin E in abruptio placentae with normal pregnancy. Int J Vitam Nutr Res 56:3–9, 1986.
128. Phelan EA, Krasinski SD, Russell RM, et al.: Dietary and supplemental vitamin A in the elderly. J Am Coll Nutr 6:435A, 1987.
129. Sato M, Lieber CS: Hepatic vitamin A depletion after chronic ethanol consumption in baboons and rats. J Nutr 111:2015–2023, 1981.
130. Grammer M, Erdman JW Jr: Effect of chronic alcohol consumption and moderate fat diet on vitamin A status in rats fed either vitamin A or beta-carotene. J Nutr 113:350–364, 1983.
131. Leo MA, Lieber CS: Hepatic fibrosis after long-term administration of ethanol and moderate vitamin A supplementation in the rat. Hepatology 3:1–11, 1983.
132. Leo MA, Lowe N, Lieber CS: Potentiation of ethanol-induced hepatic vitamin A depletion by phenobarbital and butylated hydroxytoluene. J Nutr 117:70–76, 1987.
133. Atukorala TMS, Basu TK, Dickerson JWT: Effect of corticosterone on the plasma and tissue concentrations of vitamin A in rats. Ann Nutr Metab 25:234–238, 1981.
134. Cohen BE, Cohen IK: Vitamin A: Adjuvant and steroid antagonist in the immune response. J Immunol (suppl 3): 1376–1380, 1973.
135. Leo MA, Lowe N, Lieber CS: Decreased hepatic vitamin A after drug administration in men and in rats. Am J Clin Nutr 40:1131–1136, 1984.
136. McCuaig LW, Motzok I: Excessive dietary vitamin E: Its alleviation of hypervitaminosis A and lack of toxicity. Poult Sci 49:1050–1052, 1970.
137. Jenkins MY, Mitchell GV: Influence of excess vitamin E on vitamin A toxicity in rats. J Nutr 105:1600–1606, 1975.
138. Garrett-Laster M, Oaks L, Russell RM, Oaks E: A lowering effect of a pharmacological dose of vitamin E on serum vitamin A in normal adults. Nutr Res 1:559–564, 1981.
139. Robinson WG Jr, Kuwabara T, Bieri JG: Deficiencies of vitamins E and A in the rat. Retinal damage and lipofuscin accumulation. Invest Ophthalmol Vis Sci 19:1030–1037, 1980.
140. Billitteri A, Raoul Y: Antagonism between vitamins A and D in mitochondria and lysosomes. CR Soc Biol 159:1919–1923, 1965.

141. Cho DY, Frey RA, Guffy MM, Leipold HW: Hypervitaminosis A in the dog. Am J Vet Res 36:1597–1603, 1975.
142. Veltman JR, Jensen LS, Rowland GN: Partial amelioration of vitamin A toxicosis in the chick and turkey poult by extra dietary vitamin D-3. Nutr Rep Int 35:381–392, 1987.
143. Pasantes-Morales H, Wright CE, Gaull GE: Protective effect of taurine, zinc and tocopherol on retinol-induced damage in human lymphoblastoid cells. J Nutr 114:2256–2261, 1984.

2
Vitamin D

Introduction	31
Absorption, Metabolism, and Excretion	34
Dosage Forms	36
Clinical Studies	37
Hypocalcemia and Rickets of Prematurity	37
The Breast-Fed Infant	38
Nutritional Rickets	39
Aging and Bone Loss	39
Renal Osteodystrophy	43
Malabsorption	45
Alcoholism	46
Hypoparathyroidism	46
Genetic Disease	46
Cochlear Deafness	47
Hypertension	48
Hemolytic Disorders	48
Psoriasis	48
Cancer	48
Toxicity and Side Effects	49
Interactions	51
References	52

INTRODUCTION

Vitamin D is the name used to designate a group of closely related compounds that have antirachitic activity. Chemically they belong to a class of steroid-like compounds known as seco-steroids, of which cholecalciferol (vitamin D-3) and ergocalciferol (vitamin D-2) are the most important members (Fig. 1). Vitamin D-3 occurs naturally, being synthesized in skin exposed to ultraviolet light, while vitamin D-2 is a manufactured compound obtained by irradiating ergosterol isolated from yeast or by ultraviolet treatment of foods directly. The two forms are equipotent in man and in this presentation will be

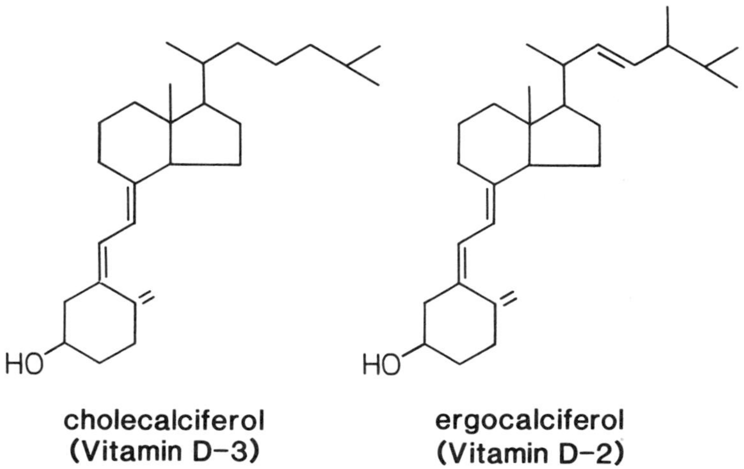

cholecalciferol
(Vitamin D-3)

ergocalciferol
(Vitamin D-2)

Fig. 1. Structural formulas of cholecalciferol (vitamin D-3) (left) and ergocalciferol (vitamin D-2) (right).

referred to simply as vitamin D with respect to their physiologic and pharmacologic properties. Both are solid at room temperature, are soluble in fats and oils, and insoluble in water. They are oxidizable, especially in aqueous suspension, and unstable to light, so that pharmaceutical preparations require coatings or solution in propylene glycol or oils with a protective antioxidant. Vitamin D is fairly stable as a food fortification agent [1–4].

Since vitamin D is synthesized in the skin when exposed to sunlight, it is technically not a vitamin, i.e., a micronutrient that must be provided in the diet, except in individuals who receive inadequate sun exposure. Research in the last two decades has established it as a prohormone, involved together with parathyroid hormone (PTH) and calcitonin in regulating calcium and bone metabolism. Vitamin D from biosynthesis in the skin or obtained from food is hydroxylated to 25-hydroxy-vitamin D (calcifediol) by the liver and then to the active hormone, 1,25-dihydroxyvitamin D (calcitriol) by the kidney (Fig. 2). Calcitriol stimulates intestinal absorption of calcium and phosphate and decreases their renal excretion. At normal circulating concentrations it is essential for mineralization of bone, but at high concentrations induced by low dietary calcium, its predominant action is to enhance resorption of bone, thereby restoring plasma calcium levels. Although the kidney is the principal source of calcitriol, this hormone also appears to be synthesized in the placenta and in bone. Other hydroxylated forms of vitamin D, e.g., the 24,25-dihydroxy and

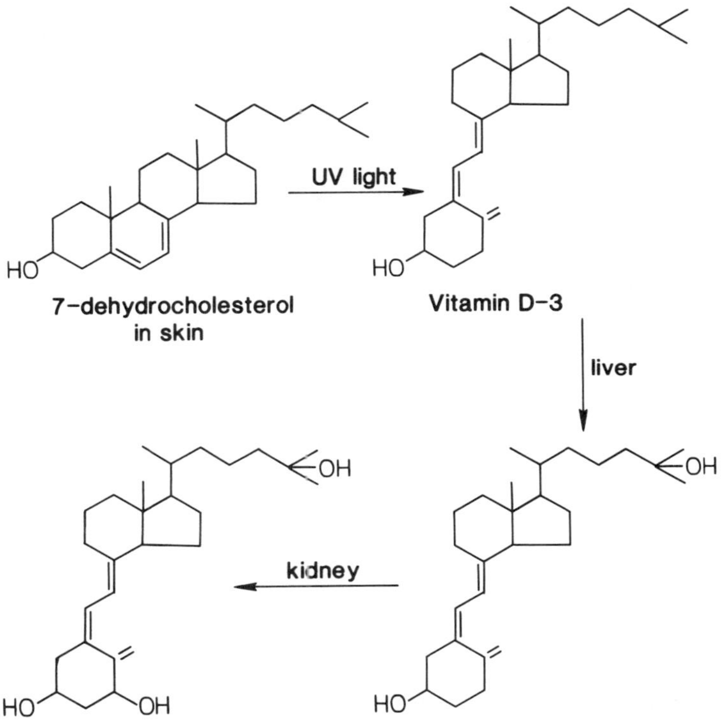

Fig. 2. Hydroxylation of vitamin D (biosynthesized in the skin — upper left) to 25-OH-D (lower right) and then to its final form, 1,25-(OH)$_2$D (lower left).

1,24,25-trihydroxy derivative, are biosynthesized, but whether they have any special functions in addition to those of calcitriol is unclear and is a subject of current research [1–4]. Recently it has been shown that macrophages are also capable of 1-hydroxylating 25-hydroxyvitamin D and that this activity is markedly stimulated by gamma-interferon [5–7].

Vitamin D appears to have immunoregulatory functions. In vitro studies showed that calcitriol at picomolar concentrations inhibited production of interleukin-2, which is important for the growth of T cells, and inhibits the proliferation of phytohemagglutinin-stimulated lymphocytes [8]. In elderly institutionalized patients in Sweden, low serum 25-OH-vitamin D levels were associated with an increased incidence of anergy (deficient delayed hypersensitivity response), and treatment with oral vitamin D or daily UV irradiation for 2–3 months normalized both serum 25-OH-D levels and the delayed hypersensitivity response in an initially vitamin D-deficient group [9]. Rook et al. [7] found that calcitriol increased the ability of cultured human myocytes to inhibit the in vitro growth of *Mycobacterium tuberculosis* and suggest that this may explain the reported effectiveness of vitamin D therapy in chronic skin tuberculosis. Calcitriol receptors and binding proteins are widespread through the body tissues, and undoubtedly additional roles for the hormone will come to light as the result of further research [1].

Because of rapid hydroxylation in the liver, circulating levels of vitamin D as such are low, of the order of 1–2 ng/ml. Normal levels of 25-OH-vitamin D-2 (derived from diet) are reported to be 2–8 ng/ml, while those of 25-OH-vitamin D-3 (derived mostly from biosynthesis in the skin) are 8–45 ng/ml. The concentration of 25-OH-vitamin D-3 is highly responsive to seasonal and occupational exposure to ultraviolet light. Circulating levels of calcitriol normally range from 19 to 70 pg/ml in children, being highest in infants and adolescents, and from 15–40 pg/ml in healthy adults [10].

The classical vitamin D deficiencies are rickets in children, which is the failure to mineralize the newly synthesized organic matrix (osteoid) of growing bone, and osteomalacia in adults, which is the demineralization of mature bone, leaving undermineralized organic matrix. Calcium absorption diminishes, plasma calcium and phosphate decline, and PTH and alkaline phosphatase levels increase [11,12]. The elevation in PTH causes a phosphaturia, resulting in hypophosphatemia that contributes to defective bone mineralization. Oral vitamin D or exposure to sunlight or ultraviolet light, together with adequate dietary calcium and phosphate, is preventive and curative [1–4].

Food sources of vitamin D include salt-water fish and artificially fortified foods, including fresh and evaporated milk, dried milk, infant formula, margarine and butter, cereals, and chocolate mixes. Milk supplies 400 IU per quart and margarine 2,000 IU per pound [2]. Fish-liver oils are especially rich sources [2]. The RDA for vitamin D is 200 IU/d for adults and 400 IU/d for infants [13]. These intakes are normally met by consumption of the fortified foods listed above, together with the small amounts contributed by unfortified animal protein foods such as meat, cheese, and eggs, but poor dietary habits and lack of sunlight exposure, as in housebound or institutionalized individuals, can lead to vitamin D deficiency [9,14,15].

ABSORPTION, METABOLISM, AND EXCRETION

Biosynthesized vitamin D-3 is formed by irradiation of 7-dehydrocholesterol, which is naturally present in skin lipids. Ultraviolet light breaks the B ring of the steroid, producing previtamin D-3, which slowly isomerizes nonenzymatically to form the vitamin [16]. Vitamin D-3 is thereby fed into the circulation fairly steadily, even when exposure to sunlight is intermittent [16,17]. Previtamin D-3 serves as a store, as it does not enter the circulation. Sunlight exposure of surgically prepared human skin sections produces a maximum conversion of 7-dehydrocholesterol to vitamin D-3 of 10–15% that does not differ much between fair and heavily pigmented skin; however, longer exposure is needed to reach this plateau in black skin, so that black children in northern cities are likely to be more susceptible than white children to vitamin D deficiency from sunlight deprivation [18]. Prolonged exposure to sunlight or UV light, even in fair-skinned individuals, does not lead to excessively high plasma 25-OH-vitamin D levels because previtamin D-3 in the skin is further isomerized by UV light to lumisterol and tachysterol [16].

Orally ingested vitamin D is absorbed throughout the small intestine, requiring the aid of bile salts. The largest proportion is absorbed in the ileum, because of the longer residence time of the digesta there. Efficiency of absorption is about 50%. The vitamin is taken up in lymph in association with chylomicrons and lipoproteins. Some is bound to vitamin D-binding protein, a plasma protein synthesized in the liver, which not only takes up biosynthesized vitamin D-3 from the skin but also binds 25-OH-vitamin D produced in the liver [1–4]. As compared with vitamin D-3 biosynthesized in the skin, which feeds into the circulation very gradually, orally ingested vitamin D reaches the liver intermittently and in relatively high concentrations after meals, so that in addition to synthesis of 25-OH-vitamin D there is considerable degradative metabolism of the vitamin in the liver to inactive metabolites that are excreted in the bile. Fraser [19] has reviewed the evidence indicating that orally administered vitamin D is less effective than sunlight in raising circulating levels of 25-OH-vitamin D. There is an enterohepatic cycling of vitamin D metabolites, but it is not of a conservative nature, as the degradation products cycled have very little vitamin D activity; whether there is any beneficial function in this cycling is doubtful [20,21].

Although fish liver often serves as a storage organ for vitamin D, mammalian liver does not store significant amounts of either vitamin D or its 25-hydroxy derivative. The largest body pool of 25-OH-vitamin D is in the circulation, with little being stored in body tissues. However, Lawson [1] reports that increasing amounts of cholecalciferol are stored in skin, adipose cells, and other tissues as plasma 25-OH-vitamin D level rises above 30 ng/ml, and release from these stores can help to maintain blood vitamin D activity during periods of deprivation.

Plasma 25-OH-vitamin D rises and falls with the vitamin D supply and serves as an index of vitamin D nutriture. The level of calcitriol is closely regulated to achieve a fairly constant level of plasma calcium. If plasma calcium falls, output of PTH increases, and this process activates increased production of calcitriol in the kidney. Calcium levels above normal depress PTH concentration, resulting in a decline of 1-hydroxylase activity in the kidney and decreased production of calcitriol. This review is a simplified description of the major aspects of control of circulating calcitriol and calcium levels. Other factors for which there is some evidence that they have roles in the regulation of 1-hydroxylase activity include plasma phosphate level and insulin and other hormones; there is continuing research in this area [1–4].

In the intestine, calcitriol acts to increase the production of calcium-binding protein and other cell proteins that stimulate uptake and transport of calcium from the intestinal lumen; uptake of phosphate also increases. In bone, calcitriol together with PTH stimulates resorption and the release of calcium and phosphate to the circulation. Whether net bone mineral content increases properly during growth, and whether in adulthood it remains constant while undergoing normal turnover, depends on the dietary mineral supply and the complex interplay between plasma mineral levels and hormonal responses [1–4].

A large number of polar catabolites of vitamin D are produced in the liver and intestine and ultimately excreted in the bile; these are devoid of antirachitic activity. Not more that 2–3% of vitamin D compounds found in bile consists of vitamin D, 25-OH-vitamin D, or the dihydroxy metabolites. Less than 5% of the body pool of vitamin D is excreted in the urine [1].

DOSAGE FORMS

The usual pharmaceutical form of vitamin D is vitamin D_2 (ergocalciferol). Ergocalciferol (Calciferol®, Kremers-Urban) is available for oral administration as 1.25 mg tablets supplying 50,000 USP units of vitamin D activity (Note: 1 mg vitamin D equals 40,000 IU or 40,000 USP units; the international and USP units are equivalent). Calciferol® is also supplied as a sesame oil solution for intramuscular injection providing 500,000 USP units/ml [4]. Generic vitamin D-2 is available over the counter (OTC) as tablets supplying 400 USP units and also, as of this writing, as tablets supplying 5,000 USP units (a highly questionable OTC form in view of potential toxicity).

25-Hydroxyvitamin D (calcifediol) is marketed as tablets containing 20 or 50 mcg for oral administration (Calderol®, Upjohn) [4].

1,25-Dihydroxyvitamin D (calcitriol) is available for oral use as capsules containing 0.25 and 0.50 mcg (Rocaltrol®, Roche Laboratories) [4].

Dihydrotachysterol, a synthetic reduction product of the vitamin D isomer, tachysterol, has calcium-regulatory properties because of the structural similarity of its 25-hydroxy derivative, formed in the liver, to calcitriol (Fig.

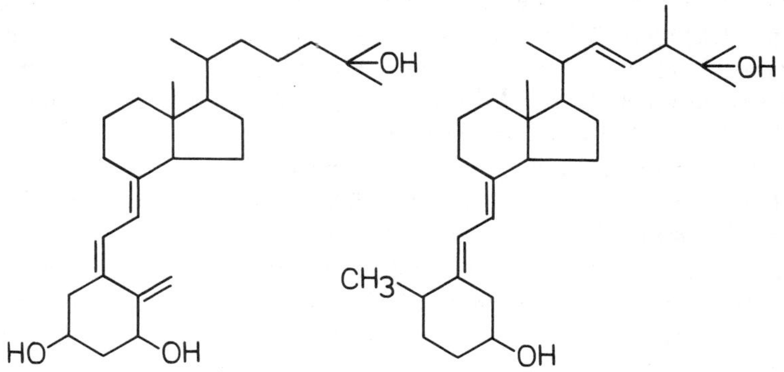

Fig. 3. Structural formulas of calcitriol (left) and 25-hydroxy-dihydrotachysterol (right), synthetic reduction product of the vitamin D isomer, tachysterol.

3). It is marketed for oral administration as 0.125, 0.2, and 0.4 mg tablets (DHT®, Roxane Laboratories) and also as 0.125 mg capsules and a solution in oil supplying 0.25 mg/ml (Hytakerol®, Winthrop) [4].

All the above drugs when used to excess can cause hypercalcemia, so that dosages must be carefully adjusted to the individual case. Titration of dosage is most easily effected with the drugs of shortest plasma half-life, i.e., calcitriol and dihydrotachysterol. The half-life of calcifediol is longer than that of these products but not nearly as long as that of vitamin D. Upon development of hypercalcemia, dosage must be terminated until plasma calcium concentration falls into the normal range, after which dosage may be resumed at a lower level. The time required for reversing hypercalcemia varies according to the calcium-regulatory drug employed: 17–60 days for ergocalciferol, 7–30 days for calcifediol, 3–14 days for dihydrotachysterol, and 2–10 days for calcitriol [21].

The effective daily dosage ranges for these drugs, when used in therapy of hypocalcemia and osteopenia that is resistant to physiologic doses of vitamin D, are as follows: ergocalciferol, 1–10 mg (40,000–400,000 IU); calcifediol, 0.05–0.5 mg (50–500mcg); calcitriol, 0.5–1mcg; and dihydrotachysterol, 0.1–1 mg [21].

CLINICAL STUDIES
Hypocalcemia and Rickets of Prematurity

In a normal pregnancy, the fetus depends on the maternal supply of 25-OH-vitamin D, as well as of calcium, magnesium, and phosphate, and the fetoplacental unit synthesizes a low level of calcitriol sufficient for its needs [22].

Studies in premature infants show that 1-hydroxylase activity is present in the kidney definitely by 32 weeks [23] and probably as early as 28 weeks of gestational age [24]. At term delivery, plasma calcitriol concentration in the infant is significantly lower than in the normal adult and in the mother, but within 24 hours it rises to normal adult values [25].

Since 80% of the bone mineral of the term infant is deposited during the third trimester [25], the very-low-birthweight infant delivered during the second trimester is prone to hypocalcemia and hypophosphatemia at birth and to the development of rickets by 2–3 months of postnatal age [24–26]. Steichen et al. [26] estimate that the amount of calcium provided by standard infant formulas, assuming 50% absorption of calcium, would be only one-third to one-half the amount normally provided to the fetus in utero during the last trimester, and similar considerations apply to phosphate. Supplementation of very-low-birthweight infants with extra calcium and phosphate, as well as vitamin D, is recommended, beginning as soon as possible after birth [26,27]. A supplementation dose of 400 IU/d of vitamin D_2 is generally advised for premature infants [4], but a large study by Hillman et al. [28] of very-low-birthweight infants (mean weight $1,178 \pm 278$ g) showed that about 55% of the group had declining or persistently low 25-OH-vitamin D levels at this vitamin D intake, suggesting that a higher supplementation level could be beneficial.

Hypocalcemia in the premature infant can be readily corrected with high doses of calcitriol, e.g., 1 mcg/d, but the resulting elevation of blood calcium concentration is probably accomplished largely at the expense of an already undermineralized skeleton, and this therapy is not recommended [23].

The Breast-Fed Infant

In a recent study, lactating women on good diets, providing mean daily intakes of 900–1,000 mg of calcium and 445–532 IU of vitamin D, maintained normal serum 25-OH-vitamin D and calcitriol levels for 6 months; thereafter, calcitriol levels gradually rose until at 12 months they were about 50% higher than in nonpregnant controls [29]. However, little of this vitamin D activity in maternal plasma is transferred to breast milk, which, measured during the 3-week postpartum period, was found to contain only about 0.4 mcg of vitamin D and its hydroxylated derivatives per liter; about three-quarters of this was 25-OH-vitamin D, and only a trace was calcitriol [30]. For this reason, it is recommended that breast-fed infants receive a prophylactic daily supplement of 400 IU of vitamin D, usually supplied in the form of an oily solution of vitamins A and D, of which there are a number of preparations on the market [4]. This is of special importance if the child's skin is dark and if there is insufficient exposure to sunlight. For fair-skinned infants breast-fed under good social and environmental conditions, vitamin D supplementation may provide somewhat higher serum 25-hydroxyvitamin D levels and better bone provides somewhat higher serum 25-hydroxyvitamin D levels and better bone mineralization at 12 weeks of age in comparison with unsupplemented controls [31], but after 16 weeks [32]and one year [33] of breast feeding, no differences were found in these parameters between supplemented and unsupplemented infants.

Nutritional Rickets

Although childhood rickets has largely been abolished by vitamin D fortification of dairy products and other foods, it is still seen occasionally in children who receive exclusively vegetarian diets, for example among those who belong to certain religious sects. Prolonged breast-feeding, black skin, veils and long garments, and urban air pollution are contributing factors. Complaints include seizures, swollen wrists, pathologic fractures, growth retardation, and developmental regression. Treatment with oral vitamin D, 400–2,000 IU/d, together with oral calcium supplements, is curative [34–36]. Subsequently, supplementation at 400 IU/d should be adequate.

Aging and Bone Loss

Within a few years after attainment of maximal bone mass around the age of 30, a slow loss of trabecular and cortical bone begins in both sexes. Superimposed on this protracted slow-phase loss is an accelerated rate of loss that occurs in women after menopause or after surgical oophorectomy and that decreases gradually over the next 8–10 years to the slow-phase rate. Incidence of Colles (distal forearm) and vertebral fractures increases in women soon after menopause, while incidence of these fractures in elderly men remains low. Hip fractures increase slowly with aging until late in life, when incidence increases exponentially in both sexes, more rapidly in women than in men. Likelihood of bone fractures in the elderly is increased by a greater tendency to suffer falls because of such factors as failing vision, muscle weakness, neurologic disorders, arthritis, and the effects of certain drugs [37].

Several physiologic changes incident to normal aging occur that affect bone homeostasis adversely. In both sexes, these include a declining efficiency of calcium absorption, decreased osteoblast function, deficient PTH response to low calcium intakes, and a reduced response of 1-hydroxylase activity in the kidney to elevations of plasma PTH concentration [37]. In postmenopausal women, there is not only a decline in plasma levels of estrogen, which has antiresorptive properties [37], but also a reduced secretory response of the potent antiresorptive hormone, calcitonin, to elevations of plasma calcium concentration [38].

Vitamin D deficiency may contribute to bone loss in the elderly, especially in Europe, where food fortification with this vitamin is not practiced. Chronic vitamin D deficiency due to low intake from food and too little exposure to sunlight, in elderly French male and female outpatients, were associated with progressive decreases in serum 25-hydroxyvitamin D and calcium levels and increases in serum PTH concentration [39]. Belgian researchers observed lower serum 25-hydroxyvitamin D levels in hip fracture patients than in healthy elderly controls and determined that in the absence of adequate sunshine exposure, at least 300 IU/d of dietary vitamin D was necessary to maintain a normal serum 25 (OH)D level [40]. Similar observations have been made in Great Britain, with findings of reduced serum 25-hydroxyvitamin D and elevations of serum alkaline phosphatase, which is an index of bone loss [14,15,41]. Longstanding vitamin D deficiency may lead to osteomalacia, either in association with osteoporosis or as the predominent type of bone loss [41]. The two types of osteopenia are definitively distinguished by bone biopsy, but also may be differentially diagnosed by the physician through careful questioning and examination of the patient. In osteoporosis, bone pain occurs only in association with fractures, while in osteomalacia, bone pain tends to be diffuse and continuous and is more commonly felt in the limbs, limb girdles, or ribs than in the spine, and muscle weakness is common [41]. Osteomalacia may be treated with oral ergocalciferol, 50,000 IU/d, usually with good biochemical and clinical responses, including reduction of serum alkaline phosphatase levels, relief of pain, increased bone density, and recovery of muscle strength within a few weeks to a few months [41]. Adequate calcium must be present in the diet or given as supplements. Serum calcium should be regularly monitored to guard against hypercalcemia. Upon recovery, vitamin D supplementation should be reduced to physiologic levels.

Nordin et al. [42] recently reported a trial of vitamin D supplementation (15,000 IU orally once a week for 2 years) in 109 ambulant healthy English women aged 65–74 years. Subjects were randomly allocated to vitamin D or placebo, and the effect on cortical bone loss was evaluated by sequential radiographic morphometry of the metacarpals. No calcium supplements were given. Compared to the controls, the vitamin D-treated women had significantly raised plasma 25-hydroxyvitamin D levels and a significantly reduced rate of cortical bone loss. Sowers et al. [43] studied midradius bone density in 324 American women aged 55–80 years and found significant positive correlations with calcium and vitamin D intakes, as well as with extended use of estrogen supplements and thiazide diuretics; mean bone density was significantly greater in those whose calcium intake was greater than 800 mg/d, consumed concurrently with vitamin D in amounts greater than 400 IU/d, than in those whose intakes were below these figures.

Because of food fortification, the incidence and severity of vitamin D deficiency is not as great among the elderly in the United States and Canada as in Europe. However, Garry et al. [44] estimate from dietary surveys that the average dietary vitamin D intake in healthy U.S. elderly (Albuquerque, NM area) is 188 IU/d, with 33% consuming less than 100 IU/d. Inadequate consumption of vitamin D-fortified milk, because of lactose intolerance or simply dislike, together with lack of sufficient sunlight exposure, may often lead to a low vitamin D status in the elderly [16]. A further report by Omdahl et al. [45] of a 5-year prospective study with 166 women and 138 men in this area (mean age 72, range 60–93 years) describes findings of significantly lower plasma 25-OH-vitamin D levels in the elderly compared with young adults (mean level, 15.5 versus 29.1 ng/ml). Plasma alkaline phosphatase levels, an index of bone loss, were inversely related to plasma 25-hydroxyvitamin D concentrations. Those who took vitamin D supplements (usually 400 IU/d in multivitamin preparations) had higher 25-hydroxyvitamin D and lower alkaline phosphatase levels than those who did not take supplements. These investigators and others [46] with extensive clinical experience in bone metabolism currently recommend that total oral intake of vitamin D, dietary and supplemental, should be 600–800 IU/d in the elderly. Earlier, Baylink et al. [41] described osteoporosis patients with mild vitamin D deficiency who had low 25-hydroxyvitamin D levels, malabsorption of calcium, elevations of plasma PTH, and bone biopsy evidence of increased resorbing surfaces without osteomalacia; these abnormalities were corrected by oral administration of 1,000 IU of vitamin D daily. Chapuy et al. [48] observed that administration for 6 months of 1,000 mg/d of calcium and 800 IU/d of vitamin D to healthy elderly French men and women reduced biochemical signs of secondary hyperparathyroidism (elevated PTH and alkaline phosphatase).

At the Mayo Clinic, encouraging results have recently been obtained in postmenopausal osteoporosis patients by treatment with calcitriol (0.5–0.75 mcg/d orally in two or three doses) [49,50]. This group had previously reported that levels of endogenously produced calcitriol in postmenopausal osteoporotics were about 30% lower than in age-matched healthy control women [51]. Calcitriol treatment raised calcium absorption, which was measured initially and at 6-, 12-, and 24-month intervals, to normal and significantly decreased urinary excretion of hydroxyproline, indicating a reduced rate of bone resorption. In 13 of 36 patients receiving 0.75 mcg of calcitriol daily, the dosage had to be reduced to 0.50 mcg daily, and in 7 of 20 patients receiving 0.50 mcg/d, the dosage was reduced to less than 0.50 mcg/d; causes were mild hypercalcemia in seven patients, hypercalciuria in eight patients, and relative increases in urinary calcium excretion in relation to glomerular filtration rate in five patients. Use of calcitriol in treatment of osteoporosis obviously has great promise, but plasma and urinary calcium levels require regular monitoring, and further clinical experience in other medical centers is needed. Annual monitoring of endogenous calcitriol levels in the elderly patient has been recommended. As of this writing, nothing has yet been published regarding the effectiveness of combinations of calcitriol with calcium and/or estrogen supplements in treating osteoporosis.

Not only adequate vitamin D nutriture, but also a generous intake of calcium is considered necessary for maintenance of good bone health. The April 1984 Consensus Conference on Osteoporosis recommended that the calcium intake of postmenopausal women should be substantially greater than the present 800 mg/d RDA, namely 1,500 mg/d [52], and it is thought that an intake of 1,000–1,500 mg/d well before menopause may reduce the incidence of osteoporosis in later life [53]. Phosphate is also required, of course, but the level of phosphate in the American diet appears ample [54]. In addition, recent evidence indicates that low intakes of magnesium, zinc, protein, and vitamin C may contribute to bone loss in the postmenopausal woman [55].

Heaney and Recker [56] have shown that efficiency of calcium absorption is highly variable by individual among middle-aged postmenopausal women (mean age 53.8 ± 6.25 years). Fractional absorption data, obtained by a double-tracer method in 273 person-studies, revealed that in these subjects (receiving no exogenous estrogen) 55% had insufficient absorption to maintain calcium balance at an intake of 800 mg/d, and nearly one-fourth would still be in negative balance at an intake of 1,500 mg/d. For the very poor absorbers it would be of particular importance to employ therapeutic measures to increase calcium absorption; estrogen and calcitriol probably exert their beneficial effects at least in part by this mechanism [56].

In long-term studies in postmenopausal osteoporotics, the vertebral fracture rate was reduced by 50% in patients treated with calcium (1,000–1,500 mg/d as calcium carbonate) plus 400 IU/d supplemental vitamin D and by 78% in those treated with calcium (1,000–3,000 mg/d), cyclical conjugated estrogen (0.625–2.5 mg/d, mean 1.3 mg/d), and 400 IU of vitamin D; additional vitamin D (50,000 IU once or twice weekly) conferred no additional benefit [57]. Fluoride administration, added to the calcium-estrogen regimen, further improved results, but this drug at effective dosages has considerable side effects and has not been approved for osteoporosis treatment by the U.S. Food and Drug Administration [37]. High calcium intakes are considered safe except when hypercalcemia is present or there is a history of nephrolithiasis [58].

Vertebral fractures were reduced by 69% in osteoporosis patients treated with 0.50–0.75 mcg/d of calcitriol [48]. Francis and Peacock [59] have made the interesting observation that in ten postmenopausal osteoporotics taking 0.25 mcg oral calcitriol b.i.d. for 7 days, their calcium absorption increased more than in ten patients taking 40 mcg oral calcifediol once daily, although both groups had similar increases in plasma calcitriol. They conclude that a major effect of oral calcitriol is a direct local action on the gut to increase calcium absorption, so that calcium status can be improved without increasing an undesirable systemic action of calcitriol on bone resorption.

Renal Osteodystrophy

Bone loss is a typical sequel of renal failure, manifesting itself clinically as osteomalacia, osteitis fibrosa, osteosclerosis, and osteoporosis. In the individual patient, these lesions may exist alone or in combinations. Contributory factors include calcium malabsorption, vitamin D resistance, hypocalcemia, phosphate retention, metabolic acidosis, and hyperparathyroidism with skeletal resistance to the calcemic action of PTH [60]. The synthesis of vitamin D in ultraviolet-exposed skin of uremic (dialysis) patients appears markedly reduced in comparison with normal controls [61]. In laboratory animals, uremia impairs the absorption of dietary vitamin D [62]. Because of deficient 1-hydroxylase activity in the diseased kidney, plasma calcitriol levels are much reduced in severe chronic renal failure and in dialysis patients [60,63].

In progressive chronic renal insufficiency, Massry and his associates [64] find that treatment with 0.5 mcg/d of calcitriol (0.25 mcg b.i.d.) is safe and beneficial in normophosphatemic patients with glomerular filtration rates (GFR) of 30–50 ml/min. Smaller amounts may be adequate if creatinine clearance is greater than 60 ml/min, and larger amounts may be needed if GFR is less than 30 ml/min [64]. Clinical improvement is manifested by a reduction in bone pain in 1–3 weeks and complete relief in a few months to 18 months, as well as by recovery of proximal muscle strength [64,65]. Calcium supplementation is also used (1,000–1,600 mg/d of elemental Ca as $CaCO_3$), as well as aluminum-hydroxide phosphate binders to maintain serum phosphate at normal levels [60,66]. Calcitriol normalizes calcium absorption, reduces plasma alkaline phosphatase, hydroxyproline, and PTH levels, and improves bone histology [64,65,67–69].

Patients with nephrotic syndrome may display marked and prolonged proteinuria, which includes losses of calcifediol (25-hydroxyvitamin D) bound to calcifediol-binding protein. This may be associated with reduction in plasma calcitriol levels even when there is normal renal function, resulting in a reduction in plasma ionized calcium concentration, elevation of PTH, and enhanced bone resorption. Administration of calcifediol (usual dose range, 25–100 mcg/d) is recommended for replacement of plasma calcifediol levels in nephrotic syndrome [64].

Calcitriol administration at dosages of 0.5–3.5 mcg/d has provided substantial clinical and biochemical improvement in most patients requiring hemodialysis, although few experience a complete reversion of bone histology to normal [65]. Hewitt et al. [70] have reported their experience in treatment of 15 children and adolescents (age range 1.3–18.6 years at start of therapy) using dosages of 0.25–1.5mcg/d of calcitriol, together with calcium supplements in eight children (330–990 mg elemental Ca as Ca gluconate) and aluminum hydroxide with meals in all patients in dosages adjusted to maintain serum phosphate at levels normal for age (900 mg to 8 g aluminum hydroxide daily). After varying periods of observation (4–28 months), usually terminated by a kidney transplant operation, it was found that seven of nine patients with hyperparathyroid bone disease experienced improvement or complete resolution of lesions by radiographic examination, and two of nine showed deterioration. Rachitic lesions, seen in three patients initially, healed completely in two and improved in the third. There was a tendency in the older patients toward continued hyperphosphatemia and elevated PTH levels. Varghese et al. [60] have commented on the tendency among dialyzed patients to "escape" from control after prolonged treatment for their osteodystrophy, sometimes requiring subtotal or total parathyroidectomy [60].

A small percentage of dialysis patients develop severe osteomalacia with multiple fractures and severe myopathy without evidence of secondary hyperparathyroidism [71]. Bone aluminum is greatly elevated in these patients, possibly deriving from aluminum contamination in the dialysate medium or from intestinal absorption of aluminum from phosphate binders [71,72]. Some of these patients show modest clinical improvement with calcitriol therapy, with reduced pain and myopathy, but no improvement in bone histology [71,72]. Those with smaller concentrations of aluminum respond best [72].

Calcifediol has been used with good results in treating juvenile renal osteodystrophy in preadolescent children with renal insufficiency (GFR 30–60 ml/min/1.73 m^2) [66]. After 1 year, mean group growth velocity had increased into the normal range and remained there in the next 2 therapy years. Dosage regimen was 1–2 mcg/kg body weight of calcifediol once daily, oral calcium at 750–1,500 mg/d, aluminum hydroxide at 50–150 mg/kg/d to maintain serum phosphate at 4–6 mg/dl, and sodium bicarbonate as and when needed to raise serum bicarbonate to greater than 20 mmole/l. Bone histology was normalized in some and improved in others, sometimes not until after 2 years, while about one-third of the group did not show improvement despite improved growth rate [66].

Others have pointed out the utility of calcifediol and dihydrotachysterol in treating osteodystrophy in both renal insufficiency and dialysis patients [21,73].

Care must be used in adjusting the dosage of vitamin D and its hydroxy derivatives to the patient's individual need and condition, as all forms of vitamin D produce hypercalcemia, generally within 1 to 2 months after initiation of therapy [21]. Calcitriol has the therapeutic advantages over other forms of vitamin D of rapid onset of action and a short half-life (3–6 hours), so that if hypercalcemia develops during its use, a return to normal serum calcium concentration will ordinarily be accomplished within 2–5 days upon cessation of calcitriol dosage and of calcium supplementation, together with use of a low-calcium diet [65]. Serum calcium should be monitored twice a month while the patient is under long-term maintenance therapy. [65].

In progressive chronic renal failure, a somewhat increased rate of decline of kidney function has been reported when dosage of hydroxylated vitamin D derivatives was sufficiently high to cause a high frequency of hypercalcemia [74–76], but others find no adverse effect on renal function if episodes of hypercalcemia are minimized by careful dosage titration [48,77,78]. Phillips et al. [79] noted an improvement in renal function with administration of calcitriol or 1-hydroxyvitamin D in terms of significantly improved reabsorption of amino acids and reduction of hyperaminoaciduria, which is common in renal insufficiency.

The manufacturer of calcitriol (Rocaltrol®, Roche Laboratories) recommends initiating dosage with 0.25 mcg/d and increasing by 0.25 mcg increments at 4–8-week intervals if satisfactory improvement in biochemical parameters and clinical manifestations is not observed. During this titration period, they recommend serum calcium determinations twice a week. If hypercalcemia occurs, drug dosage should be discontinued until serum calcium level returns within normal limits, after which calcitriol administration may be resumed at a lower dosage [see current *Physicians' Desk Reference* (PDR®)].

Malabsorption

Dietary vitamin D is absorbed with other dietary lipids and requires the presence of bile and pancreatic lipases for efficient absorption, as well as a normally functioning intestinal tract [80]. If there is inadequate exposure to sunlight or ultraviolet light, vitamin D deficiency may develop, often accompanied by decreased bone density or clinically and radiologically apparent osteomalacia, in childhood cholestasis [81], chronic cholestatic liver disease [82], cystic fibrosis [83, 84], Crohn's disease [84,85], inflammatory bowel disease [84], ulcerative colitis [84], small intestinal villous atrophy [84], scleroderma [84], small-bowel resection [86], jejunoileal bypass [87], and subtotal or total gastrectomy [80,88]. Malabsorption extends to oral calcifediol as well as vitamin D [81,86], but Sitrin and Bengoa [89] report that in chronic cholestatic liver disease, oral calcifediol is better absorbed than vitamin D-3, particularly in those with severe cholestatsis (jaundice and steatorrhea). Vitamin D orally at 4,000–50,000 IU/d, given for several months, may normalize serum calcifediol levels in Crohn's disease patients with bone disease, but bone histology may not improve. In cystic fibrosis patients, 1,000 IU/d of vitamin D appears not to be sufficient for many [83]. The importance of adequate exposure to sunlight or to artificial ultraviolet during the winter season, of for housebound patients all through the year, cannot be overemphasized [19,83,88].

Alcoholism

Lower-than-normal circulating levels of calcifediol are found in many chronic alcoholics for a variety of reasons, such as poor diet, inadequate exposure to sunlight, malabsorption (pancreatic and liver disease), and an apparent increased rate of vitamin D metabolism. Hydroxylation of vitamin D in the liver seems to be adequate even in the presence of liver disease, so that plasma calcifediol levels rise normally with seasonal increases in sunlight. Chronic alcoholism is associated with extensive bone loss and increased incidence of bone fractures in relatively young and middle-aged fully ambulatory men, sex-age goups not ordinarily subject to such phenomena. Sunlight exposure and adequate oral vitamin D and calcium intakes may be helpful in reducing the incidence of osteoporosis associated with chronic excessive use of alcohol [53,89–94].

Hypoparathyroidism

In the past, massive oral doses of vitamin D (30,000–200,000 IU/d) were used to maintain normal serum calcium levels in patients with hypoparathyroidism, either postsurgical or idiopathic [95]. Availability of calcitriol enabled rapid restoration of normal plasma calcitriol concentrations and normal calcium absorption by administration of physiologic doses, not only in hypoparathyroidism but also in pseudohypoparathyroidism, a disorder in which endogenous circulating calcitriol levels are low because kidney 1-hydroxylase is insufficiently responsive to PTH [96]. Chan et al. [76] utilized maintenance dosages of 20–40 ng/kg/d of calcitriol orally in treating children with hypoparathyroidism and pseudohypoparathyroidism. In long-term treatment of middle-aged male hypoparathyroid patients with calcitriol, Bell and Stern [97] noted a tendency for serum calcitriol level to rise, accompanied by hypercalcemia. This was correctible by temporary cessation of dosage and resumption at a lower level after serum calcium had normalized.

Dihydrotachysterol has also been effectively used for treatment of hypoparathyroidism and pseudohypoparathyroidism [4].

Genetic Disease

Congenital disorders of vitamin D metabolism, manifesting clinically as rickets, that formerly could be treated only with massive doses of vitamin D, and then sometimes with only indifferent success, can now be managed by oral calcitriol. These disorders include X-linked renal hypophosphatemic rickets and vitamin D-dependent rickets types I and II.

In X-linked renal hypophosphatemic rickets, which has a dominant mode of hereditary transmission, there is a defect in renal tubule phosphate transport that results in severe phosphate wastage and hypophosphatemia. There is defective response of renal 1-hydroxylase activity to low plasma phosphate levels and to PTH stimulation, and there seems to be partial end-organ resistance to both calcitriol and PTH. Patients manifest rickets, bone deformities, short stature, dental abscesses, and osteomalacia. Response to oral calcitriol (dosage range 20–60 ng/kg/d usually in two divided doses) and oral phosphate have given good long-term results in most patients, with healing of rickets and enhanced linear growth [98,99]. As usual with vitamin D therapy, precautions must be taken against hypercalcemia. For details of clinical and radiologic features, histologic features, laboratory values, differential diagnosis, and therapy, consult the 1985 review by Chan et al. [99].

Vitamin D-dependent rickets type I is an autosomal recessively inherited disorder characterized by profound hypocalcemia and secondary hyperparathyroidism, despite which serum calcitriol levels are low. The basic abnormality is believed to be deficient renal 1-hydroxylase activity. Patients respond well to physiologic or slightly higher doses of calcitriol (0.25–2.0 mcg/d), with correction of osteomalacia in 9–10 months [10,100].

Vitamin D-dependent rickets type II patients have hypocalcemia in the presence of high circulating levels of calcitriol and do not respond to physiologic doses of calcitriol. Studies by Eil et al. [101] in cultured skin fibroblasts indicate that a normal content of calcitriol receptors is present in the cells of these patients, but there is defective uptake in the nucleus, where calcitriol, like other steroid hormones, is thought to act. The defect, presumably general throughout the body, manifests clinically in the skin by alopecia that appears toward the end of the first year of life. Correction of serum calcium levels, rickets, and alopecia is accomplished with doses of calcitriol 20–50 times higher than the physiologic dose range of 0.25–1.0 mcg/d [10,101].

Cochlear Deafness

Brookes [102] has reported observation of bilateral cochlear deafness associated with depressed serum calcifediol levels in ten patients aged 21–53 years (mean 44 years). Various causes of vitamin D deficiency in the group were poor diet, lack of sunlight exposure, malabsorption, alcoholism, and nephritis. Of three patients given 3,000–6,000 IU/d of oral vitamin D, two improved and one did not. Subsequently, patients with less severe vitamin D deficiency were being treated with 500–1,000 IU/d of vitamin D with oral calcium supplements. Further clinical observations are needed of this apparent association between hearing impairment and vitamin D deficiency.

Hypertension

Sowers et al. [103] investigated the relation of blood pressure to calcium and vitamin D intakes in 86 women aged 20–35 years and 222 women aged 55–80 years, none of whom were taking diuretics. In the younger group, there was a significant inverse relationship between estimated dietary vitamin D intake and systolic blood pressure, which remained significant after adjusting for age, Quetelet index (wt/ht^2), alcohol consumption, and calcium intake. Older women whose consumption of both vitamin D and calcium was below the RDA of 400 IU and 800 mg, respectively, had a significantly higher systolic blood pressure than their counterparts whose estimated intake met the RDA for at least one of the two nutrients. Although a number of reports have indicated that high-calcium diets or calcium supplements exert an antihypertensive effect [104], this appears to be the first report that the calcium-regulating hormonal action of vitamin D may also be important.

Hemolytic Disorders

Subnormal serum calcifediol levels should be looked for when there is hemolysis and iron accumulation. Aloia et al. [105] report low levels of both serum calcifediol and PTH in 13 patients with thalassemia major, confirming studies of others that measured calcifediol or PTH separately. Iron deposition appeared to decrease PTH reserve and renal 1-hydroxylase activity. Severe hypocalcemia may be present and can cause tetany. Treatment with vitamin D (50,000 IU 3 times a week) with 1,000 mg/d of calcium restored serum calcium to near normal in a trial in one patient, but in several others iron chelation therapy without vitamin D supplementation produced an impressive increase in serum calcifediol concentration.

Psoriasis

Reports by Morimoto and coworkers [106,107] suggest potential usefulness of topical calcitriol in treatment of psoriasis vulgaris. The ointment they used contained 0.5 mcg of calcitriol per g of base. Of 24 patients treated, 21 responded well after 2–5 weeks of treatment.

Cancer

Abe and coworkers have determined that calcitriol at physiologic concentrations in vitro induces differentiation of mouse myeloid leukemia cells into macrophages [108] and of human myeloid leukemia cells into granulocytes [109]. In vivo, they found that injection of calcitriol intraperitoneally, or of 1-hydroxyvitamin D-3, which provides a longer and steadier maintenance of plasma calcitriol levels, significantly prolongs the survival time of mice inoculated with myeloid leukemia cells [110]; dosages were not high enough to provoke hypercalcemia. Provvedini et al. [111] have detected a calcitriol receptor molecule in peripheral mononuclear leukocytes and obtained evidence, upon culture of normal human monocytes for 4 weeks in the presence of 10^{-8}M calcitriol, of their maturation to macrophages. Frampton et al. [112] report findings that calcitriol, at concentrations somewhat above physiologic levels, inhibits replication of T-47D human breast cancer cells and human malignant melanoma MM96 cells by 30–50 percent in vitro; calcitriol stimulated their growth at lower concentrations. However, the two known metabolites of calcitriol, 1,24,25-trihydroxyvitamin D-3 and 1,25,26-trihydroxyvitamin D-3, which compete for binding to calcitriol receptors, did not stimulate but were almost equipotent with calcitriol in inhibiting the replication of both cell lines [112]. Calcitriol and its metabolites, therefore, may play an antiproliferative, prodifferentiative role in maturation of certain cell types and, potentially at least, could serve as antitumor agents in vivo [112].

Epidemiologic support for an anticancer function of vitamin D, possibly in association with calcium, is found in a recent paper by Garland et al. [113] reporting the 19-year risk of colorectal cancer in 1,954 American men who had completed detailed 28-day dietary histories during the period 1957–1959 as part of the Western Electric Health Study. Risk of colorectal cancer was inversely related to dietary intakes of vitamin D and calcium. In the quartiles of a combined index of dietary vitamin D and calcium intakes, from lowest to highest, the observed risks per 1,000 population were 38.9, 24.5, 22.5, and 14.3. In the quartile with the lowest risk, vitamin D intakes were close to or above the U.S. RDA, at 75–208 IU/1,000 kcal, and calcium intakes were at or above the RDA, at 384–906 mg/1,000 kcal. These observed risk associations remained significant after adjustment for age, daily cigarette consumption, body mass index, ethanol consumption, and percentage of calories obtained from fat.

It appears, therefore, that dietary recommendations for the protection of bone mass during aging, namely RDA intakes or higher for both vitamin D and calcium, may also have the salutary effect of reducing risk of colorectal and perhaps other types of cancer.

TOXICITY AND SIDE EFFECTS

The first signs of toxicity of any of the forms of vitamin D are hypercalcemia and hypercalciuria; these have been referred to at several places in the preceding section on Clinical Studies. Accompanying the hypercalcemia that is induced by excessive calcitriol may be a deterioration of renal function, particularly in patients with renal insufficiency [114]. Careful control of the dosage of calcitriol to minimize the number of occurrences of hypercalcemia during long-term therapy prevents any significant decline in renal function [48,77,78]. Experienced clinicians advise that all patients being treated with calcitriol must undergo regular sequential analysis of serum and urine calcium and creatinine levels [48,76,114]. Upon development of hypercalcemia, administration of calcitriol and calcium supplements must be suspended until serum calcium concentration falls below the upper limit of normal (10.1 mg/dl).

Hypercalcemia is especially hazardous to patients receiving digitalis because it enhances the toxicity of cardiac glycosides [4].

Severe vitamin D toxicity is usually observed, not with prescribed drugs such as calcifediol, calcitriol, and dihydrotachysterol, but with vitamin D itself, since it is available over the counter and may be used without a physician's supervision. Paterson [115] has reviewed 21 cases of vitamin D poisoning occurring in Scotland in patients manifesting hypercalcemia and such symptoms as anorexia, nausea or vomiting, weight loss, and headache, or mental sypmtoms such as apathy, fatigue, or confusion. All took milligram quantities (40,000 IU/d and up). In most cases, the vitamin D therapy had been prescribed by a physician initially, in several cases at too high a level, but the critical factor in bringing on toxicity was the patient's continuance of vitamin D therapy without physician follow-up. Two of the 21 patients died of their intoxication.

Other signs and sequelae of severe vitamin D intoxication include renal impairment, with polyuria, polydipsia, nocturia, decreased concentrating ability, and proteinuria; calcium deposits in the kidney, lungs, cardiovascular system, and skin; and localized or generalized osteoporosis [4].

Treatment of severe vitamin D poisoning involves the usual measures of withdrawing the vitamin and any calcium supplements and instituting a low-calcium diet [4]. Because of the long half-life of vitamin D in the body and the length of time required to reverse its toxicity [21], a glucocorticoid such as prednisone is administered [4]. Glucocorticoids appear to decrease renal synthesis of calcitriol [117], leading to reduced resorption of bone mineral and a prompt fall in serum calcium level [4,117]. Renal function may improve markedly if damage has not been too great, and abnormal calcium deposits may be slowly mobilized [4].

Although there is some controversy as to what the total oral vitamin D intake, including dietary and supplemental sources, of the normal healthy individual should be [118], official bodies have settled upon the range of 200–600 IU/d as being safe and effective in preventing vitamin D deficiency, the lowest amount being for infants and the highest for pregnant and lactating women [13]. Some clinicians advocate 600–800 IU/d for older people [44–46]. The margin of safety for vitamin D in long-term administration can be narrow for certain individuals who are unusually sensitive to its calcemic action. Thus, in a 6-month study of vitamin D supplementation at 2,000 IU/d in elderly English individuals, Johnson et al. [119] observed that 2 of 73 persons receiving this dosage developed hypercalcemia; whether sunlight exposure may have contributed more than usual to the vitamin D supplies of these individuals was not known. Sunlight exposure of itself is not known to lead to vitamin D toxicity, but of course excessive sun exposure is not desirable in terms of skin cancer risk.

Administration of vitamin D above RDA levels to pregnant women is contraindicated. Maternal hypercalcemia may contribute to supravalvular aortic stenosis in the fetus and to suppression of parathyroid function in the newborn, with resultant hypocalcemia, tetany, and seizures [4].

INTERACTIONS

Anticonvulsant drugs appear to interfere with vitamin D and calcium metabolism in some manner not well understood; serum calcitriol levels are normal, while calcifediol and calcium levels are depressed. Rickets or osteomalacia may result, usually in patients receiving high doses of anticonvulsants for a long period [120]. Phenytoin in high doses depresses plasma calcifediol levels in a dose-dependent manner [121]. During long-term, low-dosage therapy with anticonvulsants, slight skeletal changes as demonstrated by radiology may be seen during the first few years, later reverting to normal [120]; in such patients, vitamin D intakes at RDA levels and/or good sunlight exposure can maintain good bone nutrition [122-124]. If a definite diagnosis of rickets is established, treatment is the same as for nutritional rickets, i.e., supplemental calcium and vitamin D at 1,000-4,000 IU/d until hypocalcemia and elevated alkaline phosphatase are corrected, after which lower supplemental doses may be sufficient [4].

Combination of anticonvulsant therapy with a ketogenic diet for management of intractable epilepsy has been reported to enhance bone mineral loss [125]. In a group of five children with bone loss on such combined therapy, administration of 5,000 IU/d of vitamin D increased bone mass by a mean 8.1% in 12 months. The investigators recommend vitamin D treatment at this level until serum calcium, PTH, and alkaline phosphatase concentrations are normalized, with monitoring of biochemical indices every 3 months. At that point, chronic maintenance vitamin D supplementation at 1,000-2,000 IU/d can be initiated.

Glucocorticoid-induced osteopenia appears to be related to dose-dependent depression by these drugs of renal synthesis of calcitriol [116]. In addition, laboratory studies show an inhibition of bone apposition by hydrocortisone in adult male rats, which can be corrected by high-dose, intraperitoneally injected vitamin D [126]. Therapy with oral calcitriol deserves clinical investigation in patients receiving glucocorticoids long-term.

Aarskog et al. [127] reported a heparin-induced case of osteopenia, with back pain and spinal osteoporosis, in a 22-year-old woman who, because of thrombotic problems during pregnancy, was given regular heparin injections for 8 weeks before and 4 weeks after term. Heparin apparently increased bone resorption, raising plasma calcium levels, which in turn lowered PTH levels and renal calcitriol synthesis. These investigators suggest replacement therapy with calcitriol when heparin is used long-term.

Abbasi et al. [128] observed development of hypercalcemia in 22 of 79 patients undergoing therapy for pulmonary tuberculosis. Although normocalcemic on admission, they developed hypercalcemia within 4–16 weeks, possibly because of increased sensitivity to the calcemic action of vitamin D. Serum calcium levels were positively correlated with vitamin D intakes, which ranged from 400 to 1,200 IU/d. As the disease abated, serum calcium levels usually returned to normal even when vitamin D intakes did not change. The investigators recommend caution in use of vitamin D supplements or vitamin D-fortified milk during treatment of active pulmonary tuberculosis.

Rosen et al. [129] report that in lead-poisoned children, serum calcitriol is depressed to levels comparable with those seen in metabolic bone disease, while calcifediol levels remain normal, suggesting that lead impairs renal 1-hydroxylase activity. Upon chelation therapy to remove lead, serum calcitriol levels rose to normal. Mahaffey et al. [130] confirm a negative correlation between circulating lead concentrations and calcitriol levels. Earlier studies have indicated that a generous calcium intake, together with adequate vitamin D, diminishes lead absorption [131,132].

REFERENCES

1. Lawson E: Vitamin D. In: *Fat-Soluble Vitamins*, AT Diplock, (ed). Lancaster, PA: Technomic Publishing Co., 1985, pp 76–153.
2. Norman AW, Miller BE: Vitamin D. In: *Handbook of Vitamins*, LJ Machlin (ed). New York: Marcel Dekker, 1984, pp 45–97.
3. DeLuca HF: Vitamin D. In: *Modern Nutrition in Health and Disease*, RS Goodhart, ME Shils (eds). Philadelphia: Lea & Febiger, 1980, pp 160–170.
4. Haynes RC Jr, Murad F: Agents affecting calcium, parathyroid hormone, calcitonin, vitamin D, and other compounds. In: *Goodman and Gilman's The Pharmacological Basis of Therapeutics*, 7th ed., LS Goodman, TW Rall, F Murad (eds). New York: Macmillan; 1985, pp 1517–1543.
5. Adams JS, Gacad MA: Characterization of 1-alpha-hydroxylation of vitamin D-3 sterols by cultured alveolar macrophages from patients with sarcoidosis. J Exp Med 161:755–765, 1985.

6. Koeffler HP, Reichel H, Bishop JE, Norman AW: Gamma-interferon stimulates production of 1,25-dihydroxyvitamin D-3 by normal human macrophages. Biochem Biophys Res Commun 127:596–603, 1985.
7. Rook GAW, Steele J, Fraher L, et al.: Vitamin D-3, gamma-interferon, and control of *Mycobacterium tuberculosis* by human monocytes. Immunology 57:159–163, 1986.
8. Tsoukas CD, Provvedini DM, Manolagas SC: 1,25-Dihydroxyvitamin D-3: A novel immunoregulatory hormone. Science 224:1438–1440, 1984.
9. Toss G, Symreng T: Delayed hypersensitivity response and vitamin D deficiency. Int J Vitam Nutr Res 53:27–31, 1983.
10. Rosen JF Chesney RW: Circulating calcitriol concentrations in health and disease. J Pediatr 103:1–17, 1983.
11. Bilezikian JP, Canfield RE, Jacobs TP, et al.: Response of 1,25-dihydroxyvitamin D-3 to hypocalcemia in human subjects. N Engl J Med 299:437–441, 1978.
12. Adams ND, Gray RW, Lemann J Jr: The effects of oral $CaCO_3$ loading and dietary calcium deprivation on plasma 1,25-dihydroxyvitamin D concentrations in healthy adults. J Clin Endocrinol Metab 48:1008–1016, 1979.
13. Food and Nutrition Board: *Recommended Dietary Allowances*, 9th ed. Washington, DC: National Academy of Sciences, 1980.
14. McKenna MJ, Freaney R, Meade A, Muldowney FP: Hypovitaminosis D and elevated serum alkaline phosphatase in elderly Irish people. Am J Clin Nutr 41:101–109, 1985.
15. Newton HMV, Sheltawy M, Hay AWM, Morgan B: The relations between vitamin D-2 and D-3 in the diet and plasma $25OHD_2$ and $25OHD_3$ in elderly women in Great Britain. Am J Clin Nutr 41: 760–764, 1985.
16. Holick MF: Photosynthesis of vitamin D in the skin: Effect of environmental and life-style variables. Fed Proc 46:1876–1882, 1987.
17. Holick MF, MacLaughlin JA, Clark MB, et al.: Photosynthesis of previtamin D-3 in human skin and the physiologic consequences. Science 210:203–305, 1980.
18. Holick MF, MacLaughlin JA, Doppelt SH. Regulation of cutaneous previtamin D_3 photosynthesis in man: Skin pigment is not an essential regulator. Science 211:590–593, 1981.
19. Fraser DR: The physiological economy of vitamin D. Lancet 1:969–972, 1983.
20. Gascon-Barré M: Is there any physiological significance to the enterohepatic circulation of vitamin D sterols? J Am Coll Nutr 5:317–324, 1986.
21. Haussler MR, Cordy PE: Metabolites and analogues of vitamin D. Which for what? JAMA 247:841–844, 1982.
22. Delvin EE, Glorieux FH, Salle BL, et al.: Control of vitamin D metabolism in preterm infants: Feto-maternal relationships. Arch Dis Child 57:754–757, 1982.
23. Glorieux FH, Salle BL, Delvin EE, David L: Vitamin D metabolism in preterm infants: Serum calcitriol values during the first five days of life. J Pediatr 99:640–643, 1981.
24. Anon.: Vitamin D metabolism in the rickets of very-low-birth-weight, premature infants. Nutr Rev 39:234–236, 1981.
25. Steichen JJ, Tsang RC, Gratton TL, et al.: Vitamin D homeostasis in the perinatal period: 1,25-dihydroxyvitamin D in maternal, cord, and neonatal blood. N Engl J Med 302:315–319, 1980.
26. Steichen JJ, Tsang RC, Greer FR, et al.: Elevated serum 1,25-dihydroxyvitamin D concentrations in rickets of very low-birth-weight infants. J Pediatr 99:293–298, 1981.
27. Callenbach JC, Sheehan MB, Abramson SJ, Hall RT: Etiologic factors in rickets of very low-birth-weight infants. J Pediatr 98:800–805, 1981.
28. Hillman LS, Hoff N, Salmons S, et al.: Mineral homeostasis in very premature infants:

Serial evaluation of serum 25-hydroxyvitamin D, serum minerals, and bone mineralization. J Pediatr 106:970–980, 1985.
29. Greer FR, Tsang RC, Searcy JE, et al.: Mineral homeostasis during lactation—relationship to serum 1,25-dihydroxyvitamin D, 25-hydroxyvitamin D, parathyroid hormone, and calcitonin. Am J Clin Nutr 36:431–437, 1982.
30. Hollis BW, Roos BA, Draper HH, Lambert PW. Occurrence of vitamin D sulfate in human milk whey. J Nutr 111:384–390, 1981a.
31. Greer FR, Searcy JE, Levin RS, et al.: Bone mineral content and serum 25-hydroxyvitamin D concentration in breast-fed infants with and without supplemental vitamin D. J Pediatr 98:696–701, 1981.
32. Roberts CC, Chan GM, Folland D, et al.: Adequate bone mineralization in breast-fed infants. J Pediatr 99:192–196, 1981.
33. Chan GM, Roberts CC, Folland D, Jackson R: Growth and bone mineralization and the effects of lactation on maternal bone mineral status. Am J Clin Nutr 36:438–443, 1982.
34. Bachrach S, Fisher J, Parks JS: An outbreak of vitamin D deficiency rickets in a susceptible population. Pediatrics 64:871–877, 1979.
35. Edidin DV, Levitsky LL, Schey W, et al.: Resurgence of nutritional rickets associated with breast-feeding and special dietary practices. Pediatrics 65:232–235, 1980.
36. Rudolph M, Arulanantham K, Greenstein RM: Unsuspected nutritional rickets. Pediatrics 66:72–76, 1980.
37. Riggs BL, Melton JL: Involutional osteoporosis. N Engl J Med 314:1676–1686, 1986.
38. Taggart HM, Chesnut CH, Ivey JL, et al.: Deficient calcitonin response to calcium stimulation in post menopausal osteoporosis? Lancet 1:475–478, 1982.
39. Chapuy MC, Durr F, Chapuy P: Age-related changes in parathyroid hormone and 25-hydroxycholecalciferol levels. J Gerontol 38:19–22, 1983.
40. Lips P, Van Ginkel FC, Jongen MJM, et al.: Determinants of vitamin D status in patients with hip fracture and in elderly control subjects. Am J Clin Nutr 46:1005–1010, 1987.
41. Anwar M: Nutritional hypovitaminosis-D and the genesis of osteomalacia in the elderly. J Am Geriatr Soc 26:309–317, 1978.
42. Nordin BEC, Baker MR, Horsman A, Peacock M: A prospective trial of the effect of vitamin D supplementation on metacarpal bone loss in elderly women. Am J Clin Nutr 42:470–474, 1985.
43. Sowers MFR, Wallace RB, Lemke JH: Correlates of midradius bone density among postmenopausal women: A community study. Am J Clin Nutr 41:1045–1053, 1985.
44. Garry PJ, Goodwin JS, Hunt WC, et al.: Nutritional status in a healthy elderly population: Dietary and supplemental intakes. Am J Clin Nutr 36:319–331, 1982.
45. Omdahl JL, Garry PJ, Hunsaker LA, et al.: Nutritional status in a healthy elderly population: Vitamin D. Am J Clin Nutr 36:1225–1233, 1982.
46. Parfitt AM, Gallagher JC, Heaney RP, et al.: Vitamin D and bone health in the elderly. Am J Clin Nutr 36:1014–1031, 1982.
47. Baylink D, Maloney N, Morey E, et al.: Mild vitamin D deficiency ($-$D): A cause of osteoporosis in the elderly. Gerontologist 17:39, 1977.
48. Chapuy M-C, Chapuy P, Meunier PJ: Calcium and vitamin D supplements: Effects on calcium metabolism in elderly people. Am J Clin Nutr 46:324–328, 1987.
49. Gallagher JC, Jerpbak CM, Jee WSS, et al.: 1,25-Dihydroxyvitamin D-3: Short- and long-term effects on bone and calcium metabolism in patients with postmenopausal osteoporosis. Proc Natl Acad Sci USA 79:3325–3329, 1982.
50. Riggs BL, Nelson KI: Effect of long term treatment with calcitriol on calcium absorption

and mineral metabolism in postmenopausal osteoporosis. J Clin Endocrinol Metab 61:457–461, 1985.
51. Gallagher JC, Riggs BL, Eisman J, et al.: Intestinal calcium absorption and serum vitamin D metabolites in normal subjects and osteoporotic patients: Effect of age and dietary calcium. J Clin Invest 64:729, 1979.
52. Office of Medical Applications of Research, NIH: Consensus Conference—Osteoporosis. JAMA 252:799–802, 1984.
53. Spencer H, Kramer L: NIH Consensus Conference: Osteoporosis. Factors contributing to osteoporosis. J Nutr 116:316–319, 1986.
54. Draper HH, Scythes CA: Calcium, phosphorus, and osteoporosis. Fed Proc 40:2434–2438, 1981.
55. Freudenheim JL, Johnson NE, Smith EL: Relationships between usual nutrient intake and bone-mineral content of women 35–65 years of age: Longitudinal and cross-sectional analysis. Am J Clin Nutr 44:863–876, 1986.
56. Heaney RP, Recker RR: Distribution of calcium absorption in middle-aged women. Am J Clin Nutr 43:299–305, 1986.
57. Riggs BL, Seeman E, Hodgson SF, et al.: Effect of the fluoride/calcium regimen on vertebral fracture occurrence in postmenopausal osteoporosis. N Engl J Med 306:446–450, 1982.
58. Heath H III, Calloway CW: Calcium tablets for hypertension? Ann Intern Med 103:946–947, 1985.
59. Francis RM, Peacock M: Local action of 1,25-dihydroxycholecalciferol on calcium absorption in osteoporosis. Am J Clin Nutr 46:315–318, 1987.
60. Varghese Z, Farrington K, Moorhead JF: Renal osteodystrophy: Dietary influences and management. Proc Nutr Soc 38:337–350, 1979.
61. Jacob AI, Sallman A, Santiz Z, Hollis BW: Defective photoproduction of cholecalciferol in normal and uremic humans. J Nutr 114:1313–1319, 1984.
62. Vaziri ND, Hollander D, Hung EK, et al.: Impaired intestinal absorption of vitamin D-3 in azotemic rats. Am J Clin Nutr 37:403–406, 1983.
63. Chesney RW, Rosen JF, Hamstra AJ, DeLuca HF: Serum 1,25-dihydroxyvitamin D levels in normal children and in vitamin D disorders. Am J Dis Child 134:135–139, 1980.
64. Massry SG: Requirements of vitamin D metabolites in patients with renal disease. Am J Clin Nutr 33:1530–1535, 1980.
65. Voigts AL, Felsenfeld AJ, Llach F:The effects of calciferol and its metabolites on patients with chronic renal failure. II. Calcitriol, 1 alphahydroxyvitamin D-3, and 24,25-dihydroxyvitamin D-3. Arch Intern Med 143:1205–1211, 1983.
66. Langman CB, Mazur AT, Baron R, Norman ME: 25-Hydroxyvitamin D-3 (calcifediol) therapy of juvenile renal osteodystrophy: Beneficial effect on linear growth velocity. J Pediatr 100:815–820, 1982.
67. Ahmed KY, Wills MR, Varghese Z, et al.: Long-term effects of small doses of 1,25-dihydroxycholecalciferol in renal osteodystrophy. Lancet 1:629–632, 1978.
68. Goldstein DA, Malluche HM, Massry Sg: Management of renal osteodystrophy with 1,25(OH)$_2$D$_3$. I. Effects on clinical, radiographic and biochemical parameters. Min Electr Metab 2:35–47, 1979.
69. Malluche HH, Goldstein DA, Massry SG: Management of renal osteodystrophy with 1,25 (OH)$_2$D$_3$. II. Effects on histopathology of bone: Evidence for healing of osteomalacia. Min Electr Metab 2:48–55, 1979.
70. Hewitt IK, Stefanidis C, Reilly BJ, et al.: Renal osteodystrophy in children undergoing continuous ambulatory peritoneal dialysis. J Pediatr 103:729–734, 1983.

71. Hodsman AB, Sherrard DJ, Wong EGC, et al.: Vitamin D-resistant osteomalacia in hemodialysis patients lacking secondary hyperparathyroidism. Ann Intern Med 94:629–637, 1981.
72. Ott SM, Maloney NA, Coburn JW, et al.: The prevalence of bone aluminum deposition in renal osteodystrophy and its relation to the response to calcitriol therapy. N Engl J Med 307:709–713, 1982.
73. Voigts AL, Felsenfeld AJ, Llach F: The effects of calciferol and its metabolites on patients with chronic renal failure. I. Calciferol, dihydrotachysterol, and calcifediol. Arch Intern Med 143:960–963, 1983.
74. Tougaard L, Sorenson E, Brochner-Mortenson J, et al.: Controlled trial of α-hydroxycholecalciferol in chronic renal failure. Lancet 1:1044–1047, 1976.
75. Christiansen C, Rodbro P, Christensen S, et al.: Deterioration of renal function during treatment of chronic renal failure with 1,25-dihydroxy-cholecalciferol. Lancet 2:700–703, 1978.
76. Chan JCM, Young RB, Alon U, Mamunes P: Hypercalcemia in children with disorders of calcium and phosphate metabolism during long-term treatment with 1,25-dihydroxyvitamin D-3. Pediatrics 72:225–233, 1983.
77. Massry SG, Goldstein DA: Is calcitriol [1,25-$(OH)_2D_3$] harmful to renal function? JAMA 242:1875–1876, 1979.
78. Healy MD, Malluche HH, Goldstein DA, et al.: Effects of long-term therapy with calcitriol in patients with moderate renal failure. Arch Intern Med 140:1030–1033, 1980.
79. Phillips ME, Havard J, Otterud B: Aminoaciduria in chronic renal failure—its relationship to vitamins D and parathyroid status. Am J Clin Nutr 33:1541–1545, 1980.
80. Dawson-Hughes B: Osteoporosis and aging: Gastrointestinal aspects. J Am Col Nutr 5:393–398, 1986.
81. Sokol RJ, Farrell MK, Heabi JE, et al.: Comparison of vitamin E and 25-hydroxyvitamin D absorption during childhood cholestasis. J Pediatr 103:712–717, 1983.
82. Long RG, Varghese Z, Meinhard EA, et al.: Parenteral 1,25-dihydroxycholecalciferol in hepatic osteomalacia. Br Med J 1:75–77, 1978.
83. Reiter EO, Brugman SM, Pike JW, et al.: Vitamin D metabolites in adolescents and young adults with cystic fibrosis: Effects of sun and season. J Pediatr 106:21–26, 1985.
84. Lo CW, Paris PW, Clemens TL, et al.: Vitamin D absorption in healthy subjects and in patients with intestinal malabsorption syndromes. Am J Clin Nutr 42:644–649, 1985.
85. Driscoll RH, Meredith SC, Sitrin M, Rosenberg IH: Vitamin D deficiency and bone disease in patients with Crohn's disease. Gastroenterology 83:1252–1258, 1982.
86. Compston JE, Creamer B: Plasma levels and intestinal absorption of 25-hydroxyvitamin D in patients with small bowel resection. Gut 18:171–175, 1977.
87. Compston JE, Laker MF, Woodhead JS, et al.: Bone disease after jejunoileal bypass for obesity. Lancet 2:1–4, 1978.
88. Pittet PG, Davie M, Lawson DEM: Role of nutrition in the development of osteomalacia in the elderly. Nutr Metab 23:109–116, 1979.
89. Bonjour JP: Vitamins and alcoholism. X. Vitamin D. Int J Vitam Nutr Res 51:307–313, 1981.
90. Sitrin MD, Bengoa JM: Intestinal absorption of cholecalciferol and 25-hydroxycholecalciferol in chronic cholestatic liver disease. Am J Clin Nutr 46:1011–1015, 1987.
91. Saville PD: Changes in bone mass with age and alcoholism. J Bone Joint Surg 3:492–499, 1965.
92. Gascon-Barré M: Influence of chronic ethanol comsumption on the metabolism and action of vitamin D. J Am Coll Nutr 4:565–574, 1985.
93. Seeman E, Melton LJ, O'Fallon MW, et al.: Risk factors for spinal osteoporosis in men. Am J Med 75:977–983, 1983.

94. Bjorneboe G-EA, Johnsen J, Bjorneboe A, et al.: Effect of alcohol consumption on serum concentration of 25-hydroxyvitamin D-3, retinol, and retinol-binding protein. Am J Clin Nutr 44:678–682, 1986.
95. Parfitt AM: Idiopathic, surgical and other varieties of parathyroid hormone-deficient hypoparathyroidism. In: *Metabolic Basis of Endocrinology*, L DeGroot, (ed). New York, Grune & Stratton, 1979.
96. Markowitz ME, Rosen JF, Smith C, DeLuca HF: 1,25-Dihydroxyvitamin D-3-treated hypoparathyroidism: 35 patient-years in 10 children. J Clin Endocrinol Metab 55:727–733, 1982.
97. Bell NH, Stern PH: Hypercalcemia and increases in serum hormone value during prolonged administration of 1,25-dihydroxyvitamin D. N Engl J Med 298:1241–1243, 1978.
98. Chesney RW, Mazess RB, Rose P, et al.: Long-term influence of calcitriol (1,25-dihydroxyvitamin D) and supplemental phosphate in X-linked hypophosphatemic rickets. Pediatrics 71:559–567, 1983.
99. Chan JCM, Alon U, Hirschman GM: Renal hypophosphatemic rickets. J Pediatr 106:533–544, 1985.
100. Delvin EE, Glorieux FH, Maric PJ, Pettifor JM: Vitamin D dependency: Replacement therapy with calcitriol. J Pediatr 99:26–34, 1981.
101. Eil C, Liberman UA, Rosen JF, Marx SJ: A cellular defect in hereditary vitamin D-dependent rickets type II: Defective nuclear uptake of 1,25-dihydroxy-vitamin D in cultured skin fibroblasts. N Engl J Med 304:1588–1591, 1981.
102. Brookes GB: Vitamin D deficiency—a new cause of cochlear deafness. J Laryngol Otol 97:405–420, 1983.
103. Sowers MFR, Wallace RB, Lemke JH: The association of intakes of vitamin D and calcium with blood pressure among women. Am J Clin Nutr 42:135–142, 1985.
104. Weinsier RL, Norris D: Recent developments in the etiology and treatment of hypertension: Calcium, fat, and magnesium. Am J Clin Nutr 42:1331–1338, 1985.
105. Aloia JF, Ostuni JA, Yeh JK, Zaino EC: Combined vitamin D parathyroid defect in thalassemia major. Arch Intern Med 142:831–832, 1982.
106. Morimoto S, Yoshikawa K, Kozuka T, et al.: An open study of vitamin D-3 treatment in psoriasis vulgaris. Br J Dermatol 115:421–429, 1986.
107. Morimoto S, Onishi T, Imanaka S, et al.: Topical administration of 1,25-dihydroxyvitamin D-3 for psoriasis: Report of five cases. Calcif Tissue Int 38:119–122, 1986.
108. Abe E, Miyoura C, Sakagami H, et al.: Differentiation of mouse myeloid leukemia cells induced by 1 alpha, 25-dihydroxyvitamin D-3. Proc Natl Acad Sci USA 78:4990–4994, 1981.
109. Miyoura C, Abe E, Kuribashi T, et al.: 1 alpha, 25-Dihydroxyvitamin D-3 induces differentiation of human myeloid leukemia cells. Biochem Biophys Res Commun 102:937–943, 1981.
110. Honma Y, Hozumi M, Abe E, et al.: 1 alpha, 25-Dihydroxyvitamin D-3 and 1 alpha-hydroxyvitamin D-3 prolong survival time of mice inoculated with myeloid leukemia cells. Proc Natl Acad Sci USA 80:201–204, 1983.
111. Provvedini DM, Tsoukas CD, Deftos LJ, Manolagas SC: 1,25-Dihydroxyvitamin D-3 receptors in human leukocytes. Science 221:1181–1183, 1983.
112. Frampton FJ, Omond SA, Eisman JA: Inhibition of human cancer cell growth by 1,25-dihydroxyvitamin D-3 metabolites. Cancer Res 43:4443–4447, 1983.
113. Garland C, Shekelle RB, Barrett-Connor, E et al.: Dietary vitamin D and calcium and risk of colorectal cancer: A 19-year prospective study in man. Lancet 1:307–309, 1985.

114. Nielson HE, Christensen MS, Romer FK, Hansen H: Vitamin D analogues and renal function. Lancet 2:1259–1260, 1978.
115. Paterson CR: Vitamin-D poisoning: Survey of causes in 21 patients with hypercalcemia. Lancet 1:1164–1165, 1980.
116. Chesney RW, Hamstra AJ, Mazess RB, et al.: Reduction of serum 1,25-dihydroxyvitamin D-3 in children receiving glucocorticoids. Lancet 2:1123–1125, 1978.
117. Streck WF, Waterhouse C, Haddad JG: Glucocorticoid effects in vitamin D intoxication. Arch Intern Med 139:974–977, 1979.
118. Holmes RP, Kummerow FA: The relationship of adequate and excessive intake of vitamin D to health and disease. J Am Coll Nutr 2:173–199, 1983.
119. Johnson KR, Jobber J, Stonawski BJ: Prophylactic vitamin D in the elderly. Age Ageing 9:121–127, 1980.
120. Krause KH, Berlit P, Bonjour JP, et al.: Vitamin status in patients on chronic anticonvulsant therapy. Int J Vitam Nutr Res 52:375–385, 1982.
121. Gascon-Barré M, Villeneuve JP, Lebrun LH: Effect of increasing doses of phenytoin on the plasma 25-hydroxyvitamin D concentrations. J Am Coll Nutr 3:45–50, 1984.
122. Morijiri Y, Sato T: Factors causing rickets in institutionalised handicapped children on anticonvulsant therapy. Arch Dis Child 56:446–449, 1981.
123. Enneking-Ivey O, Bailey LB, Gawley L, et al.: Folic acid and vitamin D status of young children receiving minimal anticonvulsant drug therapy. Int J Vitam Nutr Res 51:349–352, 1981.
124. Marya RK, Khattar VP, Bansal RK: Occult anticonvulsant osteomalacia in North India. Nutr Metab 23:167–171, 1979.
125. Hahn RJ, Halstead LR, DeVivo DC: Disordered mineral metabolism produced by ketogenic diet therapy. Calcif Tissue Int 28:17–22, 1979.
126. Tam CS, Wilson DR, Hitchman AJ, Harrison JE: Protective effect of vitamin D_2 on bone apposition from the inhibitory action of hydrocortisone in rats. Calcif Tissue Int 33:167–172, 1981.
127. Aarskog D, Aksnes L, Lehmann V: Low 1,25-dihydroxy-vitamin D in heparin-induced osteopenia. Lancet 2: 650–651, 1980.
128. Abbasi AA, Chemplavil JK, Farah S, et al.: Hypercalcemia in active pulmonary tuberculosis. Ann Intern Med 90:324–328, 1979.
129. Rosen JF, Chesney RW, Hamstra A, et al.: Reduction in 1,25-dihydroxyvitamin D in children with increased lead absorption. N Engl J Med 302:1128–1131, 1980.
130. Mahaffey KR, Rosen JF, Chesney RW, et al.: Association between age, blood lead concentration, and serum 1,25-dihydroxycholecalciferol levels in children. Am J Clin Nutr 35:1327–1331, 1982.
131. Six KM, Goyer RA: Experimental enhancement of lead toxicity by low dietary calcium. J Lab Clin Med 77:933–942, 1970.
132. Sorrell M, Rosen JF: Interactions of lead, calcium, vitamin D and nutrition in lead-burdened children. Arch Environ Health 32:160–164, 1977.

3
Vitamin E

Introduction	60
Absorption, Metabolism, and Excretion	62
Dosage Forms	64
Clinical Studies	65
The Preterm Infant	65
Retinopathy of prematurity	65
Intraventricular hemorrhage	67
Phototherapy of hyperbilirubinemia	67
Respiratory distress and bronchopulmonary dysplasia	68
Pregnancy and Gynecologic Studies	68
Pregnancy	68
Cystic mastitis	69
Premenstrual syndrome	69
Malabsorption Disorders	70
Pediatric malabsorption	70
Adult malabsorption	71
Dosage regulation in malabsorption	72
Cardiovascular Disease	72
Intermittent claudication	72
Other circulatory actions	73
Hemolytic Diseases	74
Anti-inflammatory Action	74
Shock Therapy and Critical Care	75
Shock lung syndrome	75
Oxygen therapy	75
Surgery	75
Other Clinical Studies and Potential Uses	76
The elderly	76
Autoimmune disorders	77
Leg cramps and restless legs	77
Phrynoderma	77
Alcoholism	77
Liver disease	78

Renal failure	78
Athletics and exercise	78
Cancer prevention	79
Other clinical possibilities	80
Toxicity and Side Effects	81
Interactions	82
References	83

INTRODUCTION

Vitamin E designates a group of lipid-soluble compounds sharing certain biologic activities, namely, essentiality for reproduction in laboratory animals and, more fundamentally, antioxidant activity and stabilization of cell membranes. The most active of these compounds is a d-alpha-tocopherol, with the structural formula shown (Fig. 1). Other naturally occurring tocopherols (d-beta, gamma, and delta) differ in the number and placement of methyl groups in the benzenoid ring; the closely related tocotrienols, also naturally occurring and possessing antioxidant activity, have similar structures but with double bonds in the aliphatic side chain [1]. Synthetic alpha-tocopherol consists of a mixture of eight diastereomers, one of which is the same as naturally occurring d-alpha-tocopherol. Since the various forms of vitamin E are standardized by biologic activity, the differences in structure and steric configuration are theoretically taken account of in marketed preparations whose potency is expressed in International Units (IU). Potency may also be expressed as d-alpha-tocopherol equivalents per milligram [2], but this is more commonly seen in research papers and not in the labeling of commercial vitamin E preparations.

The richest natural sources of vitamin E are vegetable oils. In their original unprocessed state, the more unsaturated oils have higher contents of tocopherol. However, in many polyunsaturated oils as they reach the table, e.g., soybean oil, the vitamin E activity may be lower than expected because much of the tocopherol present is the beta, gamma, or delta rather than the alpha isomer, and, in addition, there may be oxidative losses of tocopherols during processing, cooking, and storage [3]. It cannot be assumed, therefore, that individuals who increase their intake of polyunsaturated fats and oils for dietary control of plasma lipids will automatically get the increased amount of vitamin E they need. Horwitt [4] points out that there is no fixed requirement for vitamin E as there is for vitamins that are components of enzyme systems, and that in his experiments with laboratory rats, the requirement varied more than fivefold depending on the diet and/or tissue composition resulting from past diets. Nevertheless, the National Research Council has published a narrow-range RDA for vitamin E of 12–15 IU/d for the adult, an intake that would probably be provided by most American diets [2], but may not be adequate

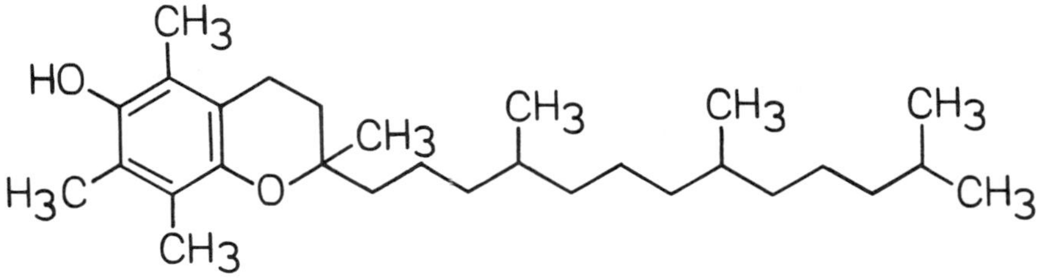

Fig. 1. Structural formula of the most active of the vitamin E compounds, d-alpha-tocopherol.

for diets high in fish oil or polyunsaturated vegetable oils. Fish oil is a particularly potent enhancer of vitamin E requirement [4,5].

Although alpha-tocopherol was isolated, identified, and synthesized in the 1930s, and although it was assumed that man requires vitamin E just as laboratory animals do, no overt clinical deficiency was reported in humans until a few decades later. Deficiencies treatable with supplemental vitamin E were found first in premature infants and later in children and adults with disorders of absorption or transport of the vitamin. In recent years, a number of medical uses for vitamin E in larger-than-nutritional quantities have been developed or are under investigation, based largely on increased understanding of the damaging effects of oxidant species (superoxide radical, hydrogen peroxide, hydroxyl radical, lipid peroxides) and free radicals in general on body tissues. There is mounting evidence that oxidant species and free radicals contribute to the etiology and pathology of many diseases, including cardiovascular disease, rheumatoid arthritis, reperfusion injury, emphysema, and cancer [6–8]. Free radicals can originate endogenously from normal metabolic reactions or exogenously as components of tobacco smoke and air pollutants and indirectly through the metabolism of certain solvents, drugs, and pesticides, as well as through exposure to radiation [8]. Tissue damage results from the reaction of free radicals with polyunsaturated lipids in cellular membranes, nucleotides in DNA, and sulfhydryl bonds in proteins [7,8]. The mammalian organism has many defenses against oxidant species and free radicals, including vitamin C and the carotenes from the diet; uric acid, ceruloplasmin, and bilirubin in the circulation; and intracellular enzymes such as superoxide dismutase, catalase, and glutathione peroxidase; but in addition to these, the lipid-soluble vitamin, tocopherol, is essential for its antioxidant and free radical quenching activity in the lipid-based membranes of cells and organelles [8].

New clinical potentials for vitamin E are constantly being explored whenever free radical damage is suspected, as for example, in a large-scale, 5-year study of vitamin E supplementation in patients with early Parkinson's disease, recently inaugurated [9], and evaluation as a suppressant of free radical damage during cardiopulmonary bypass operations [10]. Research on potential medical applications for vitamin E is one of the most active areas of present-day clinical investigation.

ABSORPTION, METABOLISM, AND EXCRETION

During digestion, the tocopherol of foods is absorbed along with other dietary lipids into the lacteals as chylomicrons [11]. Absorption efficiency is low, probably in the range of 20–40% normally. Supplemental oral intake must be tenfold or more higher than ordinary dietary intakes to double the plasma tocopherol concentration [12]. Efficiency of absorption is markedly reduced in the absence of normal flows of bile and pancreatic enzymes, so that vitamin E deficiency frequently develops in chronic malabsorption disorders [11]. Administration of oral vitamin E supplements together with medium-chain length triglycerides may improve absorption [13].

Transport of tocopherols in the blood is by the lipoproteins, the major portion being found in low-density lipoproteins (LDL) under fasting conditions. Vitamin E is delivered to cells principally via the high-affinity receptor for LDL, although a less efficient transfer mechanism probably also exists [14]. Tocopherol is also transported in erythrocytes as a constituent of the membrane, where it protects polyunsaturated lipids by its antioxidant activity. In vitamin E deficiency, erythrocytes exposed to hydrogen peroxide in vitro display increased hemolysis. Animal studies show that the unoxidized chromanol structure of alpha-tocopherol is necessary for its incorporation into the red blood cell (RBC) membrane and that oxidation to alpha-tocopherol quinone causes loss of affinity to the membrane [15].

In general, plasma tocopherol concentration is correlated with total circulating lipids (triglycerides, cholesterol, phospholipids), and individuals with high circulating lipid levels usually show above-average total tocopherol levels. Horwitt et al. [16] suggested that a ratio of plasma total tocopherols to plasma total lipids of 0.8 marks the lower limit of adequacy. In the normal North American adult's range of 0.7–2.5 mg tocopherol/dl of plasma [11], the tocopherol/lipid ratio would usually be above the minimum. Tocopherol levels of 0.5 mg/dl of plasma or less are considered unsatisfactory in the adult. The situation is radically different in premature infants, in whom deficiency (as estimated from peroxide-induced red cell hemolysis) is reported to occur when plasma total tocopherols are below 0.64 mg/dl of plasma and, since plasma lipids are low, at a ratio of total tocopherols to total lipids below 1.9 [17].

Thurnham et al. [18], based on a large-scale investigation of serum tocopherol and lipid levels in chronic alcoholics, rehabilitated alcoholics, and normal adult controls, found that the ratio of total tocopherol to total cholesterol-plus-triglycerides was almost as effective as the ratio of tocopherol: total lipid in identifying vitamin E deficiency; this ratio has the advantage that most hospitals need not set up additional laboratory methods for its use. A noninvasive and highly sophisticated method for assessing vitamin E status is based on evolution of pentane from peroxidixed omega-6-fatty acids by beta-scission as described by Evans et al. [19], a method first adapted to in vivo assessment of vitamin E status by Tappel and coworkers [20,21], using gas chromatography. Lemoyne et al. [22] report successful application of the method in adult men and women breathing air specially purified of hydrocarbon gases; they describe it as sensitive, accurate, and reproducible. It is of interest that after 10 days of supplementation with 1,000 IU/d of oral d-alpha-tocopheryl acetate in five normal subjects, plasma alpha-tocopherol levels were almost doubled and pentane exhalation levels decreased in each subject, with a significant mean group decrease of about 35%.

Experiments in rats with oral administration of isotopically labeled alpha-tocopherol show high uptake from plasma by liver, spleen, and adrenal gland per gram of tissue and lesser uptakes by heart, adipose tissue, and kidney; skeletal muscle takes up less per gram of tissue, but in total is one of the larger accumulators [11]. Distribution of uptakes is similar when the vitamin is given intravenously. Upon withdrawal of vitamin E, there is rapid mobilization from plasma and liver and slow mobilization from heart and skeletal muscle, but very slow release from fat stores, even though adipose tissue can accumulate large amounts of the vitamin [11,23]. Vitamin E is important to health of the eye, as evidenced by accumulation of a mean level of 2.9 mg/100 g wet tissue (30 human donors) in the retina and underlying pigmented layers, i.e., pigmented epithelium and choroid, accounting for over 90% of eye tocopherol, the rest being in anterior tissues [24].

The tocopherols are metabolized by oxidation to quinones and other products, some of which are excreted in the urine as such or as conjugates with glucuronic acid. Small amounts of tocopherols and their metabolites are excreted in bile. The major route of excretion of orally administered vitamin E is via the feces, primarily because the vitamin is incompletely absorbed [11].

Differences exist in the absorption and tissue uptake of the various natural and synthetic isomers of tocopherol that have not been fully worked out. gamma-Tocopherol is the isomer most abundant in the diet, but natural d-alpha-tocopherol is the one most efficiently absorbed in both man and animals, accounting for about 87% of plasma tocopherol in humans [25]. Plasma gamma-tocopherol concentration is depressed as intake of the alpha isomer increases [25–27], and Baker et al. [27] have suggested that the plasma gamma/alpha ratio might serve as a sensitive index of alpha-tocopherol ingestion. Liver cell microsomes and mitochondria preferentially bind the alpha isomer [26].

With regard to differences between natural and synthetic tocopheryl acetates, which are readily hydrolyzed in the gut to the tocopherols [11], the studies of Horwitt et al. [28] indicated better absorption in humans of natural tocopheryl acetate, while the more recent double-blind study by Baker and coworkers [27] found that in subjects taking 400 IU b.i.d. of either *all-rac-* alpha-tocopheryl acetate or d-alpha-tocopheryl acetate, there were no significant differences between the groups after 28 days in plasma alpha- or gamma-tocopherol levels, indicating equivalent biopotencies over the long term.

DOSAGE FORMS

Vitamin E is available over the counter in capsule form (100, 200, 400, 600, and 1,000 IU) as the acetate ester of either natural or synthetic tocopherol. The acetate of natural vitamin E may be prepared from d-alpha-tocopherol recovered by molecular distillation from vegetable oils or from a mixture of alpha, beta, gamma, and delta isomers from vegetable oils. Synthetic vitamin E of pharmaceutical grade is usually the acetate ester of *all-rac*-alpha-tocopherol, a mixture of four racemates including the d-alpha form. Pharmaceutical vitamin E is also manufactured as the succinate ester, which, being solid, is suitable for incorporation into multivitamin formulas. The esters do not have the antioxidant properties of the alcohol form and are resistant to oxidative destruction. They are therefore generally used in preference to the easily oxidizable alcohol in liquid and tablet preparations of vitamin E.

Aqueous solutions or dispersions of both alpha-tocopherol and alpha-tocopheryl acetate are available for oral administration and provide better absorption than oily vehicles, particularly when there is a deficiency of biliary or pancreatic secretions [13].

Preparations for *intramuscular* injection in preterm infants have been carefully studied by Hittner and coworkers [29,30], who conclude that an aqueous alcohol solution of synthetic alpha-tocoherol (Ephynal®, Roche Laboratories), but not of alpha-tocopheryl acetate in oil, is effective in rapidly elevating the plasma vitamin E level of the infant without undesirable overshoot. An alternative Ephynal® formulation by Hoffmann-LaRoche contains alpha-tocopheryl acetate instead of alpha-tocopherol. Chiswick and associates [31–33] confirm the effectiveness of Ephynal® in rapidly elevating the plasma vitamin E of preterm infants. *Intravenous injection of vitamin E is dangerous to the preterm infant, as it may produce excessively prolonged elevated plasma tocopherol levels (with slow IV infusion), which are associated with an increased incidence of necrotizing enterocolitis [34,35], or high plasma peaks (with rapid IV infusion) that are associated with increased incidence of intraventricular hemorrhage [36].*

Much additional information is needed on the pharmacokinetics of vitamin

E in its various dosage forms, comparing, for example, the alcohol and ester form of the vitamin, natural versus synthetic forms, water-dispersible and oily vehicles, oral administration on an empty stomach versus at mealtimes, etc. The failure of investigators to report the specific dosage form of vitamin E they have used causes unnecessary difficulties in the interpretation of their results.

CLINICAL STUDIES
The Preterm Infant

Retinopathy of prematurity (retrolental fibroplasia). Neonates, whether term or preterm, have low plasma tocopherol levels, usually in the range 0.2–0.4 mg/dl, largely because placental transfer of vitamin E to the fetal circulation is inefficient [17,37,38]. If vitamin E-deficient, the term as well as the preterm infant may display increased hemolysis with associated hyperbilirubinemia [39]. Normal oral feedings of breast milk or commercial infant formulas will bring the vitamin E level up to 0.5 mg/dl or higher within a week in the term infant. The small or very small premature infant, however, of gestational age < 32 weeks, is placed at risk of the retinopathy of prematurity (ROP) by this low antioxidant status [37,40,41]. Supplemental oxygen therapy for respiratory distress greatly increases the risk. Excessive light exposure in the hospital intensive care unit has also been reported to enhance the risk [42], probably by enhancing free radical production in the retina. The retina, pigmented epithelium, and choroid of the eye appear to depend on a relatively high content of vitamin E for protection against photochemically derived free radicals [24]. Vitamin E therapy, although still considered experimental as of early 1988, shows promise of preserving good retinal function and structure in many of these small preterm infants.

Although not universally accepted, the etiologic mechanism of ROP postulated by Kretzer and coworkers is supported by strong ultrastructural and biochemical evidence [43–45]. According to their findings, vascularization of the fetal retina, which is normally completed in utero under relatively hypoxic conditions, depends on an advancing apron of interdigitating spindle cells, migrating from the optic disk toward the periphery behind an advancing point of retinal development. Nascent retinal blood vessels are formed by canalization of spindle cells. Elevated oxygen tension causes formation of gap junctions between spindle cells, which activates these cells to synthesize and secrete angiogenic factors that trigger the neovascularization typical of ROP over the next 8–12 weeks. Early and sustained vitamin E supplementation, to plasma concentrations that Hittner et al. refer to as adult physiologic levels (1.1–3.3 mg/dl), suppresses gap junction formation by protecting the spindle cells from oxygen-related damage and thereby reduces the severity of ROP, although the total incidence of all grades of ROP remains constant [46]. With vitamin E supplementation, about half of the total incidence is reported to be "very transient and clinically insignificant Grade I ROP" [46].

Hittner and Kretzer [47] recommend the following dosage regimen for preterm infants at risk of ROP:

1. To rapidly raise plasma vitamin E levels, give three early IM injections of 15, 10, and 10 mg synthetic alpha-tocopherol on days 1, 2, and 4 of life, respectively. The injectable preparation (Hoffman-LaRoche Ephynal®) supplies 50 mg *all-rac*-alpha-tocopherol per ml in aqueous alcohol solution.
2. Maintain continuous oral supplementation with 100 mg/kg/day of synthetic alpha-tocopheryl acetate in medium-chain triglycerides (MCT, a low osmolality vehicle) until retinal vascularization is complete (up to 7 months) [16].
3. If the infant is non per os (NPO) for 3 days or longer, inject 10 mg/kg of tocopherol IM (Ephynal®) every third day, or employ slow IV 3 mg/kg/d of tocopherol IV (Ephynal®) or MVI Pediatric® (Armour Pharmaceuticals).
4. *Do not exceed a plasma vitamin E level of 3.5 mg/dl. An increased incidence of necrotizing enterocolitis and sepsis is associated with induction of prolonged plasma vitamin E levels of approximately 8 mg/dl or more [34,35]; increased incidence of intraventricular hemorrhage with transient high plasma peaks of vitamin E [36]; and increased occurrence of necrotizing enterocolitis with administration of hyperosmolor aqueous oral preparations [30,48].*
5. Perform a retinal screening examination at age 8 weeks and as necessary thereafter, depending on the severity of ROP observed.
6. In the very small premature infant of gestational age ≤27 weeks, vitamin E delays the initiation of gap junction activation but does not prevent the development of severe ROP, approximately 2 weeks later than in the unsupplemented infant of similar maturity. Cryotherapy should be applied to the avascular retina to ablate activated spindle cells and to shunt to destroy the source of myofibroblasts, leaving a normal quiescent vasculature between the optic nerve and the peripheral cryoablated region [44,45].

Although vitamin E therapy should begin within the first hours of life for optimal results, there is indication of benefit even if vitamin E administration is delayed until ROP has developed. In a small study involving ten premature infants diagnosed as having active grade 3-plus ROP or worse and not previously treated with vitamin E, Johnson et al.[49] instituted vigorous treatment with IM Ephynal® alpha-tocopherol (50 mg alpha-tocopherol/ml). To minimize irritation, the size of Ephynal® injections was limited to 0.4–0.5

ml, administered in thigh muscles. Simultaneous oral therapy was begun with aqueous oral vitamin E. Serum vitamin E was raised to 5–6 mg/dl as quickly as possible and maintained there until ROP had clearly begun to regress (2–4 weeks). (Whether a lower serum level would have sufficed was not determined.) Dosage was then decreased to maintain a serum level of 3–4 mg/dl for another month. Thereafter, a serum level of 2.0–2.5 mg/dl was maintained until age 1 year, using an oral dose of 25–75 mg alpha-tocopherol/day (Roche Ephynal® aqueous oral alpha-tocopherol, 0.5 ml 1 to 3 times daily). Comparison at age 1–2 years of the vitamin E-treated infants with similar premature infants (with previous active grade 3-plus ROP) who had not been treated with the vitamin revealed a 40% incidence of legal blindness in the treated group versus a 71% incidence in the untreated group. None of the 24 infants studied had been treated surgically prior to this evaluation. The probable mechanism of this late vitamin E therapy is to decrease migration and contraction of myofibroblasts that originate from the shunt [50].

There appears to be a secular downtrend in ROP incidence since 1972 (except in infants of less than 1000 g birthweight), according to data compiled by Johnson et al. [51]; they attribute this largely to the increasing vitamin E content of nursery formulas, greater use of maternal and infant multivitamins, and, from 1981 on, specific oral vitamin E supplements to achieve plasma tocopherol levels in the infant of 1–2 mg/dl if possible.

Intraventricular hemorrhage (IVH). Chiswick et al. [31,32] demonstrated in a controlled study a significant reduction of IVH in premature infants of 26–32 weeks gestational age by use of IM Ephynal® alpha-tocopheryl acetate. Their dosage schedule was 20 mg/kg of alpha-tocopheryl acetate within 24 hours of birth and thereafter at daily intervals for a total of 80 mg/kg. Individual doses of alpha-tocopheryl acetate greater than 25 mg (0.5 ml Ephynal®) were divided and given into two different injection sites to avoid local inflammatory reactions. Incidental to their studies of ROP, Hittner et al. [8,15] and Speer et al. [52] also observed a lower incidence of IVH in vitamin E-treated premature infants when the vitamin E (Ephynal®, Roche) was given within the first hours of life by IM injection. Similarly, Sinha et al.[53] report a significant reduction in incidence of brain hemorrhage in very preterm infants given vitamin E by IM injection during the first 3 days of life, compared with controls.

Phototherapy of hyperbilirubinemia. Gross [54], in a study involving 20 preterm infants of birthweight 1,000–1,500 g, treated one-half the group with intramuscular alpha-tocopheryl acetate at 50 mg/kg and compared response of their serum bilirubin with phototherapy. Reduction to acceptable levels took an average of 48 hours in the vitamin E group and 107 hours in the control group. A similar study with preterm infants weighing 1,501–2,000 g at birth showed smaller and statistically insignificant benefits for vitamin E. Further studies are needed to confirm usefulness of this therapy.

Respiratory distress, bronchopulmonary dysplasia (BPD). An early controlled trial by Ehrenkranz et al. [55] indicated a reduction in severity of BPD by vitamin E supplementation in premature infants placed on ventilation for respiratory distress syndrome (RDS), but a later study by the same group [56] and evaluations by others [57] did not confirm benefit. However, Finer et al. [40] found that BPD was associated with a significantly increased risk of active ROP, so that vitamin E therapy appears called for on this basis irrespective of doubtful benefit for BPD. It is worth mentioning, also, that in rabbit neonates injected subcutaneously at 1 hour and 24 hours of life with vitamin E (100 mg/kg of synthetic alpha-tocopherol), the adverse pulmonary effects of hyperoxia (reduction in lung lavage phospholipids, decreased maximum distensibility, altered compliance) were abolished [58].

Huijbers et al.[59] compared plasma tocopherol levels in premature infants who survived RDS and premature infants without RDS who received no supplemental oxygen. At similar vitamin E intakes, the plasma tocopherol levels of the infants who had not developed RDS gradually increased, whereas those who had survived RDS maintained below-normal tocopherol concentrations, suggesting that preterm infants with RDS may need more supplemental vitamin on a long-term basis than those without RDS [59].

Pregnancy and Gynecologic Studies

Pregnancy. There are no studies indicating benefit to mother or fetus of maternal vitamin E intakes above the RDA, which will be met if the diet is good or a multivitamin supplement taken. A marginal nutritional state with respect to vitamin E, as well as other vitamins, has been observed in abruptio placentae [60]. Yoshioka et al. [61] observe that blood lipid peroxide levels increase markedly in late pregnancy, both in humans and in laboratory rats on standard diet; blood lipid peroxide levels were further elevated in the pregnant rat by a vitamin E-deficient diet. High blood lipid peroxide concentrations are usually observed in toxemia of pregnancy [61]. Mino and Nagamatu [62] report that although plasma tocopherol increases during human pregnancy, the ratio of tocopherol to total lipids remains almost constant while RBC tocopherol levels decline markedly during the second and third trimesters. Inasmuch as changes in RBC tocopherol levels closely reflect changes in liver tocopherol concentrations, these investigators suggest that the hyperlipemia of late pregnancy diminishes biologic availability of tocopherol to other tissues including the placenta. Further clinical investigations are needed in pregnancy of the benefit, if any, of vitamin E supplements above the RDA level.

Cystic mastitis (mammary dysplasia—benign breast disease). Although vitamin E acetate has been used for treatment of cystic mastitis on an empirical basis for more than two decades [63,64], it is only in recent years that double-blind placebo-controlled evaluations have been carried out. Conflicting results have been obtained. In studies reported in the 1978–1982 period [65–70], in which synthetic alpha-tocopheryl acetate was the supplement used, London and his coworkers obtained significant subjective and objective responses in 70–80% of cystic mastitis patients. Dosage regimen was 300 IU orally b.i.d. Elevated levels of luteinizing and follicle-stimulating hormones on day 21 of the menstrual cycle were lowered to normal [67], and reduced ratios of progesterone to estradiol were increased to normal [66]. Ratios of HDL to total cholesterol, which were typically low in patients before treatment, were increased in patients but not in normal controls [67,68].

When London et al. [71] repeated their investigation of vitamin E therapy in cystic breast disease, using this time synthetic alpha-tocopherol (the alcohol) instead of the acetate ester, negative results were obtained. The trial was randomized, placebo-controlled, and double-blind; subjects were 128 women with confirmed mammary dysplasia. Dosages of vitamin E were 150, 300, or 600 IU/d for 2 months. No significant effects were noted in breast examinations, sonography, or thermography, and there were no significant changes in hormone levels. The investigators speculate that the change in form of vitamin E employed may have had something to do with the different results, although serum tocopherol levels rose in the patients. Differences in absorption [25,26,28] and tissue binding [26] between different tocopherol isomers and between the alcohol and acetate forms have been reported.

Ernster et al. [72] also obtained negative results in a double-blind, randomized clinical trial in which, like London et al., they used synthetic d-alpha-tocopherol (not the acetate). The dosage regimen was 600 IU/d (300 IU b.i.d.) for 2 months. There were no significant changes in the breast examinations and no changes in blood lipids except for serum tocopherol levels. The high tocopherol intake raised mean serum tocopherol concentration only about one-third (from an initial 1.5 mg/dl to a final 2.0 mg/dl). An initially high vitamin E status and/or limited transport/uptake of the form of vitamin E employed may have reduced the likelihood of observing a response. Further clinical studies should be initiated to explore further the possible benefits from appropriate vitamin E therapy of this commonly occurring, benign but distressing disorder.

Premenstrual syndrome (PMS). London et al. have also examined the effect of vitamin E on the severity of PMS symptoms in two studies, the first of which involved 75 women with documented benign breast disease [73]. A double-blind randomized dose-response methodology was used, involving 2 months of oral treatment with placebo or with 150, 300, or 600 IU synthetic alpha-tocopheryl acetate per day. Adjusted mean PMS symptom scores were significantly improved in three of the four symptom categories listed by Abraham [74], i.e., in PMT-A (nervous tension, mood swings, irritability, anxiety), PMT-C (headache, craving for sweets, increased appetite, heart pounding, fatigue, dizziness, fainting), and PMT-D (depression, forgetfulness, crying, confusion, insomnia). There was no improvement in category PMT-H (weight gain, swelling of extremities, breast tenderness, abdominal bloating). A dose of 300 IU of vitamin E per day seemed optimal, both 150 and 600 IU/day being somewhat less effective.

The second study involved 41 women with PMS but without breast disease, assigned randomly to 1 capsule per day of 400 IU natural d-alpha tocopherol (rather than synthetic) or placebo, taken for three menstrual cycles [75]. Vitamin E therapy resulted in significant improvement of all major PMS symptom categories, including PMT-H of the Abraham classification. There were no demonstrable side effects.

The mechanism of action of vitamin E on PMS symptomatology is unclear, but does not involve change in serum levels of testosterone, dehydroepiandrosterone sulfate, estradiol, or progesterone [76]. London et al. suggest that the PMS effects of vitamin E therapy may be mediated by the action of alpha-tocopherol on prostaglandin metabolism [75,76]. Further clinical studies, with and without cotherapy with prostaglandin-inhibiting drugs, would be desirable. The physical and chemical form in which the vitamin E is administered, as well as time of dosage, may be important factors in the response.

Malabsorption Disorders

Pediatric malabsorption. Low serum vitamin E levels and/or low ratios of alpha-tocopherol to total serum lipids have been reported in children with chronic cholestasis (biliary atresia, congenital or neonatal hepatitis, arteriohepatic dysplasia, Alagille syndrome) [77–80] and cystic fibrosis [81]. Chronic vitamin E deficiency in children leads to progressive neuraxonal changes and demyelination of posterior columns and brainstem [82], as well as peripheral nerve dysfunction [78,83,84]. The spectrum and progression rates of neurologic symptoms in clinical vitamin E deficiency, including gait disturbances, loss of tendon reflexes, muscle weakness, and visual disturbances, have been reviewed by Satya-Murti et al. [85]. Progression to fullblown nutritional muscular dystrophy, together with severe neurologic symptoms, was reported by Tomas [86] in a 7-year-old child who had had severe malabsorption from birth; the defects were considerably, but not completely, reversed by 16 months of therapy with 400 IU/day of oral aqueous alpha-tocopheryl acetate. In children with chronic cholestasis corrected with Kasai-type enterostomy, vitamin E deficiency and progression of neurologic complications may persist, requiring vigorous vitamin E therapy to elevate serum vitamin E levels [77,87]. Tazawa et al. found that oral supplements of alpha-tocopheryl acetate in doses of 5–10 mg/kg/day were needed in postoperative infants, and they suggest IM administration in poorly responding patients [87].

Sokol et al. [88,89] describe use of IM injection as well as oral vitamin E in long-term treatment of children with chronic cholestatic liver disease. Oral dosage of aqueous synthetic alpha-tocopheryl acetate was 120 IU/kg body weight per day. Injection dosage was 1–2 mg/kg/d (Ephynal®, 1 ml equals 50 mg or 55 IU). Correction of vitamin E deficiency did not improve the malabsorption of fats and tocopherol, but appeared to improve liver function, as indicated by a reduction in serum bile acid levels [89]. The neurologic syndrome associated with this disease was effectively treated by vitamin E therapy. With institution of treatment early in life, dramatic effects were seen, but the treatment was less effective in restoring previously damaged nerve function.

Runge et al. [90] report that in patients with abetalipoproteinemia, the progressive retinal deterioration associated with the untreated disease was arrested by large oral doses of water-dispersed tocopheryl acetate (100 mg/kg/d), used in conjunction with a low-fat diet and supplements of the other fat-soluble vitamins. This therapy also prevented development of neurologic dysfunction in abetalipoproteinemia patients if begun early in life, and if begun later, ameliorated certain neurologic abnormalities that had already developed [90].

Early and aggressive treatment with oral and/or IM vitamin E is advised whenever malabsorption of any cause is present and the serum tocopherol:total lipid ratio is low [88,90].

Adult malabsorption. Adults with malabsorption due to chronic liver disease or pancreatitis frequently have low serum vitamin E levels [91–93]. Vitamin E deficiency associated with mild-to-severe neurologic symptoms has also been reported in adults with cystic fibrosis [94,95], Crohn's disease [96], and short-bowel syndrome [96,97].

Traber et al. [98] have shown that the peripheral nerves (biopsy specimens) of adult malabsorption patients with peripheral neuropathy contain a significantly lower concentration of tocopherol than the nerves of patients with other neurologic disorders without vitamin E deficiency. They suggest that peripheral nerve damage during vitamin E deficiency is caused by insufficient antioxidant tocopherol to protect nerve cell membranes and that destruction of the nerves is caused by free radical damage. They strongly recommend measurement of serum vitamin E in patients with symptoms of peripheral neuropathy.

Arria et al. [99] found 18 of 42 adult female patients with primary biliary cirrhosis (PBC) to be deficient in vitamin E by blood analysis and to perform more poorly on neuropsychologic tests of psychomotor capacity than PBC patients without vitamin E deficiency. They recommend routine evaluation of vitamin E status in this patient population.

Relief of back pain and epigastric pain in patients with chronic pancreatitis has been reported upon administration of 6 100-IU capsules/d of synthetic alpha-tocopheryl acetate [100].

Dosage regulation in malabsorption. Dosage of vitamin E in malabsorption is on an individual case-by-case basis, with monitoring of the serum vitamin E level in adults and of the serum vitamin E-to-total serum lipids ratio in children and hyperlipemic adults. Water-dispersed vitamin E preparations usually are better absorbed than oily preparations. Most children with chronic cholestasis respond to oral aqueous alpha-tocopheryl acetate administered in doses of 100–400 IU/day, while cystic fibrosis patients may do well on 50 IU/d [101]. Children who do not respond to massive oral doses may require IM injection with Ephynal®. Adults with malabsorption may require only modest supplements of oily or aqueous forms of alpha-tocopherol or alpha-tocopheryl acetate to bring serum vitamin E levels to normal, but megadoses are safe if response to lower dosages is poor. However, Jeffrey et al. [102] found it necessary to resort to IM injections of vitamin E in some adult patients with chronic cholestatic liver disease and severe vitamin E deficiency associated with peripheral neuropathy.

Cardiovascular Disease

Intermittent claudication. The possible value of high-dose vitamin E in cardiovascular disease has been a subject of debate since the first favorable report by Shute et al. in 1948 [103], but it is now generally agreed that benefit in cardiac conditions remains to be proved by controlled studies [104]. However, there appears to be clinical benefit in the peripheral vascular disorder, intermittent claudication. Haeger [105] has summarized extensive studies carried out at his clinic in Sweden involving single-blind, long-term trials of supplementation with 100 IU alpha-tocopheryl acetate 3 times daily in 122 patients; there were 36 clinically comparable controls. Observation time ranged from 1 to 16 years.

Vitamin E-treated patients showed statistically significant gains over controls in walking distance accomplishment and in arterial blood flow as determined by lower leg plethysmography, as well as a marked reduction in the number of patients requiring surgical reconstruction or amputation. Active exercise was part of the treatment of both the control and experimental groups and appeared to have benefits of its own. Continuous long-term use of the vitamin E supplement is necessary. For example, there was no difference in arterial blood flow between controls and vitamin E-treated subjects initially or after 12 months, but significantly greater flow rates in the vitamin E group after 18 and 36 months.

Other circulatory actions. alpha-Tocopheryl acetate supplements of 600–800 IU/d have been shown to increase the ratio of HDL cholesterol to total cholesterol (without lowering the latter) in individuals with low initial values of this ratio, including healthy men and women without clinical pathology [106,107], spinal cord injury patients [107], and women with mammary dysplasia [67,68]. Individuals with normal or relatively high HDL-C levels show no change. Short-term (4-week) administration of 600 IU of vitamin E per day to renal patients on maintenance hemodialysis produced no significant change in their HDL-C levels [108].

Szczeklik et al. [109] report that daily supplementation with 300 mg vitamin E for 2 weeks in 46 patients with hyperlipoproteinemia effectively reduced their elevated plasma lipid peroxide levels to the levels of 18 healthy adult controls receiving placebo. A daily intake of 600 mg vitamin E resulted in a mild suppressant effect on platelet clumping in the patients with hyperlipoproteinemia; the 300 mg/d dose did not produce this effect [109].

Vitamin E has been demonstrated to inhibit biosynthesis of thromboxane A_2, the potent proaggregatory and prothrombotic metabolite of arachidonic acid, both in vitro [110] and, at a dosage of 1,600 IU/d, in vivo, in man [111]. Unlike aspirin, which at a dosage of 2,600 mg/day strongly inhibits not only thromboxane synthesis in man, but also that of prostacyclin (PGI_2), vitamin E at 1,600 IU/day did not depress prostacyclin biosynthesis [111]. Even lower aspirin dosages (80 mg, about one-fourth of a 5-grain aspirin tablet) have been shown to substantially inhibit both thromboxane and prostacyclin synthesis [112]. Rabbits fed an atherogenic diet for 1 week were found to have an elevated plasma lipid peroxide level and a 90% decrease in arterial biosynthesis of prostacyclin; 100 mg vitamin E daily prevented the increase in plasma peroxides and protected the prostacyclin synthesis system [109].

Since prostacyclin is antiaggregatory, antithrombotic, and vasodilative, these effects of vitamin E on prostaglandin metabolism, together with the above-described actions on blood lipids, suggest potential value for this vitamin in cardiovascular disorders. However, long-term controlled studies remain to be carried out to determine whether these experimental observations can be translated into practical clinical use.

Stuart [110] has suggested that vitamin E administration may be beneficial in diabetes mellitus, as diabetics manifest an increase in platelet proaggregatory thromboxane together with a decrease in prostacyclin biosynthesis. A recent preliminary report of Bierenbaum et al. [113], indicating possible benefits in lowering fasting blood glucose levels and blood pressure in adult-onset (type

II) diabetics, provides strong justification for vigorous investigation of the usefulness of high-dose vitamin E in this disease. Pritchard et al. [114] found that when vitamin E level was increased about fivefold (to 200 IU/kg) in diets of streptozocin-induced diabetic rats, lipid peroxide levels were reduced in plasma and liver, plasma triglycerides were lowered toward normal, and hepatic lipoprotein lipase activity was increased.

Hemolytic Diseases

Erythrocyte disorders. Modest clinical benefits have been reported for high-dose vitamin E treatment of certain genetic hemolytic diseases, e.g., sickle-cell anemia, beta-thalassemia, and Mediterranean-type glucose-6-phosphate dehydrogenase (G6PD) deficiency. Patients with these disorders tend to have low serum tocopherol levels despite normal intakes [115–117]. Some, but not all, sickle-cell anemia patients show a reduction in the percentage of irreversibly sickled cells upon vitamin E therapy (150 IU alpha-tocopherol 3 times daily), with greatest response in those with initially highest percentages [116,118]. In beta-thalassemic patients, serum vitamin E levels can be normalized by administration of 750–1,000 IU synthetic alpha-tocopheryl acetate per day, with decreased RBC membrane peroxidation, but no significant changes in hemoglobin levels or transfusion requirements [117]. Variable results have been obtained in treatment of G6PD-deficiency patients, some showing modest increases in RBC half-life and in hemoglobin level, others showing no significant response at dosages up to 2,400 IU/day [119–121]. It may be assumed that it is desirable to normalize low serum vitamin E levels in such patients, whether or not other beneficial changes are observed.

Anti-inflammatory Action

Osteoarthritis. Israeli rheumatologists reported a single-blind, cross-over study of the "analgesic" effect of vitamin E (presumably alpha-tocopheryl acetate) in osteoarthritis [122]. Most of the patients had spondylosis, other diagnoses being gonarthrosis, symptomatic Heberden's nodes, or osteoarthritis at other sites. Most were women, average age was 56.5 years, and average duration of symptoms was 9.3 years. Patients were randomly assigned to vitamin E (600 IU/day) or identically appearing placebo; after 10 days' treatment, the groups were switched. Fifteen of the 29 patients experienced marked relief of pain while on vitamin E and only one while on placebo. Weak antiinflammatory activity for vitamin E, administered orally to laboratory rats, has been reported in experimentally induced inflammation [123]. Additional clinical studies are needed.

Shock Therapy and Critical Care

Shock lung syndrome (adult respiratory distress syndrome, ARDS). Shock lung syndrome, or ARDS, is considered to be the main cause of death after severe shock states. Wolf and Steeger [124,125] have demonstrated the value of high-dose enteral vitamin E, both in experimental animals and clinically, in reducing the severity of ARDS. They describe the pulmonary lesions as progressive deterioration of gas exchange and compliance, increased pulmonary vascular resistance, and severe interstitial edema. The syndrome may develop in the late course of traumatic, hypovolemic, and septic shock, as well as in severe intoxications with drugs and poisons [124]. Comparative studies were carried out during 1977–1982 in 176 shock patients aged 16–82 years, of whom 87 received enteral vitamin E therapy and 89 did not. The studies were not randomly controlled, as the vitamin E-treated patients were those who in general were more severely ill at beginning of treatment. Oxygen-supplemented ventilation and other standard supportive measures were used, including heparin treatment.

Vitamin E therapy was via gastric tube (heparinization ruling out IM administration) as *all-rac*-alpha-tocopheryl acetate (6–8 x 300–500 mg doses/d). Survivorship in the tocopherol-treated group was 39%, while in the non-vitamin E group it was 17%. Measured serum tocopherol response to enteral vitamin E was less in shock patients than in healthy subjects [125].

The promising results of this investigation warrant further trials in other acute care centers.

Oxygen therapy. More broadly, as suggested by the work of Wolf and Seeger [124,125] and supported by the animal studies of Taniguchi et al. [126], a good vitamin E status is important for any patient requiring oxygen-assisted ventilation because of reduced lung capacity from any cause, in order to prevent or delay the onset of lung injury. In rats exposed to 60% or 80% oxygen, by the second day there was a significant increase in lung peroxides, degradation of lung microsomal phospholipids, and a reduction of lung microsomal electron transport activity; treatment with tocopheryl acetate injections prevented these early changes [126]. If oxygen therapy was prolonged, vitamin E therapy delayed the appearance of chemical and functional changes until 4 days at 80% oxygen and 8 days at 60% oxygen [126].

Surgery

Contracture inhibition after breast implant. Baker [127] compared incidence of contracture after breast implant surgery in 100 patients receiving 1,000 IU synthetic vitamin E (presumably alpha-tocopheryl acetate) by mouth twice daily with the incidence in a comparable control group not receiving supplemental vitamin E. Incidence of contracture was 19% in the treated group and 30% in the untreated group. Baker recommends that vitamin E therapy begin 1 week before surgery and that it be maintained for 2 years. No harmful side effects were noted.

Reduction of peritoneal adhesions. Kagoma et al. [128] compared postsurgical adhesions in a group of mice receiving vitamin E at 300 IU/kg of diet with those of a control group receiving 65 IU/kg diet. They found a statistically significant decrease in the incidence and degree of abdominal adhesions in the group receiving the higher vitamin E intake (equivalent to about 150 IU/d for a human consuming 500 g of food, dry weight). The results take on added significance in view of the modest level of supplementation. Comparable studies in human abdominal surgery have not been reported.

Skin flap survival. Hayden et al. [129] report that pretreatment, orally or IV, with an antioxidant mixture of vitamin E, vitamin C, beta-carotene, and glutathione significantly prolonged mean skin flap survival at 1 week after surgery in laboratory rats. The increased survival was attributed to inhibition of free radical production during reperfusion of the skin flaps. The investigators recommend clinical evaluation of "these inexpensive, readily available and realtively nontoxic agents" in reconstructive surgery.

Cardiopulmonary bypass. Cavarocchi et al. [10] compared blood vitamin E levels in ten cardiopulmonary bypass patients given 2,000 IU oral vitamin E 12 hours before surgery with 20 control bypass patients who did not receive supplemental vitamin E. Blood vitamin E levels after surgery were significantly reduced in controls but not in the supplemented patients, and blood free radical activity increased progressively during surgery in control patients whereas they did not increase in the vitamin E-treated group. The authors conclude that vitamin E, by trapping free radicals, breaks the chain reaction of membrane lipid peroxidation that can lead to endothelial injury and capillary leakage [10].

Other Clinical Studies and Potential Uses

The elderly. Considerable evidence has accumulated that free radical damage may contribute to the pathogenesis of diseases associated with aging: cancer, atherosclerosis, hypertension, senile dementia, and immune disorders [130]. Animal studies show that vitamin E is important to inhibition of lipid peroxidation, determined ex vivo, in the cerebrum [131] and cerebellum and brainstem [132]. In old rats maintained on a low-tocopherol diet and then repleted, long-term supplementation (20 weeks) was required to inhibit peroxidation capacity in the brain [131]. The vitamin E minimum requirement level in the diet, which maintained steady tocopherol levels in cerebellum and brain stem of young rats, was not adequate for old rats but resulted in declining levels of brain tocopherol [132]. Similarly, in laboratory mice, a dietary vitamin E level that was adequate to keep liver lipid peroxidation at a low (normal) level in young mice was not adequate to do so in old animals; when dietary vitamin E was increased from 40 IU/kg to 140 IU/kg, liver peroxidation was lowered by one-half in young animals and by about two-thirds in old animals [133].

In human studies of plasma tocopherol concentration versus age, either no significant change was seen [134], or some degree of rise up to age 60 and a gradual small decline thereafter [135]. The ratio of tocopherol to total lipid does not change [135], suggesting unaltered vitamin E status, but platelet vitamin E levels decline with age, indicating less tissue activity of vitamin E [134].

Wartanowicz et al. [136] report that administration of 200 mg/d of alpha-tocopheryl acetate to a group of individuals aged 60–100 years lowered serum peroxide levels 14% in 4 months and 26% in 12 months.

Since immune capacity declines in old age in association with a steady increase in death rate [137], it is of interest and importance that vitamin E supplementation above ordinary dietary intakes improves immune response in experimental animals [138,139], probably by suppressing prostaglandin E_2 synthesis [139], with greater effect in old than in young animals [130,139,140].

Porphyria, lupus erythematosus and autoimmune disorders. Nair et al. [141] noted clinical improvement of porphyria in four patients treated with 100 IU water-miscible alpha-tocopheryl acetate per day; urinary excretion of coproporphorins and uroporphorins declined to normal levels. With higher dosages, usually 800–1,600 IU/day, Ayres and Mihan [142] reported marked improvement in two cases of porphyria cutanea tarda, four of discoid lupus erythematosus, three of morphea, 20 of Raynaud's disease, five of vasculitis, and one of severe polymyositis; they give instances of relapse upon withdrawal of vitamin E and lack of effect of doses of 300 IU/day. Another report by these physicians describes favorable responses in seven cases of chronic discoid lupus erythematosus at dosages of 800–2,000 IU of d-alpha-tocopheryl acetate per day [143].

Leg cramps, restless legs syndrome. Benefits in treating idiopathic nocturnal leg cramps, rectal cramps, and the "restless legs" syndrome have been reported at dosages of 300–800 IU of d-alpha-tocopheryl acetate per day [144,145].

Phrynoderma. Nadiger [146] reported results of treating 50 children with phrynoderma (follicular keratosis) with 100 IU synthetic alpha-tocopheryl acetate t.i.d., with and without simultaneous B-complex supplementation. Vitamin E alone was partially effective, while vitamin E plus B-complex was curative.

Alcoholism. Clinical studies have demonstrated that a low vitamin E and selenium status is common in chronic alcoholics [147]. Vitamin E deficiency

[148], and particularly combined vitamin E-selenium deficiency [147,149], contribute to clinical liver disease and fibrosis; the mechanism leading to hepatic biochemical and structural damage is postulated to be free radical chain reactions and peroxidation of unsaturated lipids in cell membranes [147–149]. Increased free radical activity induced by alcohol consumption has been demonstrated in human alcoholism [150] and in laboratory animals [151]. An association between chronic alcohol use and heart disease is well established. In rats, chronic alcohol intake elevates the level of free peroxide radicals in heart muscle; this increase could be prevented by simultaneous administration of vitamin E [151]. The data indicate desirability of vitamin E and selenium supplementation in chronic alcoholics if biochemical evidence of deficiency is present.

Liver disease. Studies in the laboratory rat have demonstrated that vitamin E deficiency causes a progressive rise in lipid peroxide free radical production [152] and that supplementation with vitamin E above ordinary dietary levels can protect the liver to a considerable degree from free radical activity induced by chemical agents [153] or induced by reperfusion following an ischemic period [154]. As in the case of cirrhosis in alcoholics [147,148], liver damage in rats is greater when there is a deficiency of both vitamin E and selenium [152].

Renal failure. Accelerated breakdown of red blood cells with resultant anemia occurs commonly in hemodialysis patients. Ono et al. [155], investigating the potential value of vitamin E supplementation in this condition, noted that although plasma vitamin E levels appeared normal, the tocopherol concentration in packed red cells was significantly lower than that of normal controls. Daily supplementation with 600 IU of oral vitamin E for 30 days caused elevations in plasma and RBC tocopherol concentrations, reduced susceptibility of RBC to osmotic hemolysis, and, in 12 of 15 patients, a significant increase in hematocrit. Lubrano et al. [156] report similar favorable results with therapeutic oral vitamin E in nine hemodialysis patients, observing increased RBC membrane stability and a significant rise in hemoglobin level.

Athletics and exercise. There are hints from animal studies that exercise training and regular strenuous exercise may increase the vitamin E requirement. Packer and coworkers [157,158] report that endurance capacity of vitamin E-deficient rats is markedly decreased and that vitamin C supplementation did not counteract this effect. A subsequent report of this group [159] shows that the mitochondrial content of trained muscle of rats increases but without an increase in total muscle tocopherol, indicating that there is a reduction in tocopherol concentration in the proliferating mitochondrial membranes. Vitamin E is of course the major lipid-soluble, chain-breaking antioxidant in membranes of the mitochondria, where free radicals are likely to be produced by mitochondrial electron transport activity mediated by ubiquinones.

Cancer prevention. Cancer-preventive activity has been proposed for vitamin E based on its antioxidant and free radical-quenching properties inasmuch as prooxidant states (i.e., increased tissue concentrations of active oxygen, organic peroxides, and free radicals) can promote initiated cells into neoplastic growth [6]. Free radical damage to liver DNA in rats on high-polyunsaturated-fat diets is inhibited by vitamin E [160]. Furthermore, in experimental animals vitamin E supplementation above ordinary dietary intakes significantly improves immune response [138,139], tending to offset age-associated declines in immune function [130,139,140], which are believed to play a significant role in the increased incidence of cancer seen in old age.

Several recent publications establish an epidemiologic correlation between low serum or plasma tocopherol levels and subsequent mortality from stomach and colon cancer [161], lung cancer [162,163], breast cancer [164], and cancer of all types [165,166]. If both vitamin E and selenium status are low, risk of cancer is enhanced [163,165]. Miyamoto et al. [163] suggest that there may be familial factors predisposing to low serum vitamin E and selenium levels, since healthy relatives of lung cancer patients were found to have significantly lower serum tocopherol levels than control subjects, as well as a trend toward lower selenium levels.

Both vitamin E and selenium have been shown to inhibit radiogenic and chemically induced transformation of mouse embryo cells in vitro [167]. Tocopherols, like vitamin C, reduce nitrite present in cured meats and render it unavailable or less available for reaction with amine compounds occurring in food, intestinal secretions, and feces that may produce carcinogenic nitrosamines; this has been demonstrated in animals [168] and in humans [169]. Bruce and Dion [170] reported isolation from human feces of a naturally occurring, organic solvent-extractable mutagen, quantifiable by UV absorbance. In volunteers taking 120, 400, or 1,200 IU/d of synthetic alpha-tocopherol, mutagen formation was markedly reduced in an apparent dose-dependent manner. A later study [171] of fecal mutagenicity in 20 healthy volunteers (2 males, 18 females) showed that combined supplemental ingestion of 400 mg/d of synthetic alpha-tocopherol (440 IU) and 400 mg/d of vitamin C lowered fecal mutagenicity by 74%. Addition of vitamins E and C directly to feces did not reduce mutagenicity, indicating that the vitamins act to reduce formation of the mutagen, rather than inactivating it in the stool.

In experimental animal cancer, Shklar and coworkers have shown, using topical 0.1% dimethylbenz(a)anthracene (DMBA) as cancer-inducing agent in the buccal pouch of the hamster, that not only could carcinogenesis (at observation 28 weeks later) be entirely prevented by oral administration of synthetic alpha-tocopherol (10 mg on alternate days) [172], but considerable tumor regression in established DMBA-induced buccal pouch tumors could be obtained by injecting natural d-alpha-tocopheryl succinate into the buccal tissues twice weekly for 4 weeks [173]. Topical alpha-tocopheryl acetate reduced the rate of formation of skin tumors in mice treated topically with DMBA [174].

Cigarette smoking has been established as the cause of most cases of lung cancer. A connection between smoking-induced lung cancer and vitamin E status of the lung is indicated by the report of Pacht et al. [175] that alveolar fluid samples obtained from cigarette smokers by bronchoalveolar lavage contained only about 15% as high a concentration of vitamin E as comparable samples from nonsmokers. Three weeks' administration of synthetic alpha-tocopheryl acetate (2,400 IU/d orally) increased the vitamin E level of smokers' alveolar fluid to only about 45% of unsupplemented nonsmokers' aveolar fluid. There was increased oxidation of tocopherol to tocopherol quinone in the smokers' lungs. Low vitamin E level in alveolar fluid of smokers suggests the likelihood of increased oxidant damage to their lung tissues, with diminished resistance to free radical-promoted carcinogenesis.

There appears to be a place for vitamin E supplementation as an adjunct to other cancer therapy. Animal studies have shown that vitamin E protects against side effects of radiation therapy and reduces toxicity of Adriamycin® in mice [176]. Oral vitamin E (40 mg/d of d-alpha-tocopheryl acetate) markedly reduced the cardiotoxicity of Adriamycin® in rabbits by reducing build-up of Adriamycin®-induced peroxide free radical generation [177]. In vitro, d-alpha-tocopheryl acid succinate enhanced the cytotoxic action of Adriamycin® against human prostatic carcinoma cells while decreasing its toxicity to normal cells [178]. Wood [179] reports that in 11 of 16 patients receiving therapy with this antitumor drug, simultaneous administration of 1,600 IU synthetic alpha-tocopheryl acetate per day prevented development of significant alopecia; vitamin E administration should be started 5–7 days before the first dose of Adriamycin®.

Other clinical possibilities. In the older literature, which was critically reviewed by Marks in 1962 [180], claims have been made for usefulness in other conditions, including senile vaginitis, infertility, muscular dystrophy, Dupuytren's contracture, fibrositis, Peyronie's disease, interstitial keratitis, nephritis, purpura, scleroderma, burns (topically), wound healing, and diabetes.

Confirmatory clinical studies, preferably under controlled conditions, are obviously needed for these miscellaneous proposed uses of vitamin E before credence can be given to their claimed effectiveness. A recent animal study by Till et al. [181] provides strong encouragement for clinical evaluation of oral vitamin E in burn patients. They found that in rats suffering burns over 30% of body area, vitamin E treatment protected their lungs from fluid accumulation, hemmorrhage, and structural damage.

TOXICITY AND SIDE EFFECTS

Oral alpha-tocopherol and its esters have very low toxicity according to controlled studies performed with healthy normal adults. Farrell and Bieri [182] carried out 20 standard clinical blood tests in a group of 28 adults who had taken vitamin E supplements of 100–800 IU/day for an average of 3 years; no disturbances were revealed in liver, kidney, muscle, thyroid gland, erythrocytes, leukocytes, coagulation parameters, or blood glucose. Tsai et al. [183] examined blood parameters in 202 college students who ingested 600 IU of synthetic alpha-tocopheryl acetate per day for 4 weeks. They found no effect on prothrombin time, total leukocyte count, or serum creatine phosphokinase. There was no muscle weakness or gastrointestinal disturbance. However, they noted a slight but statistically significant decrease in thyroid hormone (T-3 and T-4) levels. Bierenbaum et al. [113] administered 2,000 IU/day of synthetic alpha-tocopheryl acetate to 65 adults (25 normal, 15 diabetic, 15 postcoronary) in a double-blind cross-over study involving 6 weeks on vitamin E and 6 weeks on placebo; no significant side effects were noted, and laboratory tests showed no change in thyroid function.

There are numerous reports of rare side effects associated with high-dose administration of vitamin E, reviewed recently by Roberts [184]. These include breast tenderness, elevation of blood pressure, fatigue, myopathy, intestinal cramps, urticaria, and possible aggravation of diabetes. Roberts associates precipitation of thrombophlebitis in some of his patients with high-dose vitamin E [185] despite the known platelet antiaggregatory properties of the vitamin [109–111]. As evidence for his point of view, he cites general improvement in these patients after withdrawal of vitamin E, but it must be noted that they were receiving other treatment that could be credited with the improvement. The most that can be concluded from such evidence is that in thrombophlebitis-prone patients, vitamin E is probably an ineffective antithrombotic agent. Most of the other rare side effects that have been reported are probably coincidental occurrences, but some may reflect individual idiosyncratic reactions, and the physician should be aware of such possibilities.

In vitamin E treatment of the preterm infant, oral dosage with 100 mg/kg/day of synthetic alpha-tocopheryl acetate in medium-chain triglycerides for several months has been found safe [29,46], but intramuscular injection of alpha-tocopherol to produce a rapid initial elevation of serum vitamin E should be closely monitored to keep serum alpha-tocopherol from exceeding 3.5 mg/dl. *Rapid intravenous injection of vitamin E is absolutely contraindicated, as it may elevate serum tocopherol to levels associated with an increased incidence of intraventricular hemorrhage [36]. Prolonged elevation of serum levels by slow IV infusion of vitamin E is associated with necrotizing enterocolitis and sepsis [34,35], while oral administration of 200 mg/kg of vitamin E in hyperosmolar solutions produces an increased incidence of necrotizing enterocolitis* [48].

Several years ago the vitamin E product E-Ferol, introduced in 1983 for IV use in premature infants, was found to produce a syndrome of pulmonary deterioration, liver and kidney failure, and death. This led to recall of the product in 1984. Its composition was 25 IU of vitamin E as synthetic alpha-tocopheryl acetate per ml in an aqueous vehicle containing 9% polysorbate 80 and 1% polysorbate 20. Investigations in human cell systems by Alade et al. [186] revealed that addition of polysorbate 80, but not of tocopheryl acetate, to the culture media had adverse effects on lymphocytes and on liver and kidney cells. The "E-Ferol syndrome" is therefore believed to be primarily caused by the polysorbate 80 component of the preparation [186,187].

INTERACTIONS

A lengthened prothrombin time has been reported in a patient taking high-dose vitamin E while on therapy with warfarin and clofibrate [188]. Experimentally, administration of vitamin E enhances the coagulopathy of vitamin K-deficient laboratory animals [189]; therefore, in patients receiving anticoagulant therapy, high-dose vitamin E should be looked upon as a possible potentiator of anticoagulant activity, although in one study of 12 patients receiving warfarin therapy, vitamin E at dosages of 100 or 400 IU/d for 4 weeks did not result in any significant change in prothrombin time [190].

Studies of serum vitamin E in epileptic children suggest that anticonvulsant drugs may increase the vitamin E requirement [191]. When vitamin E-deficient rats were given high-dose phenobarbital for 5 days, hepatic vitamin E levels were further reduced and lipid peroxide levels increased [192]. Pall et al. [193] report that patients receiving phenothiazine have higher concentrations of lipid peroxidation products in their cerebrospinal fluid than controls. Side effects of many commonly used drugs may reflect induced increases in peroxide free radical production, and vitamin E supplementation may prove useful in controlling such side effects [192,193].

Excessive amounts of iron in infant formulas or administration of therapeutic iron can interfere with vitamin E absorption, probably by catalyzing oxidation of the vitamin in the GI tract [194]. Present-day formulas contain more vitamin E and less iron in order to prevent vitamin E deficiency from this cause [104]. Administration of B vitamins, especially riboflavin and vitamin B-6, simultaneously with vitamin E has been reported to reduce the absorption of tocopheryl acetate to some extent; no mechanism has been suggested [146,195]. This may be only a short-term effect, as individuals who had been taking "therapeutic" supplements of vitamin B-6 and other B vitamins for years, along with 400–800 IU of vitamin E daily, had plasma tocopherol levels 3 times as high as those of unsupplemented control subjects [196].

REFERENCES

1. Kasparek S: Chemistry of tocopherols and tocotrienols. In: *Vitamin E: A Comprehensive Treatise*, LJ Machlin (ed). New York: Marcel Dekker, Inc., 1980, pp. 7–65.
2. National Research Council: *Recommended Dietary Allowances*, 9th ed. Washington, DC: National Academy of Sciences, 1980.
3. Lehmann J, Martin HL, Lashley EL, et al.: Vitamin E in foods from high and low linoleic acid diets. J Am Diet Assoc 86:1208–1216, 1986.
4. Horwitt MK: The promotion of vitamin E. J Nutr 116:1371–1377, 1986.
5. Meydani SN, Shapiro AC, Meydani M, et al.: Effect of age and dietary fat (fish, corn and coconut oils) on tocopherol status of C57BL/6NIa mice. Lipids 22:345–350, 1987.
6. Cerutti PA: Prooxidant states and tumor promotion. Science 227:375–381, 1985.
7. Halliwell B: Oxidants and human disease: Some new concepts. FASEB J 1:358–364, 1987.
8. Machlin LJ, Bendich A: Free radical tissue damage: Protective role of antioxidant nutrients. FASEB J 1:441–445, 1987.
9. Lewin R: Drug trial for Parkinson's. Science 236:1420, 1987.
10. Cavarocchi NC, England MD, O'Brien JF, et al.: Superoxide generation during cardiopulmonary bypass: Is there a role for vitamin E? J Surg Res 40:519–527, 1986.
11. Gallo-Torres HE: Absorption, blood transport and metabolism. In: *Vitamin E: A Comprehensive Treatise*, LJ Machlin (ed). New York: Marcel Dekker, 1980, pp 170–267.
12. Chan AC, Raynor C, Douglas C, et al.: Transitory stimulation of human platelet 12-lipo oxygenase by vitamin E supplementation. Am J Clin Nutr 44:278–282, 1986.
13. Machlin LJ: Vitamin E. In: *Handbook of Vitamins*, LJ Machlin (ed). New York: Marcel Dekker, 1984, pp 99–145.
14. Traber MG, Kayden HJ: Vitamin E is delivered to cells vita the high affinity receptor for low-density lipoprotein. Am J Clin Nutr 40:747–751, 1984.
15. Nakamura T, Masugi F: Transfer of α-tocopherol from plasma to erythrocytes in vitamin E-deficient rats. Int J Vitam Nutr Res 49:364–369, 1979.
16. Horwitt MK, Harvey CC, Dahm Jr CH, Searcy MT: Relationship between tocopherol and serum lipid levels for determination of nutritional adequacy. Ann NY Acad Sci 203:223–235, 1972.
17. Gutcher GR, Raynar WJ, Farrell PM: An evaluation of vitamin E status in premature infants. Am J Clin Nutr 40:1078–1089, 1984.
18. Thurnham DI, Davies JA, Crump BJ, et al.: The use of different lipids to express serum tocopherol:lipid ratios for the measurement of vitamin E status. Am Clin Biochem 23:514–520, 1986.
19. Evans CD, List GR, Dolev A, et al.: Pentane from thermal decomposition of lipoxidase-derived products. Lipids 2:432–434, 1967.
20. Dillard CJ, Litov RE, Tappel AL: Effects of dietary vitamin E, selenium, and polyunsaturated fats on in vivo lipid peroxidation in the rat as measured by pentane production. Lipids 13:396–402, 1978.
21. Downey JE, Irving DH, Tappel AL: Effects of dietary antioxidants on in vivo lipid peroxidation in the rat as measured by pentane production. Lipids 13:403–407, 1978.

22. Lemoyne M, Van Gossum A, Jurian R, et al.: Breath pentane analysis as an index of lipid peroxidation; a functional test of vitamin E status. Am J Clin Nutr 46:267–272, 1987.
23. Traber MG, Kayden HJ: Tocopherol distribution and intracellular localization in human adipose tissue. Am J Clin Nutr 46:488–495, 1987.
24. Alvarez RA, Liou GI, Fong S-L, Bridges CDB: Levels of alpha- and gamma-tocopherol in human eyes: Evaluation of the possible role of IRBP in intraocular alpha-tocopherol transport. Am J Clin Nutr 46:481–487, 1987.
25. Behrens WA, Madere R: Alpha- and gamma-tocopherol concentrations in human serum. J Am Coll Nutr 5:91–96, 1986.
26. Behrens WA, Madere R: Mechanisms of absorption, transport and tissue uptake of RRR-alpha-tocopherol and d-gamma-tocopherol in the rat. J Nutr 117:1562–1569, 1987.
27. Baker H, Handelman GJ, Short S, et al.: Comparison of plasma alpha and gamma tocopherol levels following chronic oral administration of either *all-rac*-alpha-tocopheryl acetate or RRR-alpha-tocopheryl acetate in normal adult male subjects. Am J Clin Nutr 43:382–387, 1986.
28. Horwitt MK, Elliott WH, Kanjananggulpany P, Fitch CD: Serum concentrations of alpha-tocopherol after ingestion of various vitamin E preparations. Am J Clin Nutr 40:240–245, 1984.
29. Hittner HM, Speer ME, Rudolph AJ, et, al.: Retrolental fibroplasia and vitamin E in the preterm infant—comparison of oral versus intramuscular: oral administration. Pediatrics 73:238–249, 1984.
30. Hittner HM, Kretzer FL: Toxicity of vitamin E in preterm infants. In: *Retinopathy of Prematurity: Current Concepts and Controversies*, AR McPherson, HM Hittner, FL Kretzer (eds). Toronto; B.C. Decker Inc., 1986, pp 111–116.
31. Chiswick ML, Wynn J, Toner N: Vitamin E and intraventricular hemorrhage in the newborn. Ann NY Acad Sci 393:109–120, 1982.
32. Chiswick ML, Sinha S, Davies J, Sims DG: Vitamin E protects against periventricular haemorrhage in preterm babies (abstract). British Pediatric Association meeting, York, UK, April 15–19, 1986.
33. Sinha S, Davies J, Sims DG, Chiswick ML: Muscle paralysis and vitamin E in the prevention of periventricular haemorrhage (PVH) (abstract). British Pediatric Association meeting, York, UK, April 15–19, 1986.
34. Sobel S, Guerigian J, Troendle G, et al.: Vitamin E in retrolental fibroplasia (letter). N Engl J Med 306:867, 1982.
35. Johnson L, Bowen F, Herrmann N, et al.: The relationship of prolonged elevation of serum vitamin E levels to neonatal bacterial sepsis (SEP) and necrotizing enterocolitis (NEC). Pediatr Res 17:319A, 1983.
36. Rosenbaum AL, Phelps DL, Isenberg DJ, et al.: Retinal hemorrhage in retinopathy of prematurity associated with tocopherol treatment. Ophthalmology 92:1012–1014, 1985.
37. Hittner HM, Godio LB, Rudolph AJ, et al.: Retrolental fibroplasia: Efficacy of vitamin E in a double-blind clinical study of preterm infants. N Engl J Med 305:1365–1371, 1981.
38. Abbasi S, Ludomirski A, Bhutani V, et al.: Maternal and fetal serum vitamin E to total lipid ratio during gestational development as determined by percutaneous umbilical blood sampling (PUBS). J Am Coll Nutr 6:445A, 1987.
39. Ojo CO, Dawodu AH, Osifo BOA: Vitamin E deficiency in the pathogenesis of hemolysis and hyperbilirubinemia of neonatal jaundice. J Trop Pediatr 32:251–254, 1986.
40. Finer NN, Grant G, Schindler RF, et al.: Effect of intramuscular vitamin E on frequency and severity of retrolental fibroplasia: A controlled trial. Lancet 1:1087–1091, 1982.

41. Schaffer DL, Johnson L, Quinn GE, et al.: Vitamin E and retinopathy of prematurity: The ophthalmologist's perspective. In: *Retinopathy of prematurity: Update 1985*, JT Flynn, DL Phelps (eds). New York: Alan R. Liss Inc., 1986.
42. Glass P, Avery GB, Subramanian KNS, et al.: Effect of bright light in the hospital nursery on the incidence of retinopathy of prematurity. N Engl J Med 313:401–404, 1985.
43. Kretzer FL, Mehta RS, Johnson AT, et al.: Vitamin E protects against retinopathy of prematurity through action on spindle cells. Nature 309:793–795, 1984.
44. Kretzer FL, McPherson AR, Hittner HM: An interpretation of retinopathy of prematurity in terms of spindle cells: Relationship to vitamin E prophylaxis and cryotherapy. Graefes Arch Clin Exp Ophthalmol 224:205–214, 1986.
45. Kretzer FL, McPherson AR, Rudolph AJ, Hittner HM: Pathogenic mechanism of retinopathy of prematurity: a controversial explanation for the efficacy of oral and intramuscular vitamin E supplementation and cryotherapy. Bull NY Acad Med 61:883–902, 1985.
46. Hittner HM, Rudolph AJ, Kretzer FL: Suppression of severe retinopathy of prematurity with vitamin E supplementation. Ultrastructural mechanism of clinical efficacy. Ophthalmology 91:1512–1523, 1984.
47. Hittner HM, Kretzer FL: Efficacy of vitamin E in retinopathy of prematurity. In: *Retinopathy of Prematurity: Current Concepts and Controversies*, AR McPherson, HM Hittner, FL Kretzer (eds). Toronto: B.C. Decker Inc., 1986, pp 89–103.
48. Finer NN, Peters KL, Hayek Z, Merkel CL: Vitamin E and necrotizing enterocolitis. Pediatrics 73:387–393, 1984.
49. Johnson L, Schaffer D, Quinn G, et al.: Vitamin E supplementation and the retinopathy of prematurity. Ann NY Acad Sci 393:473–495, 1982.
50. F.L. Kretzer and H.M. Hittner, private communication.
51. Johnson LH, Quinn GE, Abbasi S, Bowen FW: Vitamin E and retinopathy of prematurity (ROP). J Am Coll Nutr 6:85A, 1987.
52. Speer ME, Blifeld C, Rudolph AJ, et al.: Intraventricular hemorrhage and vitamin E in the very low birth weight infant: Evidence for efficacy of early intramuscular administration. Pediatrics 74:1107–1112, 1984.
53. Sinha S, Toner N, Davies J, et al.: Vitamin E supplementation reduces frequency of periventricular hemorrhage in very preterm babies. Lancet 1:446–470, 1987.
54. Gross SJ: Vitamin E and neonatal bilirubinemia. Pediatrics 64:321–323, 1979.
55. Ehrenkranz RA, Bonta BW, Ablow RC, Warshaw JB: Amelioration of bronchopulmonary dysplasia after vitamin E administration. N Engl J Med 299:564–569, 1978.
56. Ehrendranz RA, Ablow RC, Warshaw JB: Prevention of bronchopulmonary dysplasia with vitamin E administration during the acute stages of respiratory distress syndrome. J Pediatr 95:873–878, 1979.
57. Saldanha RL, Cepeda EE, Poland RL: The effect of vitamin E prophylaxis on the incidence and severity of bronchopulmonary dysplasia. J Pediatr 101:89–93, 1982.
58. Ward JA, Roberts RJ: Vitamin E inhibition of the effects of hyperoxia on the pulmonary surfactant system of the newborn rabbit. Pediatr Res 18:329–334, 1984.
59. Huijbers WA, Schrijver J, Speek AJ, et al.: Persistent low plasma vitamin E levels in premature infants surviving respiratory distress syndrome. Eur J Pediatr 145:170–171, 1986.
60. Sharma S, et al.: Comparison of blood levels of vitamin A, beta-carotene, and vitamin E in abruptio placentae with normal pregnancy. Int J Vitam Nutr Res 56:3–9, 1986.
61. Yoshioka T, Motoyama H, Yamasaki F, et al.: Lipid peroxidation and vitamin E levels during pregnancy in rats. Biol Neonate 52:223–231, 1987.
62. Mino M, Nagamatu M: An evaluation of nutritional status of vitamin E in pregnant women with respect to red blood cell tocopherol level. Int J Vitam Nutr Res 56:149–153, 1986.

63. Abrams AA: Use of vitamin E in chronic cystic mastitis (letter). N Engl J Med 272:1080–1081, 1965.
64. Solomon D, Strummer D, Nair PP: Relationship between vitamin E and urinary excretion of ketosteroid fractions in cystic mastitis. Ann NY Acad Sci 203:103, 1972.
65. London RS, Solomon D, London ED, et al.: Mammary dysplasia: Clinical response and urinary excretion of 11-desoxy-17-ketosteroids and pregnanediol following alpha tocopherol therapy. Breast 4:19, 1978.
66. London RS, Sundaram GS, Schultz M, et al.: Endocrine parameters and α-tocopherol therapy of patients with mammary dysplasia. Cancer Res 41:3811–3814, 1981.
67. Sundaram GS, London R, Margolis S, et al.: Serum hormones and lipoproteins in benign breast disease. Cancer Res 41:3814–3816, 1981.
68. Sundaram GS, London R, Manimekalai S, et al.: Alpha tocopherol and serum lipoproteins. Lipids 16:223–227, 1981.
69. London R, Sundaram G, Manimekalai S, et al.: Mammary dysplasia: Endocrine parameters and tocopherol therapy. Nutr Res 2:243–247, 1982.
70. London RS, Sundaram GS, Goldstein PJ: Medical management of mammary dysplasia. Obstet Gynecol 59:519–523, 1982.
71. London RS, Sundaram GS, Murphy L, et al.: The effect of vitamin E on mammary dysplasia: A double-blind study. Obstet Gynecol 65:104–106, 1985.
72. Ernster VL, Goodson WH, Hunt TK, et al.: Vitamin E and benign breast "disease": A double-blind, randomized clinical trial. Surgery 97:490–494, 1985.
73. London RS, Sundaram GS, Murphy L, Goldstein PJ: The effect of α-tocopherol on premenstrual symptomatology: A double-blind study. J Am Coll Nutr 2:115–122, 1983.
74. Abraham G: Premenstrual tension. In: *Current Problems in Obstetrics and Gynecology.* Chicago: Year Book Medical Publishers, 1981, pp 1–39.
75. London RS, Murphy L, Kitlowski KE, Reynolds MA: Efficacy of alpha-tocopherol in the treatment of premenstrual syndrome. J Reprod Med 32:400–404, 1987.
76. London RS, Sundaram G, Manimekalai, et al.: The effect of alpha-tocopherol in premenstrual symptomatology. II. Endocrine correlates. J Am Coll Nutr 3:351–356, 1984.
77. Guggenheim MA, Jackson V, Lilly J, Silverman A: Vitamin E deficiency and neurologic disease in children with cholestasis: A prospective study. J Pediatr 102:577–579, 1983.
78. Sokol RJ, Bove KE, Heubi JE, Iannaconne ST: Vitamin E deficiency during chronic childhood cholestasis: Presence of sural nerve lesion prior to 2-1/2 years of age. J Pediatr 103:197–204, 1983.
79. Sokol RJ, Farrell MK, Heubi JE, et al.: Comparison of vitamin E and 25-hydroxyvitamin D absorption during childhood cholestasis. J Pediatr 103: 712–717, 1983.
80. Sokol RJ, Heubi JE, Iannaccone ST, et al.: Vitamin E deficiency with normal serum vitamin E concentrations in children with chronic cholestasis. N Engl J Med 310:1209–1212, 1984.
81. Congden PJ, Bruce G, Rothburn MM, et al.: Vitamin status in treated patients with cystic fibrosis. Arch Dis Child 56:708–714, 1981.
82. Sung JG, Park SH, Mastri AR, Warwick WJ: Axonal dystrophy in the gracile nucleus in congenital biliary atresia and cystic fibrosis (mucoviscidosis): Beneficial effect of vitamin E therapy. J Neuropathol Exp Neurol 39:584–597, 1980.
83. Rosenblum JL, Keating JP, Prensky AP, Nelson TL: A progressive, disabling, neurologic syndrome in children with chronic liver disease: A possible result of vitamin E deficiency. N Engl J Med 304:503, 1981.
84. Muller D: Vitamin E—its role in neurological function. Postgrad Med J 62:107–112, 1986.
85. Satya-Murti S, et al.: The spectrum of neurological disorders from vitamin E deficiency. Neurology 36:917–921, 1986.

86. Tomas LG: Reversibility of human myopathy caused by vitamin E deficiency. Neurology 29:1182–1186, 1979.
87. Tazawa Y, Nakagawa M, Yamada M, et al.: Serum vitamin E levels in children with corrected biliary artresia. Am J Clin Nutr 40:246–250, 1984.
88. Sokol RJ, Guggenheim M, Iannaccone S, et al.: Improved neurologic function after long-term correction of vitamin E deficiency in children with chronic cholestasis. N Engl J Med 313:1580–1586, 1985.
89. Sokol RJ, Heubi JE, McGraw C, Balistreri WF: Correction of vitamin E deficiency in children with chronic cholestasis. II. Effect on gastrointestinal and hepatic function. Hepatology 6:1263–1269, 1986.
90. Runge P, Muller DPR, McAllister J, et al.: Oral vitamin E supplements can prevent the retinopathy of abetalipoproteinemia. Br J Ophthalmol 70:166–173, 1986.
91. Vatassery GT, Chiang T:Serum α-tocopherol, lipids, potassium, and creatine phosphokinase in normal and malabsorption patients. Am J Clin Nutr 32:2061–2064, 1979.
92. Dutta SK, Bustin MP, Russell RM, Costa BS: Deficiency of fat-soluble vitamins in treated patients with pancreatic insufficiency. Ann Intern Med 97:549–552, 1982.
93. Sokol RJ, Balistreri WF, Hoofnagle JH, Jones EA: Vitamin E deficiency in adults with chronic liver disease. Am J Clin Nutr 41:66–72, 1985.
94. Elias E, Muller DPR, Scott J: Association of spinocerebellar disorders with cystic fibrosis or chronic childhood cholestasis and very low serum vitamin E. Lancet 2:1319–1321, 1981.
95. Stead RJ, Muller DPR, Mathews S, et al.: Effect of abnormal liver function on vitamin E status and supplementation in adults with cystic fibrosis. Gut 27:714–718, 1986.
96. Harding AE, Muller DPR, Thomas PK, Willison HJ: Spinocerebellar degeneration secondary to chronic intestinal malabsorption: A vitamin E deficiency syndrome. Ann Neurol 12:419–424, 1982.
97. Howard L, Ovesen L, Satya-Murti S, Chu R: Reversible neurological symptoms caused by vitamin E deficiency in a patient with short bowel syndrome. Am J Clin Nutr 36:1243–1249, 1982.
98. Traber MG, Sokol RJ, Ringel SP, et al.: Lack of tocopherol in peripheral nerves of vitamin E-deficient patients with peripheral neuropathy. N Engl J Med 317:262–265, 1987.
99. Arria AM, Tarter RE, Warty V, Van Thiel DH: Vitamin E and neurologic function in adults with chronic cholestatic liver disease. J Am Coll Nutr 6:442A, 1987.
100. Tanimura H, Kato H, Hikasa Y: The role of vitamin E in the etiology and treatment of pancreatitis. Ann NY Acad Sci 393:214–216, 1982.
101. Kelleher J, Miller MG, Littlewood JM, et al.: The clinical effect of correction of vitamin E depletion in cystic fibrosis. Int J Vitam Nutr Res 57:253–259, 1987.
102. Jeffrey GP, Muller DPR, Burroughs AK, et al.: Vitamin E deficiency and its clinical significance in adults with primary biliary cirrhosis and other forms of chronic liver disease. J Hepatol 4:307–317, 1987.
103. Shute EV, Vogelsang AB, Skelton FR, Shute WE: The influence of vitamin E on vascular disease. Surg Gynecol Obstet 86:1–8, 1948.
104. Bieri JG, Corash L, Hubbard VS: Medical uses of vitamin E. N Engl J Med 308:1063–1071, 1983.
105. Haeger K: Long-term study of α-tocopherol in intermittent claudication. Ann NY Acad Sci 393:369–375, 1982.
106. Herman WJ Jr: The effect of vitamin E on lipoprotein cholesterol distribution. Ann NY Acad Sci 393:467–472, 1982.

107. Barboriak JJ, Shetty KR, El-Ghatit AZ, Kalbfleisch JH: Plasma high-density lipoprotein cholesterol and vitamin E supplements. Ann NY Acad Soc 393:174, 1982.
108. Chapkin RS, Haberstroh B, Liu T, Holub BJ: Effect of vitamin E supplementation on serum and high-density lipoprotein-cholesterol in renal patients on maintenance hemodialysis. Am J Clin Nutr 38:253–256, 1983.
109. Szczeklik A, et al.: Dietary supplementation with vitamin E in hyperlipoproteinemias: Effects on plasma lipid peroxides, antioxidant activity, prostacyclin generation and platelet aggregability. Thromb Haemost 54:425–430, 1985.
110. Stuart MJ: Vitamin E deficiency: Its effect on platelet-vascular interaction in various pathologic states. Ann NY Acad Sci 393:277–288, 1982.
111. FitzGerald GA, Brash AR: Endogenous prostacyclin and thromboxane biosynthesis during chronic vitamin E therapy in man. Ann NY Acad Sci 393:209–211, 1982.
112. Editorial: Aspirin: what dose? Lancet 1:592–593, 1986.
113. Bierenbaum ML, Noonan FJ, Machlin LJ, et al.: The effect of supplemental vitamin E on serum parameters in diabetics, post-coronary and normal subjects. Nutr Rep Int 31:1171–1180, 1985.
114. Pritchard KA, et al.: Triglyceride-lowering effect of dietary vitamin E in streptozotocin-induced diabetic rats. Diabetes 35:278–281, 1986.
115. Natta C, Machlin L: Plasma levels of tocopherol in sickle cell anemia. Am J Clin Nutr 32:1359–1362, 1979.
116. Chiu D, Vichinsky E, Yee M, et al.: Peroxidation, vitamin E, and sickle-cell anemia. Ann NY Acad Sci 393:323–335, 1982.
117. Rachmilewitz EA, Kornberg A, Acker M: Vitamin E deficiency due to increased consumption in β-thalassemia and Gaucher's disease. Ann NY Acad Sci 393:336–347, 1982.
118. Natta CL, Machlin, LJ, Brin M: A decrease in irreversibly sickled erythrocytes in sickle cell anemia patients given vitamin E. Am J Clin Nutr 33:968–971, 1980.
119. Corash L, Spielberg S, Bartsocas C, et al.: Reduced chronic hemolysis during high-dose vitamin E administration in Mediterranean-type-glucose-6-phosphate dehydrogenase deficiency. N Engl J Med 303:416–420, 1980.
120. Corash LM, Sheetz M, Bieri JG, et al.: Chronic hemolytic anemia due to glucose-6-phosphate dehydrogenase deficiency of glutathione synthetase deficiency: The role of vitamin E in its treatment. Ann NY Acad Sci 393:348–360, 1982.
121. Johnson GJ, Vatassery GT, Finkel B, Allen DW: High-dose vitamin E does not decrease the rate of chronic hemolysis in glucose-6-phosphate dehydrogenase deficiency. N Engl J Med 308:1014–1017, 1983.
122. Machtey I, Ouaknine L: Tocopherol in osteoarthritis: A controlled pilot study. J Am Geriat Soc 26:328–330, 1978.
123. Levy L: The antiinflammatory action of some compounds with antioxidant properties. Inflammation 1:333–345, 1976.
124. Wolf H, Seeger HW: Prevention of respiratory distress syndrome by high dose α-tocopherol: Experimental results and clinical aspects. Int J Vit Nutr Res 51:181–183, 1981.
125. Wolf HRD, Seeger HW: Experimental and clinical results in shock lung treatment with vitamin E. Ann NY Acad Sci 393:392–410, 1982.
126. Taniguchi H, Tagaki K, Satake T, et al.: Mechanism of rat lung microsomal dysfunction induced by oxygen breathing in relation to the effect of alpha-tocopherol. Arzneimittelforsch/Drug Res 36:924–927, 1986.
127. Baker JL: The effectiveness of alpha-tocopherol (vitamin E) in reducing the incidence of spherical contracture around breast implants. Plast Reconstruct Surg 68:696–698, 1981.

128. Kagoma P, et al.: The effect of vitamin E on experimentally induced peritoneal adhesions in mice. Arch Surg 120:949–951, 1985.
129. Hayden RE, Yeung CST, Paniello RC, et al.: The effect of glutathione and vitamins A, C and E on acute skin flap survival. Laryngoscope 97:1176–1179, 1987.
130. Harman D: Nutritional implications of the free-radical theory of aging. J Am Coll Nutr 1:27–34, 1982.
131. Meydani M, et al.: Effect of vitamin E, selenium and age on lipid peroxidation events in rat cerebrum. Nutr Res 5:1227–1236, 1985.
132. Meydani M, Macauley JB, Blumberg JB: Influence of dietary vitamin E, selenium and age on regional distribution of alpha-tocopherol in the rat brain. Lipids 21:786–791, 1986.
133. Chen LH: The effect of age and dietary vitamin E on the tissue lipid peroxidation of mice. Nutr Rep Int 10:339–344, 1974.
134. Vatassery GT, Johnson GJ, Krezowski AM: Changes in vitamin E concentrations in human plasma and platelets with age. J Am Coll Nutr 4:369–375, 1983.
135. Vandewoude MFJ, Vandewoude MG: Vitamin E status in a normal population: The influence of age. J Am Coll Nutr 6:307–311, 1987.
136. Wartanowicz M, et al.: The effect of alpha-tocopherol and ascorbic acid on the serum peroxide level in elderly people. Ann Nutr Metab 28:186–191, 1984.
137. Makinodan T: Mechanism of senescence of immune response. Fed Proc 37:1239–1240, 1978.
138. Bendich A, Gabriel E, Machlin LJ: Dietary vitamin E requirement for optimum immune responses in the rat. J Nutr 116:675–681, 1986.
139. Meydani SN, Meydani M, Verdon CP, et al.: Vitamin E supplementation suppresses prostaglandin E_2 synthesis and enhances the immune response of aged mice. Mech Ageing Dev 34:191–201, 1986.
140. Harman D, Heidrick ML, Eddy DE: Free radical theory of aging: Effect of free-radical-reaction inhibitors on the immune response. J Am Geriatr Soc 25:400–407, 1977.
141. Nair PP, Mezey E, Murty HS, et al.: Vitamin E and porphorin metabolism in man. Arch Intern Med 128:411–415, 1971.
142. Ayres S Jr, Mihan R: Is vitamin E involved in the autoimmune mechanism? Cutis 21:321–325, 1978.
143. Ayres S Jr, Mihan R: Lupus erythematosus and vitamin E: An effective and nontoxic therapy. Cutis 23:49–54, 1979.
144. Catchcart RF: Leg cramps and vitamin E (letter). JAMA 219:216–217, 1972.
145. Ayres S Jr, Mihan R: Leg cramps and vitamin E (letter). JAMA 219:217, 1972.
146. Nadiger HA: Role of vitamin E in the aetiology of phrynoderma (follicular hyperkatosis) and its interrelationship with B-complex vitamins. Br J Nutr 44:211–214, 1980.
147. Tanner AR, Bantock I, Hinks L et al.: Depressed selenium and vitamin E levels in an alcoholic population. Dig Dis Sci 31:1307–1312, 1986.
148. Bjorneboe G-E, Bjorneboe A, Hagen B, et al.: Reduced hepatic alpha-tocopherol content after long-term administration of ethanol to rats. Biochim Biophys Acta 918:236–241, 1987.
149. Lu W, Bantok I, Desai S, et al.: Aminoterminal procollagen III peptide elevation in alcoholics who are selenium and vitamin E deficient. Clin Chim Acta 154:165–170, 1986.
150. Fink R, et al.: Increased free-radical activity in alcoholics. Lancet 2:291–294, 1985.
151. Edes I, et al.: Myocardial lipid peroxidation in rats after chronic alcohol ingestion and effects of different antioxidants. Cardiovasc Res 20:542–548, 1986.
152. Fraga CG, ARias RF, Llesuy SF, et al.: Effect of vitamin E and seleniumn deficiency on rat liver chemiluminescence. Biochem J 242:383–386, 1987.

153. Sclafani L, et al.: Protective effect of vitamin E in rats with acute liver injury. J Paren Ent Nutr 10:184–187, 1986.
154. Marubayashi S, et al.: Role of free radicals in ischemic rat liver cell injury—prevention of damage by vitamin E, coenzyme Q-10 or reduced glutathione administration. Surg Forum 36:136–138, 1985.
155. Ono K, et al.: Effects of large dose vitamin E supplementation on anemia in hemodialysis patients. Nephron 40:440–445, 1985.
156. Lubrano R, et al.: Relationship between red blood cell lipid peroxidation, plasma hemoglobin, and red blood cell osmotic resistance before and after vitamin E supplementation in hemodialysis patients. Artif Organs 10:245–250, 1986.
157. Packer L: Vitamin E, physical exercise and tissue damage in animals. Med Biol 62:105–109, 1984.
158. Gohil K, et al.: Vitamin E deficiency and vitamin supplements: Exercise and mitochondrial oxidation. J Appl Physiol 60:1986–1991, 1986.
159. Gohil K, Rothfuss L, Lang J, Packer L: Effect of exercise training on tissue vitamin E and ubiquinone content. J Appl Physiol 63:1638–1641, 1987.
160. Summerfield FW, Tappel AL: Effects of dietary polyunsaturated fats and vitamin E on aging and peroxidative damage to DNA. Arch Biochem Biophys 233:408–416, 1984.
161. Stahelin HB, Rosel F, Buess E, Brubacher G: Cancer, vitamins and plasma lipids: Prospective Basel study. J Natl Cancer Inst 73:1463–1468, 1984.
162. Menkes MS, Comstock GW, Vuilleumier JP, et al.: Serum beta-carotene, vitamins A and E, selenium, and the risk of lung cancer. N Engl J Med 315:1250–1254, 1986.
163. Miyamoto H, Aroya Y, Ito M, et al.: Serum selenium and vitamin E concentrations in families of lung cancer patients. Cancer 60:1159–1162, 1987.
164. Wald NJ, Boreham J, Hayward JL, Bulbrook RD: Plasma retinol, beta-carotene and vitamin E levels in relation to the future risk of breast cancer. Br J Cancer 49:321–324, 1984.
165. Salonen JT, Salonen R, Lappetelainen R, et al.: Risk of cancer in relation to serum concentrations of selenium and vitamins A and E: Matched case-control analysis of prospective data. Br Med J 290:417–420, 1985.
166. Kok FJ, Van Duijn CM, Hofman A, et al.: Micronutrients and the risk of lung cancer. N Engl J Med 316:1416, 1987.
167. Borek C, Ong A, Mason H, et al.: Selenium and vitamin E inhibit radiogenic and chemically induced transformation in vitro via different mechanisms. Proc Natl Acad Sci USA 83:1490–1494, 1986.
168. Mergens WJ: Efficacy of vitamin E to prevent nitrosamine formation. Ann NY Acad Sci 393:61–69, 1982.
169. Wagner DA, Shuker DEG, Bilmazes C, et al.: Effect of vitamins C and E on endogenous synthesis of N-nitrosamino acids in humans: Precursor-product studies with [^{15}N]nitrate. Cancer Res 45:6519–6522, 1985.
170. Bruce WR, Dion PW: Studies relating to a fecal mutagen. Am J Clin Nutr 33:2511–2512, 1980.
171. Dion PW, Bright-See EB, Smith CC, Bruce WR: The effect of dietary ascorbic acid and alpha-tocopherol on fecal mutagenicity. Mutat Res 102:27–37, 1982.
172. Trickler D, Shklar G: Prevention by vitamin E of experimental oral carcinogenesis. J Natl Cancer Inst 78:165–169, 1987.
173. Shklar G, Schwartz J, Trickler DP, Niukian K: Regression by vitamin E of experimental oral cancer. J Natl Cancer Inst 78:987–992, 1987.
174. Slaga TJ, Bracken WM: The effects of antioxidants on skin tumor initiation and aryl hydrocarbon hydroxylase. Cancer Res 37:1631–1635, 1977.

175. Pacht ER, Kasaki H, Mohammed JR, et al.: Deficiency of vitamin E in the alveolar fluid of cigarette smokers: Influence on alveolar macrophage cytotoxicity. J Clin Invest 77:789-796, 1986.
176. Tanigawa N, Katoh H, Kan N, et al. Effect of vitamin E on toxicity and antitumor activity of Adriamycin in mice. Cancer Res 77:1249-1255, 1986.
177. Milei J, Boveris A, Llesuy, et al.: Amelioration of Adriamycin-induced cardiotoxicity in rabbits by prenylamine and vitamins A and E. Am Heart J 111:95-102, 1986.
178. Ripoll EAP, Rama BN, Webber MM: Vitamin E enhances the chemotherapeutic effects of Adriamycin on human prostatic cancer cells in vitro. J Urol 136:529-531, 1986.
179. Wood LA: Vitamin E in advance of administration of chemotherapy prevents or reduces hair loss. N Engl J Med 312:1060, 1985.
180. Marks J: Critical appraisal of the therapeutic value of α-tocopherol. Vitam Horm 20:573-598, 1962.
181. Till GO, Hatherill JR, Tourtelotte WW, et al.: Lipid peroxidation and acute lung injury after thermal trauma to skin. Evidence of a role for hydroxyl radical. Am J Pathol 119:376-384,1985.
182. Farrell PM, Bieri JG. Megavitamin E supplementation in man. Am J Clin Nutr 28:1381-6, 1975.
183. Tsai AC, Kelley JJ, Pen G, Cook N: Study on the effect of megavitamin E supplementation in man. Am J Clin Nutr 31:831-837, 1978.
184. Roberts HJ: Perspective on vitamin E as therapy. JAMA 246:129-131, 1981.
185. Roberts HJ: Thrombophlebitis associated with vitamin E therapy. Angiology 30:169-177, 1979.
186. Alade SL, Brown RE, Paquet A Jr: Polysorbate 80 and E-Ferol toxicity. Pediatrics 77:593-597, 1986.
187. Anon: Mystery of the E-Ferol syndrome. Nutr Rev 45:76-77, 1987.
188. Corrigan JJ Jr, Marcus FL: Coagulopathy associated with vitamin E ingestion. JAMA 230:1300-1301, 1974.
189. Corrigan JJ Jr: The effect of vitamin E on warfarin-induced vitamin K deficiency. Ann NY Acad Sci 393:361-367, 1982.
190. Corrigan JJ Jr, Ulfers LL: Effect of vitamin E on prothrombin levels in warfarin-induced vitamin K deficiency. Am J Clin Nutr 34:1701-1705, 1981.
191. Ogunmekan AO: Relationship between age and vitamin E level in epileptic and normal children. Am J Clin Nutr 32:2269-2271, 1979.
192. Ono J, Mimaki T, Yabuuchi H: Effects of phenobarbital on lipid peroxidation in vitamin E-deficient rats. Pediatr Pharmacol 5:223-227, 1986.
193. Pall HS, Blake DR, Williams AC, Lunec J: Evidence of enhanced lipid peroxidation in cerebrospinal fluid of patients taking phenothiazines. Lancet 2:596-599, 1987.
194. Draper HH: Nutrient interrelationships. In: *Vitamin E: A Comprehensive Treatise*, LJ Machlin (ed) New York: Marcel Dekker, 1980, pp 272-288.
195. Nadiger HA: Studies on interrelationship between vitamins E and B-complex. Ann Nutr Metab 24:352-356, 1980.
196. Baker H, Pauling L, Frank O: Mega-ascorbate taken with other vitamins permits elevation of circulating vitamins including B-12 in humans. Nutr Rep Int 23:669-677, 1981.

4
Vitamin K

Introduction	93
Absorption, Metabolism, and Excretion	95
Dosage Forms	96
Clinical Uses	96
Hypoprothrombinemia of Infancy	96
Acute Care Patients	98
The Elderly	98
Malabsorption	99
Genetic Vitamin K Dependency	99
Toxicity and Side Effects	100
Interactions	100
References	101

INTRODUCTION

The term vitamin K is used generically to designate 2-methyl-1, 4-naphthoquinone (also called menadione, menaquinone, or vitamin K-3) and all 3-substituted derivatives of this compound that show antihemorrhagic activity in vitamin K-deficient animals. The form that occurs naturally in plants, 2-methyl-3-phytyl-1, 4-naphthoquinone, has been given the designation vitamin K-1 and is also known as phylloquinone or phytonodione. (See Fig. 1 for structural formulas.) Other naturally occurring forms of vitamin K have longer side chains at the 3-position [1–5].

Phylloquinone is a viscous liquid. The synthesized K-3 form, menadione, is a bright yellow powder. Both of these are lipid-soluble. A synthetic water-soluble derivative, menadiol sodium diphosphate, is available, which in the body is converted to menadione and alkylated at the 3-position with a polyisoprenoid chain [4]. All forms of vitamin K are light-sensitive; solutions should be stored in the dark [1].

Vitamin K is necessary for the production of normal plasma clotting factors, specifically factors II (prothrombin), VII, IX, and X. The activity of

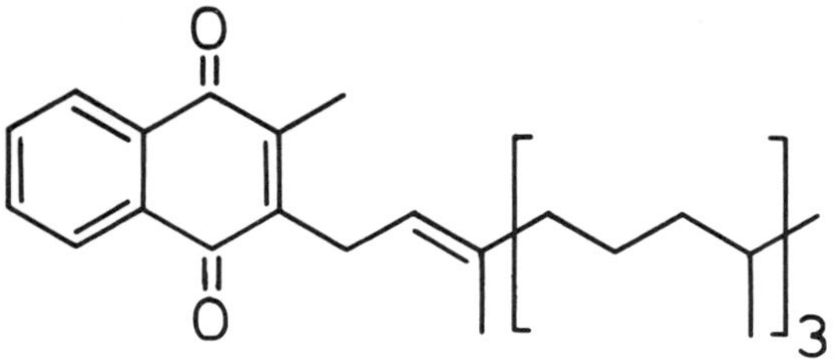

**phylloquinone
(phytonodione, Vitamin K-1)**

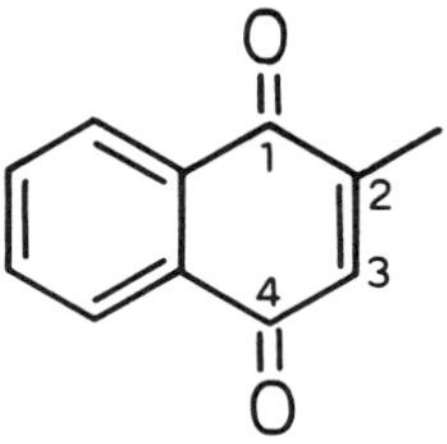

**menadione
(Vitamin K-3)**

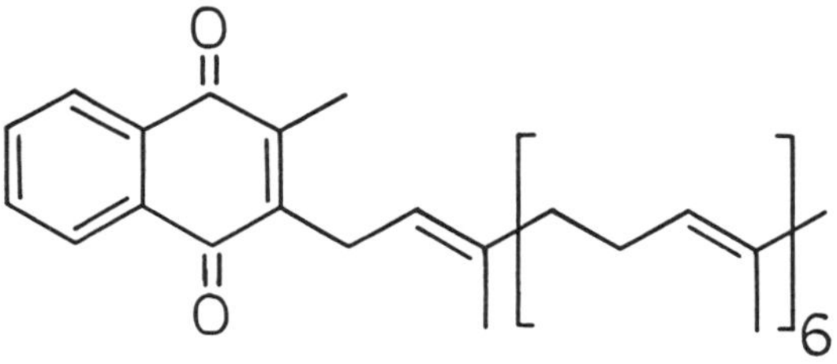

**menaquinone-7
(Vitamin K-2 series)**

Fig. 1. Structural formulas of vitamins K-1, K-2, and K-3.

these proteins depends on their ability to bind calcium ions, which in turn depends on the vitamin K-dependent introduction of gamma-carboxyl groups in glutamyl residues near the amino-terminal ends of their polypeptide chains. Other proteins containing gamma-carboxylated glutamyl residues have been found in plasma, bone, kidney, and other tissues whose function is not yet understood, except for plasma protein C, which acts to shut down the overall coagulation reaction. The gamma-carboxylation of clotting factors is carried out by a hepatic carboxylase through a molecular mechanism that is not fully understood but that is closely coupled with simultaneous epoxidation of vitamin K hydroquinone, the active form of the vitamin. There is an epoxide reductase in liver microsomes that reduces vitamin K epoxide back to the hydroquinone, thus conserving the vitamin. In vitamin K deficiency, the liver continues to synthesize clotting factor precursors that circulate in inactive undercarboxylated forms. The presence of abnormal plasma prothrombin during vitamin K deficiency has been demonstrated by both immunologic and chemical means [1].

Vitamin K is widely distributed in foods, the best sources being green and leafy vegetables. Naturally occurring vitamin K in foods is stable to food processing and meal preparation methods [1]. Some vitamin K is synthesized by intestinal bacteria; just how much of this is absorbable is uncertain. The latest estimate of the adult human requirement is 0.15–0.4 mcg/kg/d, an amount widely exceeded by normal dietary intakes, even without counting bacterial synthesis in the intestine [2].

Using high-performance liquid chromatography to analyze phylloquinone levels in serum samples obtained from 95 healthy Red Cross volunteer blood donors, Mummah-Schendel and Suttie [6] determined that the median concentration was 1.1 ng/ml, with a mean of 1.3 ± 0.64 (SD) ng/ml. Sex, age, smoking habits, alcohol intake, and consumption of foods high in vitamin K were not found to influence serum phylloquinone concentrations.

Vitamin K deficiency, with hypothrombinemia and hemorrhagic tendencies, is observed in neonates due to inefficient placental transfer of the vitamin to the fetus as well as lack of intestinal bacterial synthesis and has also been seen in adults and children with malabsorption disorders, in hospitalized elderly patients, and in individuals on antibiotic medication whose dietary intakes of vitamin K are low. A state of marginal vitamin K deficiency, resulting in prolonged clotting time, is purposely induced by use of the coumarin anticoagulant drugs, which inhibit the epoxide reductase enzyme that regenerates vitamin K hydroquinone [1].

ABSORPTION, METABOLISM, AND EXCRETION

Phylloquinone of plant foods is absorbed in the proximal small intestine by a saturable, energy-dependent process, according to animal studies. However, other forms of vitamin K, including menadione, appear to be absorbed in the distal intestine and colon by passive diffusion. As in the case of other fat-soluble vitamins, efficient absorption depends on the presence of bile and pancreatic juice. Vitamin K, together with other lipids, is absorbed as micelles into the luteals. In the lymph, it is associated with chylomicrons, and in plasma with lipoproteins. Phylloquinone is rapidly concentrated in liver, as shown by administration of radiolabeled vitamin. Other organs also concentrate it, including the adrenal glands, lungs, bone marrow, kidneys, and lymph nodes. Turnover is rapid and total storage low, so that the difficulty of inducing deficiency appears attributable more to ready accessibility from the diet, and possibly intestinal synthesis, rather than to buildup of any reserve [3].

Menadione enters the blood stream directly and is rapidly distributed through the body, but is not well retained by the tissues. It is devoid of vitamin K activity until alkylated with an isoprenoid side chain. The water-soluble menadiol sodium diphosphate is converted to menadione and alkylated [1].

Phylloquinone is metabolized to various oxygenated derivatives, mostly by shortening the side chain to five or seven carbon atoms, yielding carboxylic acids that are conjugated with glucuronic acid. Only a few of the oxidation products have been identified. After a single injected dose of radioactive phylloquinone in man, about 20% of the radioactivity was recovered in the urine in 3 days, and 40–50% was excreted in the feces via the bile. Menadione is rapidly conjugated with sulfate, phosphate, and glucuronide, and the conjugates are excreted in both bile and urine; only a small fraction is converted to active vitamin K [1–4].

DOSAGE FORMS

Phylloquinone (phytonodione, vitamin K-1) is marketed in 5-mg tablets (Mephyton®, Merck Sharp & Dohme) for oral administration. Injectable forms are supplied in ampules containing a dispersion of 2 or 10 mg/ml of phylloquinone in a solution of buffered polysorbate and propylene glycol (Konakion®, Roche Laboratories) or of polyoxyethylated fatty acid derivatives and dextrose (Aquamephyton®, Merck Sharp & Dohme). Konakion® is administered only intramuscularly, while Aquamephyton® may be given by any parenteral route; *however, occasional severe reactions, including anaphylaxis, have occurred after intravenous administration, so that subcutaneous or intramuscular injection is preferred* [4].

Menadione is supplied as 5-mg tablets for oral administration. Menadiol sodium diphosphate (Synkayvite®, Roche Laboratories) is marketed for injection by any parenteral route and is supplied in 1-ml ampules at 5 or 10 mg/ml and in 2-ml ampules at 75 mg/2 ml; 5 mg tablets are also available [4, current PDR®].

CLINICAL USES

Hypoprothrombinemia of Infancy

The infant, both term and preterm, is born with low serum and hepatic levels of vitamin K and low circulating levels of plasma clotting factors, probably because of inefficient transfer of vitamin K across the placenta as well as lack of intestinal bacterial synthesis [4,8]. These levels continue to decline for the first few days of life, but thereafter, in the normal term infant, a gradual rise toward adult values begins if there is an adequate oral vitamin K intake, such as supplied by commercial formulas; this is aided by establishment of a normal intestinal bacterial flora [2,4]. However, in some cases hypoprothrombinemia may persist, particularly in the low-birthweight infant. It may also persist during the breast-feeding of any infant, as mother's milk is low in vitamin K compared with cow's milk or formula [9]. Since this condition may lead to the serious complication of hemorrhagic disease of the newborn (HDN), the American Academy of Pediatrics recommends the routine administration parenterally of 0.5–1.0 mg vitamin K-1 (phylloquinone, phytonodione) to all infants at birth [10]. Alternatively, oral administration of a colloidal suspension of vitamin K-1 (2 mg) to low-birthweight infants being fed by nasogastric tube has been reported by Sann et al. to produce a good early rise in serum vitamin K at 6 hours, which remains at satisfactory levels for up to 7 days [8]. Phylloquinone is preferred over menadione and its water-soluble derivatives for administration to infants because of greater safety and efficacy [4].

Lane and Hathaway [11] classify HDN as "early," occurring in the first 24 hours of life; "classic," occurring during days 1–7; and "late," occurring from 1 to 12 months of age. The early form, in which the infant may manifest intracranial bleeding, cephalohepatoma, or intrathoracic or intra-abdominal bleeding at birth or during the first 24 hours, is usually caused by maternal therapy with drugs that interfere with blood clotting, including warfarin, anticonvulsants, and antituberculosis drugs; idiopathic cases also occur. Most are not responsive to vitamin K therapy, though some may be, so that administration of the vitamin should be tried. For example, Dreyfus et al. [12] describe two cases of early HDN manifested on delivery at term; one infant responded nicely to administration of 5 mg phylloquinone IM, while the other showed no response and had to be treated by exchange transfusion. Deblay et al. [13] reported that oral administration of vitamin K to epileptic mothers under treatment with anticonvulsants prevented hypoprothrombinemia and hemorrhages in their infants; vitamin K was given daily during the 2 weeks before delivery.

Classic HDN usually appears at 2–5 days of age, manifested by ecchymoses, nasal or gastrointestinal bleeding, or excessive bleeding at circumcision. Incidence is enhanced by breast feeding. Classic HDN is virtually abolished by the routine intramuscular administration of phylloquinone at birth [12].

Late HDN usually involves acute intracranial hemorrhage, which may be intracerebral, intracerebellar, subarachnoid, subdural, or epidural. The first signs may be severe CNS dysfunction with vascular collapse. Other signs are widespread deep ecchymoses or "nodular purpura," gastrointestinal bleeding, and excessive bleeding at puncture sites or surgical incisions [11]. Factors increasing the likelihood of late HDN are breast-feeding, diarrhea, treatment of the infant with antibiotics, and malabsorptive conditions [4,11]. In treating the hemorrhagic patient, transfusion of fresh blood or reconstituted fresh plasma may be necessary in addition to administration of vitamin K [4]. Payne and Hasegawa [14] reported successful emergency treatment of a 4-week-old breast-fed female infant presenting in a comatose condition with a left-sided intracerebral hematoma, markedly prolonged prothrombin time and partial thromboplastin time, and jaundice. Intravenous administration of injectable phylloquinone (5 mg) and 10 ml/kg of fresh frozen plasma returned clotting parameters to normal in 18 hours. The intracerebral hematoma was then evacuated, with good clinical outcome. The acute vitamin K deficiency was attributed to a combination of factors: no prophylactic vitamin K at birth, breast-feeding, and cholestatic liver disease associated with congenital alpha-1-antitrypsin deficiency.

To prevent vitamin K deficiency in the breast-fed infant, it is recommended that they receive 1 mg phylloquinone intramuscularly once a month [1]. Alternatively, it has been suggested that the vitamin K content of human milk may be raised to satisfactory levels by administering 20 mg oral phylloquinone twice weekly to the lactating mother [7].

In vitamin K deficiency secondary to chronic cholestasis of infancy, Balistreri [15] recommends replacement with 2.5–5.0 mg orally every other day as water-soluble derivative of menadione (e.g. Synkayvite®).

Acute Care Patients

A number of investigators have reported development of vitamin K deficiency with coagulation defects in hospitalized patients, caused usually by a combination of circumstances. Among these are injury, gastrointestinal surgery, parenteral nutrition, cancer, renal failure, and use of antibiotics [1,16,17]. Blood factor assays may occasionally be necessary to make a diagnosis, but a therapeutic trial with parenteral vitamin K is often enough to make the right diagnosis [17].

Phylloquinone suspension can be added to TPN solutions for once weekly administration if intramuscular injection is inconvenient, as for example for home TPN patients or for those with diminished muscle mass. The solutions appear visually and chemically compatible for 24 hours [18]. *Without dilution in TPN solutions, however, intravenous administration of injectable phylloquinone, at least in large doses (5–10 mg), is inadvisable, as severe decreases in blood pressure and pulse rate may occur* [18].

The Elderly

Although it has generally been assumed that dietary deficiency of vitamin K does not occur, Hazell and Baloch [19] noted a relatively high percentage of hypoprothrombinemia in older patients on admission to the hospital; the condition was responsive to oral vitamin K. The tendency to be easily bruised and ecchymoses in the elderly can also be due to vitamin C deficiency; if these conditions do not respond to vitamin K administration, vitamin C supplements should be tried.

Malabsorption

Using an immunologic method to measure the plasma concentration of abnormal (undercarboxylated) prothrombin, Krasinski et al. [20] found vitamin K deficiency in 18 of 58 patients with chronic gastrointestinal disease and/or resection. All patients with vitamin K deficiency had either Crohn's disease involving the ileum or ulcerative colitis that was being treated with sulfasalazine or antibiotics. No bleeding tendencies were noted. Administration intramuscularly of 10 mg of aqueous collooidal phylloquinone (Aqua-Mephyton®, Merck Sharp & Dohme) reduced the concentration of abnormal prothrombin to normal.

Malabsorption conditions leading to hypoprothrombinemia, often with bleeding tendencies, include intrahepatic or extrahepatic biliary obstruction with resultant jaundice, biliary fistula, mucoviscidosis, celiac sprue, regional enteritis and enterocolitis, and dysentery [1,4]. The usual treatment is 10 mg/d of vitamin K-1 or menadione orally. If oral administration is not feasible, a smaller dose of phylloquinone may be given intramuscularly. If there is obstructive liver disease, coadministration of bile salts is helpful for absorption of oral vitamin K-1. Vitamin K administration is not efficacious for treatment of hypoprothrombinemia due to severe hepatitis or cirrhosis [4,21].

Corrigan et al. [22] found inadequate vitamin K status, measured as the ratio of coagulant activity to total normal and abnormal prothrombin, in 14 of 24 cystic fibrosis patients. Oral vitamin K (5 mg) given two or three times weekly improved some but not all patients, suggesting that daily supplements and perhaps larger doses may be needed for some patients.

Caballero and Buchanan [23] describe hemorrhagic disease and vitamin K deficiency in a 6-week-old infant with abetalipoproteinemia. Hemorrhaging was initially controlled with parenteral vitamin K. Following diagnosis of the underlying disorder, management has involved a diet rich in medium-chain triglycerides supplemented with large oral doses of vitamins A, E, and K.

Genetic Vitamin K Dependency

A few cases have been reported of individuals with congenital deficiency of vitamin K-dependent clotting factors. They have responded to massive oral doses of vitamin K [1,17,24,25].

TOXICITY AND SIDE EFFECTS

Oral administration of usually recommended dosages of vitamin K, either as phylloquinone (phytonodione) or as menadiol sodium diphosphate, has not been reported to cause side effects other than occasional allergic reactions such as skin rash and urticaria. Incidence of sensitivity reactions is greater and more severe with injectable forms of these drugs. *With intramuscular or subcutaneous injection of phylloquinone or menadiol sodium diphosphate, occasional individual hypersensitivity and anaphylactoid reactions have been reported* [4, current *Physician's Desk Reference*].

Intravenous injection of any form of vitamin K is a hazardous mode of administration, although sometimes considered necessary for emergency treatment of severe hemorrhagic episodes due to the hypoprothrombinemia of acute vitamin K deficiency. Dosages of 10 mg phylloquinone administered in 10 minutes or less have been reported to occasionally produce severe decreases in blood pressure and pulse rate, and in one case a fatality [26,27]. The intravenous route should be undertaken only with careful consideration of the possible risks. On the other hand, no adverse results have been reported for the administration to TPN patients of a weekly dose of 10 mg of injectable phylloquinone after its incorporation into TPN solution to be infused over 8–12 hours [18].

In newborns, large doses of injectable phylloquinone (5 mg or more) given IM or subcutaneously may induce hyperbilirubinemia and kernicterus, particularly in the premature infant. The usual routine dosage of 0.5–1.0 mg is considered safe [10]. Neither menadione nor menadiol sodium diphosphate is considered safe for administration to newborns, as they tend to induce hemolytic anemia, hyperbilirubinemia, and kernicterus, particularly in the premature infant [4]. The same is true if these forms of vitamin K are administered to the mother prior to delivery (see current *Physician's Desk Reference*).

Menadione and its derivatives can also induce hemolysis in individuals who are genetically deficient in glucose-6-phosphate dehydrogenase [4].

All forms of vitamin K are ineffective in treating hypoprothrombinemia resulting primarily from advanced liver disease, and administration of large doses of vitamin K in this situation may have the paradoxical effect of further decreasing prothrombin levels. The mechanism is unknown, but does not seem to involve further suppression of liver function [4].

Menadione and its derivatives may interfere with the modified Reddy-Jenkins-Thorn procedure for determining urinary 17-hydroxycorticosteroids, producing falsely elevated values (see current PDR®.)

INTERACTIONS

The coumarin and indanedione anticoagulants at recommended dosages induce a moderate degree of hypoprothrombinemia by antagonizing enzymatic reduction of vitamin K epoxide to vitamin K hydroquinone [1]. Phylloquinone, but not menadione or menadiol sodium diphosphate, is effective in treating overdosage of anticoagulant drugs. Mild overdosage may be treated with a single IM injection of 2.5–10 mg of phylloquinone. Larger doses may interfere with the effectiveness of subsequent oral anticoagulant therapy for several days. However, if bleeding is severe, 20–40 mg of phylloquinone should be administered, and further dosage may be needed at 4-hour intervals. Additional therapy for the severe case may include transfusion of fresh whole blood, frozen plasma, or clotting factor concentrates [4].

Aspirin and salicylates may also induce a vitamin K-responsive hypoprothrombinemia with hemorrhagic tendencies. Treatment, after withdrawal of salicylate, is similar to that for anticoagulant overdosage [4]. Hemorrhagic problems have also been associated with administration of antihypertensive drugs [1].

Antibiotics, especially those of the cephalosporin series, may induce vitamin K-responsive coagulopathies, and prophylactic administration of parenteral vitamin K has been recommended when these antibiotics are used [28].

Administration of anticonvulsant and antituberculosis drugs, as well as anticoagulants, to pregnant women has been reported to induce a vitamin K-responsive hypoprothrombinemia in their newborns [1,11].

REFERENCES

1. Suttie JW: Vitamin K. In: *Fat-Soluble Vitamins*, AT Diplock (ed). Lancaster, PA: Technomic Publishing Co., 1985, pp 225–311.
2. Olson JA: Recommended dietary intakes (RDI) of vitamin K in humans. Am J Clin Nutr 45:687–692, 1987.
3. Suttie JW: Vitamin K. In: *Handbook of Vitamins*, LJ Machlin (ed). New York: Marcel Dekker, 1984, pp 147–198.
4. Mendel HG, Cohn VH: Fat-soluble vitamins. Vitamins A, K, and E. In: *Goodman and Gilman's The Pharmacologic Basis of Therapeutics*, 7th ed., LS Goodman, TW Rall, F Murad (eds). New York: Macmillan, 1985, pp 1573–1591.
5. Olson RE: Vitamin K. In: *Modern Nutrition in Health and Disease*, 6th ed., RS Goodhart, ME Shils (eds). Philadelphia: Lea & Febiger, 1980, pp 170–180.
6. Mummah-Schendel LL, Suttie JW: Serum phylloquinone concentrations in a normal adult population. Am J Clin Nutr 44:686–689, 1986.
7. Shearer MJ, Rahim S, Barkhan P, Stimmler L: Plasma vitamin K-1 in mothers and their newborn babies. Lancet 2:460–463, 1982.
8. Sann L, Leclercq M, Guillaumont M, et al.: Serum vitamin K-1 concentrations after oral administration of vitamin K-1 in low birth weight infants. J Pediatr 107:608–611, 1985.
9. Haroon Y, Shearer MJ, Rahim S, et al.: The content of phylloquinone (vitamin K-1) in human milk, cows' milk and infant formula foods determined by high-performance liquid chromatography. J Nutr 112:1105–1117, 1982.

10. American Academy of Pediatrics: Vitamin K supplementation for infants receiving milk substitute infant formulas and for those with fat malabsorption. Pediatrics 48:483–487, 1971.
11. Lane PA, Hathaway WE: Vitamin K in infancy. J Pediatr 106:351–359, 1985.
12. Dreyfus M, Lelong-Tissier MC, Lombard C, Tchernia G: Vitamin K deficiency in the newborn. Lancet 1:1351, 1979.
13. Deblay MF, Vert P, Andre M, Marchal F: Transplacental vitamin K prevents hemorrhagic disease of infant and epileptic mother. Lancet 1:1247, 1982.
14. Payne NR, Hasegawa DK: Vitamin K deficiency in newborns: A case report in alpha-1-antitrypsin deficiency and a review of factors predisposing to hemorrhage. Pediatrics 73:712–716, 1984.
15. Balistreri WF: Neonatal cholestasis. J Pediatr 73:712–716, 1984.
16. Pineo GF, Gallus AS, Hirsh J: Unexpected vitamin K deficiency in hospitalized patients. Can Med Assoc J 109:880–883, 1973.
17. Ansell JE, Kumar R, Deykin D: The spectrum of vitamin K deficiency. JAMA 238:40–42, 1977.
18. Demers R, Schneider PJ: Vitamin K administration to patients receiving TPN. Infusion 8:191–192, 1984.
19. Hazell K, Baloch KH: Vitamin K deficiency in the elderly. Gerontol Clin 12:10–17, 1970.
20. Krasinski SD, Russell RM, Furie BC, et al.: The prevalence of vitamin K deficiency in chronic gastrointestinal disorders. Am J Clin Nutr 41:639–643, 1985.
21. Blanchard RA, Furie BC, Jorgensen M, et al.: Acquired vitamin K-dependent carboxylation deficiency in liver disease. N Engl J Med 305:242–248, 1981.
22. Corrigan JJ, Taussig LM, Beckerman R, Wagener JS: Factor II (prothrombin) coagulant activity and immunoreactive protein: Detection of vitamin K deficiency and liver disease in patients with cystic fibrosis. J Pediatr 99:254–257, 1981.
23. Caballero FM, Buchanan GR: Abetalipoproteinemia presenting as severe vitamin K deficiency. Pediatrics 65:161–163, 1980.
24. Gallop PM, Lian JB, Hauschka PV: Carboxylated calcium-binding proteins and vitamin K. N Engl J Med 302:1460–1466, 1980.
25. Chung KS, Bezeaud A, Goldsmith JC, et al.: Congenital deficiency of blood clotting factors II, VII, IX, and X. Blood 53:776–787, 1979.
26. Barash P, Kitohata LM, Mandel S: Acute cardiovascular collapse after intravenous phytonodione. Anesth Analg 55:304–306, 1976.
27. Rich EC, Drage CW: Severe complications of intravenous phytonodione therapy. Postgrad Med 72:303–306, 1982.
28. Clinical Nutrition Case: New examples of vitamin K-drug interaction. Nutr Rev 42:161–163, 1984.

5
Thiamin (Vitamin B-1)

Introduction	103
Absorption, Metabolism, and Excretion	105
Dosage Forms	105
Clinical Studies	106
Alcoholism	106
Neurologic Disorders	107
Nonalcoholic Fulminant Hepatic Failure	108
Diabetes Mellitus	108
The Elderly	108
Pregnancy	109
Malabsorption	109
Optic Neuropathy During Ketogenic Diet	109
Optic Atrophy in Glaucoma	110
"Neurotic Tension"	110
Exercise and Athletic Training	111
Genetic Disorders	111
Recurrent Febrile Lymphadenopathy	112
Other Possible Uses	112
Toxicity and Side Effects	113
Interactions	113
References	114

INTRODUCTION

Thiamin is a double-ring compound in which a substituted pyrimidine ring and a substituted thiazole ring are connected by a methylene group. The most common commercially available form is thiamin chloride hydrochloride (see Fig. 1), usually called simply thiamin hydrochloride. It is a colorless crystalline material, very soluble in water, with a characteristic odor and a slightly bitter taste [1].

Thiamin is phosphorylated in the body to the diphosphate, which functions as a coenzyme in oxidative decarboxylation of pyruvate to give acetylcoenzyme A; the oxidative decarboxylation of alpha-ketoglutarate to succinyl-coenzyme A;

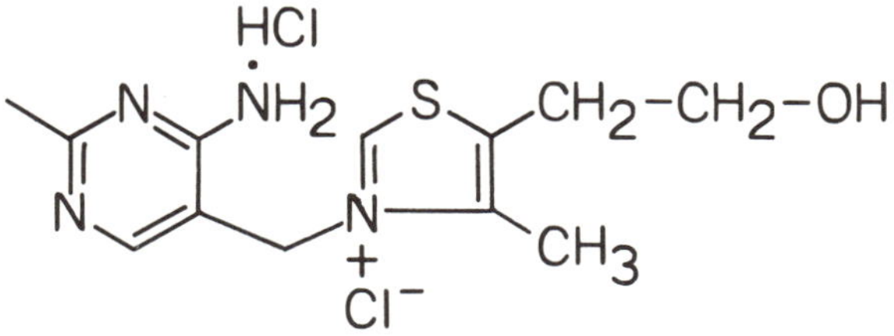

Fig. 1. Structural formula of thiamin hydrochloride, the most common commercially available form of thiamin.

and oxidative decarboxylation of the branched-chain alpha-ketoacids derived from the deamination of leucine, isoleucine, and valine. Without thiamin diphosphate, the citric acid (Krebs) cycle cannot function to generate energy. Also dependent on this coenzyme are reactions catalyzed by transketolase in the hexose monophosphate shunt pathway, which produces pentoses for RNA and DNA synthesis and NADPH for biosynthesis of fatty acids and other products. Apart from these enzymatic roles of thiamin, the vitamin also appears to have an essential function in nerve transmission in the form of its triphosphate ester [1].

Chronic deprivation of thiamin leads to the classical deficiency disease, beriberi, which is still seen in areas where the inhabitants live mostly on polished rice that has not been enriched with synthetic thiamin. General symptoms include anorexia, weight loss, cardiac enlargement with tachycardia, muscle weakness, and neurologic symptoms, including loss of knee and ankle jerks. The disease was almost completely abolished in Japan by dietary improvement and rice enrichment after World War II, but a few cases have recently been reported among teenagers consuming excessive amounts of sugar-sweetened soft drinks and refined carbohydrate foods [2]. In the United States and other western countries, frank thiamin deficiency is seen mostly in chronic alcoholics in association with Wernicke-Korsakoff syndrome [1].

Thiamin is widely distributed in the body tissues, mostly in coenzyme form bonded firmly but not covalently to enzyme proteins. Its concentration is particularly high in metabolically active tissues such as heart, liver, brain, kidney, and digestive organs [1].

Good food sources include lean pork, wheat germ, liver and other organ meats, lean meats, poultry, eggs, fish, beans and peas, nuts, whole grains, and enriched bread, pasta, and breakfast cereals. Dairy products, fruit, and vegetables are generally not good sources. The RDA for thiamin is 0.5 mg/1,000 kcal, or 1.0–1.5 mg/d for the average adult; this amount is available in most American diets. However, considerable losses of thiamin can occur during cooking or other heat-processing of food and also if cooking water is discarded [1]. Polyphenolic compounds in coffee and tea, whether or not decaffeinated, can inactivate thiamin, so that heavy use of these beverages could compromise thiamin nutrition [3]. Thiaminases present in raw fish and shellfish can also destroy the vitamin [4].

ABSORPTION, METABOLISM, AND EXCRETION

Thiamin occurs almost entirely in phosphorylated forms in animal protein foods, but is mostly nonphosphorylated in plant foods and enriched cereal products. Dephosphorylation occurs in the intestinal lumen. In animal studies, free thiamin was found to be absorbed in the upper small intestine by an active saturable process that is important in utilization of the low concentrations ordinarily found in food. At higher luminal concentrations of thiamin attainable by administering supplements, the vitamin is also absorbed by passive diffusion. In humans, thiamin hydrochloride absorption was found to follow Michaelis-Menten kinetics, i.e., the rate of absorption approaches a limit (V_{max}) as the dosage increases; this limits the amount absorbed after single large doses (up to 50 mg) to about 8 mg [5]. Phosphorylated forms of thiamin are present in the enterocyte, but it is not known whether phosphorylation-dephosphorylation plays an active role in thiamin absorption. Free thiamin appears on the serosal side of the intestinal wall and is transported to the body tissues, where it is taken up by active saturable transport systems and phosphorylated intracellularly [1,5–9]. Thiamin triphosphate, constituting 10% of thiamin in nerve tissue, has a noncoenzymatic function in that tissue; nerve action causes a decline in thiamin triphosphate level and some decline in diphosphate level with a rise in free thiamin concentration [1,9].

Thiamin over and above tissue requirements for coenzyme and noncoenzyme functions is rapidly excreted in the urine, together with small amounts of thiamin metabolites, a few of which have been identified as substituted pyrimidine and thiazole compounds [1].

DOSAGE FORMS

Thiamin hydrochloride is available over the counter as tablets in 25-, 50-, 100-, and 250-mg strengths; a timed-release 500-mg tablet is also available. Thiamin mononitrate is often found in multivitamin preparations and is preferred over the hydrochloride in food enrichment because of its greater stability in processing and storage. Injectable thiamin·HCl (100 mg/ml in single-dose vials) is available by prescription (see current PDR®); *occasional instances have been reported of sensitization after repeated parenteral administration with resultant risk of anaphylactic shock* [1].

CLINICAL STUDIES
Alcoholism

Chronic alcoholism often leads to deficiency of thiamin because of poor dietary habits and, metabolically, because of the interference of ethanol with gastrointestinal absorption of thiamin and its utilization for the synthesis of coenzyme and thiamin-dependent holoenzymes [1,5,10]. The frequency of deficiency increases with gastritis or pancreatitis and especially with liver disease (fatty liver, alcoholic hepatitis, cirrhosis) [10]. Alcoholics tend to be deficient in many nutrients simultaneously and to require multinutrient therapy for rehabilitation, but thiamin deficiency is considered specifically causative of neurologic disorders in the alcoholic because of thiamin's essentiality for nerve function and for metabolism of glucose in nervous tissue. Neurologic malfunction may manifest as Wernicke's encephalopathy, Korsakoff's psychosis, or combined Wernicke-Korsakoff syndrome [1,10]. Thiamin deficiency is also a characteristic finding in the peripheral neuropathy of alcoholism [10]. Bonjour [11] has written a comprehensive review on the clinical and biochemical effects of thiamin deficiency in the alcoholic, methods of assessing thiamin status, and treatment of alcohol withdrawal.

Leevy [10] recommends for treatment of alcohol addiction, after withdrawal of alcohol, the administration parenterally of all vitamins in therapeutic amounts, as well as zinc, magnesium, and amino acids. Ten mg/d of thiamin is given intravenously along with other vitamins for 4 or 5 days or until oral administration becomes feasible. Larger IV doses may be beneficial if there is severe malnutrition or hepatic dysfunction. Various other investigators have recommended dosages of 50–100 mg thiamin daily orally, intravenously, or intramuscularly in the general management of alcohol withdrawal symptoms [11]. Patients with cardiac symptoms of the beriberi type may require prolonged (up to 3 weeks) intravenous or intramuscular thiamin administration for full recovery [11].

The malnourished alcoholic with gastritis, neuropathy, or liver disease may not respond well to oral thiamin·HCl [10]. Absorption should improve gradually as thiamin receptor and absorption sites in the gut increase with nutritional rehabilitation [5,10,12]. High-dose thiamin·HCl (50–100 mg/d) should be continued until the patient is well recovered.

For the malnourished alcoholic patient with gastritis, neuropathy, or liver disease, Leevy [10] prefers to administer oral thiamin in the form of thiamin alkyl disulfides, which are lipid-soluble and absorbed through the lymphatics. For example, 50 mg of thiamin tetrahydrofurfuryl disulfide gave a far superior plasma thiamin response in alcoholics in comparison with 50 mg of thiamin·HCl [10]. Other alkyl disulfide derivatives include allithiamin and thiamin propyl disulfide. These are readily converted to thiamin in the body

[1]. Although widely used in Japan and Europe, these thiamin derivatives had not yet received, as of this writing, the approval of the United States Food and Drug Administration.

Some Wernicke-Korsakoff patients appear to have a genetically altered apotransketolase with a reduced affinity for thiamin diphosphate; clinical improvements of greater or lesser degree have been obtained by treatment with oral thiamin tetrahydrofurfuryl disulfide [13].

A fulminating or pernicious beriberi of alcoholism has been reported from Australia by Campbell [14], occurring mostly in aborigines and accompanied by severe lacticacidosis. The symptoms appear suddenly in heavy drinkers, manifested as hyperpnea and tachypnea, hypotension, weak or impalpable pulse, and in severe cases confusion, agitation, cold limbs, and oliguria or anuria. Cardiomegaly is present. This acute disorder, previously invariably fatal, was found to be rapidly responsive to high-dose thiamin·HCl administered intravenously; Campbell recommends 100 mg IV twice daily for several days [14]. Supportive therapy, which without high-dose thiamin had previously been unsuccessful in salvaging these patients, included administration of other vitamins, sodium bicarbonate IV to treat the acidosis, and frusemide IV to stimulate diuresis.

It is of interest that adult male rats, given equal access to tap water or a dilute alcohol solution, consumed voluntarily only one-fifth the amount of alcohol as control rats when they received 20 mg thiamin·HCl/kg diet, compared with 4 mg/kg diet for the controls [15]. The 4-mg/kg level is considered optimal for the adult rat. For an adult human consuming roughly one-half a kilogram of food (dry weight) per day, the corresponding amount of thiamin·HCl needed as a supplement, if a similar phenomenon occurs in humans, would be 10 mg/d.

Neurologic Disorders

Langohr et al. [16] have reported on the thiamin, riboflavin, and vitamin B-6 status of 176 patients with various neurologic disorders. Incidence of biochemical thiamin deficiency (estimated by erythrocyte transketolase activation) was 31%, of riboflavin deficiency 22% (by erythrocyte glutathione reductase activation), and of B-6 deficiency 6% (by erythrocyte transaminase activation). More than half of the thiamin-deficient patients had been addicted to alcohol, the remainder being diagnosed as follows: polyneuropathy of malabsorption, polyneuropathy of diabetes mellitus, optic atrophy of unknown eitology, cerebellar ataxia, neoplasm-related neurologic disorders, B-12 neuromyelopathy, myelopathy of unknown origin, and Thévenard's syndrome. Alcoholics with and without deficiencies of the B vitamins showed a similar frequency of polyneuritis, but the vitamin-deficient patients had a much higher incidence of cerebellar and/or brainstem lesions. It thus appears that deficiencies of thiamin and other B vitamins must be suspected in neurologic disorders of many types and origins, and they should be corrected by therapeutic dosages of these vitamins.

Nonalcoholic Fulminant Hepatic Failure

Labadarios et al. [17] found that 9 of 24 patients with acute hepatocellular necrosis leading to fulminant hepatic failure were biochemically deficient in thiamin early in the course of their illness, probably as a result of inadequate intake of the vitamin. Of the 24, 15 had overdosed on paracetamol, 7 had acute viral hepatitis, and 2 had halothane-associated hepatic necrosis. All patients received intensive supportive therapy for hepatic failure. Those with thiamin deficiency received 100 mg thiamin·HCl intravenously b.i.d. plus 500 mg ascorbic acid b.i.d. and 50 mg riboflavin b.i.d. intravenously. Despite the severe hepatocellular necrosis, there was adequate conversion of thiamin to the diphosphate coenzyme form and restoration of normal erythrocyte transketolase activation values. The authors recommend that high-dose intravenous thiamin be included in the therapy of hepatic failure.

Diabetes Mellitus

Kjosen and Seim [18] reported finding low erythrocyte transketolase activities due apparently to low apoenzyme levels in diabetics (7 type I, 10 type II); enzyme levels were most markedly reduced in the type I cases. It is not known whether the neuropathies of diabetes are primarily related to marginal thiamin status, but these investigators suggest that such complications, if due to this cause, might be reduced by giving the diabetic sufficient thiamin to ensure optimal utilization of the patient's reduced level of transketolase apoenzymes [18]. A supplement of 10 mg/d of thiamin·HCl should be sufficient in most cases. Further clinical studies are needed in this area.

The Elderly

The triad of old age, alcoholism, and malabsorption may lead to poor thiamin nutriture and is associated with such senile disorders as dizziness, gait disturbances, anorexia, weight loss, irritability, and altered electrocardiograms [19,20]. Apart from the adverse influences of alcoholism and malabsorption, Chen et al. [21] have documented an age-related decline in plasma and red cell thiamin levels in healthy, free-living Americans. The cause is unknown, as the diets appeared to provide adequate thiamin. Mean plasma and red cell thiamin levels in 33 individuals aged 60–91 years were significantly (20–25%) lower than those of 18 people aged 20–40 years. Hoorn et al. [22], based on TPP (thiamin pyrophosphate) activation of erythrocyte transketolase, found 23% of geriatric patients, aged 65–93, to be biochemically thiamin-deficient. Oral supplementation with 20 mg/d of thiamin·HCl for 12 days lowered their ETK (erythrocyte transketolase) activation values to the normal range. Similarly, Griffiths et al. [23] found 40% of hospitalized elderly patients to be biochemically deficient in thiamin, compared with a control group of healthy young hospital employees. A 3-week period of supplementation with 25 mg/d of thiamin·HCl lowered the ETK activation values of the deficient patients into the range of the control group.

Pregnancy

Pregnant women may be in a marginal state of thiamin nutriture if dietary intake of the vitamin is low or if there is excessive consumption of "empty-calorie" high-carbohydrate foods and beverages. Vir et al. [24], studying a group of 60 pregnant women in Northern Ireland, all of whom were receiving at least two-thirds of the RDA in their diets, found about one-third of them to be biochemically deficient by the erythrocyte transketolase activation test. Surprisingly, the same was true for nonpregnant controls. Multiparity increased the incidence of thiamin deficiency. Thiamin supplements of 5 or 10 mg/d are commonly included in multivitamin products for pregnant and lactating women.

Malabsorption

A variety of malabsorptive disorders can jeopardize nutritional status with respect to thiamin, as well as other vitamins, for example celiac disease and short-bowel syndrome [5]. Glad et al. [25] reported a case of severe polyneuropathy that developed in a 37-year-old woman 4 months after jejunoileal bypass surgery for morbid obesity; high-dose thiamin produced gradual improvement. Other malabsorptive disorders that may initiate or maintain thiamin deficiency include diarrhea, dysentery, sprue, ulcerative colitis, cancer, achlorhydria, and biliary disease [1]. Depending on the severity and duration of the disorder, 10–50 mg/d of oral thiamin·HCl may be used for supportive therapy or maintenance.

Optic Neuropathy During Ketogenic Diet

The high-protein, high-fat ketogenic diet sometimes used to aid in control of seizure disorders is less demanding of dietary thiamin than a high-carbohydrate diet. Nevertheless, thiamin deficiency can occur, as illustrated by a report of Hoyt and Billson [26] on development of a symmetrical bilateral optic neuropathy in two children aged 5 and 7 years who had been maintained on such a diet. Serum folate and vitamin B-12 levels were normal, and urinary tests for toxic heavy metals were negative, but biochemical thiamin deficiency was present. The parents of these children had neglected to provide the multivitamin supplements recommended by their physician. Treatment with 50 mg/d of oral thiamin·HCl plus a routine B-complex supplement restored visual function to normal in several weeks.

Optic Atrophy in Glaucoma

Asregadoo [27] found, in comparing thiamin and vitamin C levels in blood of 38 chronic open-angle glaucoma patients (mean age 63 years, range 45–86 years) and 12 controls in California, that thiamin levels in the glaucoma patients were significantly lower than in the controls, while vitamin C levels were similar. Dietary intakes of both vitamins were the same in patients and controls, suggesting that glaucoma patients absorb thiamin less efficiently than healthy normals. The author cites epidemiologic impressions (supporting data not given) that incidence of chronic open-angle glaucoma is greater in population groups and areas where B-vitamin intake is low. In treating Guyanan patients with optic atrophy resulting from chronic open-angle glaucoma, Asregadoo obtained good clinical response of visual function to daily intramuscular injection of 100 mg thiamin·HCl and therapeutic B complex for 10 days, followed by oral supplements. In subsequent California practice for 12 years, he has routinely prescribed a vitamin C and B-complex therapeutic formulation that, taken once daily, provides 15 mg/d of thiamin; none of these patients have required surgery to control intraocular pressure [27]. Confirmatory studies in other ophthalmology centers are needed.

"Neurotic Tension"

Lonsdale and Shamberger [28] describe a group of symptoms observed in 20 patients with biochemical thiamin deficiency that were classifiable as "neurotic dysfunction, conversions or hysterical," conditions that are frequently treated by sedatives and psychological counseling. The patients were mostly teenagers who had been consuming excessive amounts of low-thiamin "junk" foods, sweet beverages, candy, and between-meal "empty-calorie" snacks. Symptoms included abdominal and/or chest pain, sleep disturbances, fatigue, anorexia, nausea, and personality changes, often of a hostile or aggressive nature. Seen in 8 of the 20 patients was intermittent fever of unknown cause, and 6 of 20 experienced diaphoresis, usually nocturnal. Eight of 20 had intermittent diarrhea, often alternating with constipation. Some of these symptoms resemble those of adult human volunteers in the early stages of thiamin deficiency induced experimentally by a low-thiamin diet, while others appear to be nonspecific symptoms accompanying early or mild deficiency of any one of several different vitamins, as tabulated by Brin [29].

Symptoms were cleared slowly over weeks and months of treatment with high-dose thiamin together with dietary instruction [28]. Normal erythrocyte transketolase activities and activation coefficients, however, were rapidly restored. Dosages of thiamin·HCl were 150–600 mg/d for periods of 11 days to 5 months. The need for dosages this large has been questioned by Leevy [30], who points out that 10 mg/d is sufficient to correct a more serious thiamin-deficiency disorder, extraocular palsy, in uncomplicated Wernicke's encephalopathy [31].

Exercise and Athletic Training

High rates of metabolism and sweat losses in athletic training appear to increase the need for thiamin as well as other water-soluble vitamins. Thiamin deficiency (transketolase activation test) has been shown to be common among male and female high-performance fencers [32,33] and teenage girls training for competitive skiing, gymnastics, and swimming [34]. A therapeutic-type multivitamin supplement providing 5–10 mg/d of thiamin·HCl may be indicated for those engaging in strenuous sports activities.

Genetic Disorders

Deficiency of pyruvate decarboxylase, first described by Lonsdale in 1968 [35], has since been noted a number of times, though considered a very rare disease [36,37]. Elevated levels of lactic acid, pyruvic acid, and alanine are found in the urine, and there are movement disorders of the type called cerebellar ataxia. Responsiveness to pharmacologic thiamin has not been reported. The disease may be partially controlled by a low-carbohydrate ketogenic diet [37].

Congenital deficiency of branched-chain alpha-ketoacid dehydrogenase, resulting in maple syrup urine disease, is less rare. Branched-chain ketoacids in the urine impart the distinctive odor. If the mutant enzyme has a moderate degree of activity, the disease responds to pharmacologic doses of thiamin·HCl orally, but if activity is very low or absent, there is no response to thiamin supplements. Some patients have responded to oral thiamin·HCl doses as low as 10 mg/d with marked reductions of urinary branched-chain ketoacids, but most require doses of 100–200 mg/d, and some physicians have used as much as 1,000 mg oral thiamin·HCl per day [38,39]. Elsas and Danner employed 200 mg/d in treating a 10-kg child [39]. Dietary restriction of branched-chain amino acids is important in controlling the disease.

Reduced affinity of transketolase for thiamin diphosphate, possibly of genetic origin, is found in some Wernicke-Korsakoff patients and may make these individuals more susceptible to development of the syndrome upon abuse of alcohol. Response to pharmacologic oral thiamin·HCl is fair-to-good [13].

Leigh's disease, or subacute necrotizing encephalopathy (SNE), is an autosomal recessive disease with fatal outcome that usually manifests before age 2 years in a variety of symptoms, including feeding problems, weakness, visual disturbances, ataxia, convulsions, and peripheral neuropathy [40]. Patients lack thiamin triphosphate in brain tissues, and there appears to be an inhibitor of thiamin triphosphate synthesis from the diphosphate [1,38]. Cooper and Pincus [40] noted improvement in clinical status of Leigh's disease patients when given large daily doses of thiamin·HCl or thiamin tetrahydrofurfuryl disulfide for about 5 weeks. The improvement could be maintained for variable periods ranging from 3 months to 2 years, but the patients then became refractory to thiamin, no matter how large the dose.

A very rare thiamin-responsive megaloblastic anemia has been reported, accompanied by sensorineural deafness and diabetes mellitus. The disease responds to high-dose thiamin·HCl, and one patient has been maintained in good health but without remission of deafness on a daily supplement of 25 mg [41].

Recurrent Febrile Lymphadenopathy

Lonsdale [42] has reported successful treatment of recurrent febrile lymphadenopathy in two boys, aged 4 and 5 years, with 150 mg/d of oral thiamin hydrochloride. In neither case was an infectious agent demonstrated, and biopsy of a pathologically enlarged lymph gland revealed only reactive hyperplasias in each case. There was no response to antibiotics. Abnormal thiamin status was suggested in one child by detection of a substance in urine reported to be diagnostic of Leigh's subacute necrotizing encephalopathy, a disease in which there is a congenital failure of conversion of TPP to thiamin triphosphate in the brain. The other child had deficient activity of red cell transketolase, a thiamin-dependent enzyme. Recurrences occurred if thiamin supplementation was interrupted. Confirmatory reports have not been published, but it appears worthwhile to try thiamin supplementation if recurrent febrile lymphadenopathy fails to resolve spontaneously or to respond to antibiotics.

Other Possible Uses

In Japan, amyotrophic lateral sclerosis (ALS) has been treated with 50–100 mg/d of allithiamin, 500 mg/d of magnesium oxide, and 1 g/d of calcium carbonate, with resultant improvement in strength [43]. There appears to have been no follow-up of this lead.

Thiamin deficiency or impaired thiamin metabolism has been proposed as the responsible primary cause for respiratory arrest in sudden infant death syndrome (SIDS or crib-death) [44,45]. Lonsdale and coworkers [46,47] found in the urine of two threatened SIDS infants a factor that inhibits the phosphotransferase reaction that forms thiamin triphosphate in the brain. Davis et al. [48], however, found serum thiamin levels in SIDS infants to be higher than in healthy controls, an observation that appears to rule out deficiency as a cause of SIDS, but does not negate the possibility that impaired phosphorylation of the vitamin is the underlying cause. High-dose thiamin abolished recurrent attacks of respiratory arrest in one threatened SIDS infant [46]. Clearly much more investigation is needed of the relation between SIDS and thiamin nutriture, but adequate thiamin supplementation of expectant and lactating mothers as well as of infants is likely to be of nutritional benefit in many cases and may possibly decrease the likelihood of SIDS.

Alvarez and Gilbreath [49,50] studied the effects of thiamin on collagen formation and wound healing in young adult rats fed a thiamin-deficient diet and then given by gavage either 1 ml distilled water (deficient group), 1 ml water containing 1 mg thiamin·HCl, or 1 ml water containing 3 mg thiamin·HCl on alternate days after wounding the skin. Thiamin deficiency reduced the amount of type III collagen in the healing wound. Breaking strength measurements after 7 days showed only 55% of the wound strength in the deficient rats as in the supplemented (1 mg) rats. The higher thiamin·HCl supplement (3 mg) strengthened the wound by another 14%. Since 1 mg every other day is many times higher than the rat's optimal dietary requirement (4 mg/kg diet), the additional increase in strength with 3 mg every other day appears to represent a true pharmacologic effect on wound healing. There does not appear to have been any follow-up of these observations clinically.

TOXICITY AND SIDE EFFECTS

Animal studies indicate that thiamin·HCl is an extraordinarily safe substance. For example, in monkeys, up to 600 mg/kg of body weight is required to produce toxic symptoms (respiratory depression). In humans, no toxic effects have been reported with high-dose oral supplements, e.g., chronic dosage at 100–200 mg/d, except possibly for some gastric upset [1]. Likewise, administration of similar amounts by subcutaneous, intramuscular, or intravenous administration has been well tolerated, *except for rare instances of sensitization after repeated parenteral administration, which could place the patient in jeopardy of anaphylactic shock* [1].

INTERACTIONS

Reduced blood thiamin levels have been reported in epileptics treated with phenytoin [51]. In general, drugs that cause nausea or anorexia, induce diuresis, or increase intestinal motility can decrease the availability of dietary thiamin [1].

REFERENCES

1. Gubler CJ: Thiamin. In: *Handbook of Vitamins*, L.J. Machlin (ed). New York: Marcel Dekker, 1984, pp 245–297.
2. Kawai C, Wakabayashi A, Matsumura T, Yui Y: Reappearance of beriberi heart disease in Japan. A study of 23 cases. Am J Med 69:383–386, 1980.
3. Hilker, DM, Somogyi JC: Antithiamins of plant origin: Their chemical nature and mode of action. Ann NY Acad Sci 378:137–145, 1982.
4. Murata K: Actions of two types of thiaminase on thiamin and its analogues. Ann NY Acad Sci 378:146–156, 1982.
5. Thomson AD, Leevy CM: Observations on the mechanism of thiamine hydrochloride absorption in man. Clin Sci 43:153–163, 1972.
6. Nose Y, Iwashima I, Nishino A: Thiamine uptake by rat brain slices. In: *Thiamine*, CJ Gubler, M Fujiwara, PM Dreyfus (eds). New York: John Wiley & Sons, 1976, pp 157–168.
7. Sklan D, Trostler N: Site and extent of thiamin absorption in the rat. J Nutr 107:353–356, 1977.
8. Hoyumpa AM Jr: Characterization of normal intestinal thiamin transport in animals and man. NY Acad Sci 378:337–343, 1982.
9. Spector R: Thiamin homeostasis in the central nervous system. Ann NY Acad Sci 378:344–354, 1982.
10. Leevy CM: Thiamin deficiency and alcoholism. Ann NY Acad Sci 378:316–326, 1982.
11. Bonjour JP: Vitamins and alcoholism. IV. Thiamin. Int J Vitam Nutr Res 50:321–338, 1980.
12. Dart R, Howard L: Alcoholic peripheral neuropathy. Nutr Rev 39:237–239, 1981.
13. Blass JP, Gibson GE: Abnormality of a thiamin-requiring enzyme in patients with Wernicke-Korsakoff syndrome. N Engl J Med 297:1367–1370, 1977.
14. Campbell CH: The severe lacticacidosis of thiamine deficiency: Acute pernicious or fulminating beriberi. Lancet 2:446–449, 1984.
15. Eriksson K, Pekkanen L, Rusi M: The effects of dietary thiamin on voluntary ethanol drinking and ethanol metabolism in the rat. Br J Nutr 43:1–13, 1980.
16. Langohr HD, Petruch F, Schroth G: Vitamin B-1, B-2 and B-6 deficiency in neurological disorders. J Neurol 225:95–108, 1981.
17. Labadarios D, Rossouw JE, McConnell JB, et al.: Thiamine deficiency in fulminant hepatic failure and effects of supplementation. Int J Vitam Nutr Res 47:17–22, 1977.
18. Kjosen B, Seim SH: The transketolase assay of thiamine in some diseases. Am J Clin Nutr 30:1591–1596, 1977.
19. Baum RA, Iber FL: Thiamin—the interaction of aging, alcoholism, and malabsorption in various populations. World Rev Nutr Diet 44:85, 1984.
20. Iber FL, Blass JP, Brin M, et al.: Thiamin in the elderly—relation to alcoholism and to neurological degenerative disease. Am J Clin Nutr 36(suppl):1067, 1982.
21. Chen MF, Boyce HW Jr, Barry PP, Amontree JS: Blood levels of thiamin in older Americans in relation to the magnesium concentration. Nutr Rep Int 33:559–564, 1986.
22. Hoorn RKJ, Flikweert JP, Westerink D: Vitamin B-1, B-2, and B-6 deficiencies in geriatric patients, measured by coenzyme stimulation of enzyme activities. Clin Chim Acta 61:151, 1975.
23. Griffiths LL, Brocklehurst JC, Scott DL, et al.: Thiamin and ascorbic acid levels in the elderly. Gerontol Clin 9:1, 1967.
24. Vir SC, Love AHG, Thompson W: Thiamin status during pregnancy. Int J Vitam Nutr Res 50:131–140, 1980.

25. Glad BW, Hodges RE, Michas CA, et al.: Atrophic beri-beri. A complication of jejunoileal bypass surgery for morbid obesity. Am J Med 65:69–74, 1978.
26. Hoyt CS, Billson FA: Optic neuropathy in ketogenic diet. Br J Ophthalmol 63:191–194, 1979.
27. Asregadoo ER: Blood levels of thiamine and ascorbic acid in chronic open-angle glaucoma. Ann Ophthalmol 11:1095–1100, 1979.
28. Lonsdale D, Shamberger RJ: Red cell transketolase as an indicator of nutritional deficiency. Am J Clin Nutr 33:205–211, 1980.
29. Brin M: Red cell transketolase as an indicator of nutritional deficiency. Am J Clin Nutr 33:169–171, 1980.
30. Leevy CM: Red cell transketolase as an indicator of nutritional deficiency. Am J Clin Nutr 33:172–173, 1980.
31. Cole M, Turner A, Frank O, et al.: Extraocular palsy and thiamine therapy in Wernicke's encephalopathy. Am J Clin Nutr 22:44, 1969.
32. Haralambie G, Van Dam B: Untersuchungen über den Vitamin-Elektrolyt-Status bei Spitzenfechterinnen. Leistungssport 7:214–219, 1977.
33. Van Dam B: Vitamins and sport. Br J Sports Med 12:74–79, 1978.
34. Haralambie G: Recherches sur les carences minerales et vitaminiques chez l'enfant pratiquant le sport. In: *Le Sport et L'Enfant*, JM Bourgeois (ed). Montpellier: Euromed, 1980, pp 233–242.
35. Lonsdale D: Hyperalaninemia with pyruvicemia (letter). N Engl J Med 278:1235, 1968.
36. Blass JP, Gibson GE, Kark RAP: Pyruvate decarboxylase deficiency. In: *Thiamine*, CJ Gubler, M Fujiwara, PM Dreyfus (eds). New York: John Wiley & Sons, 1976, pp 321–334.
37. Koike M, Koike K: Biochemical properties of mammalian 2-oxo acid dehydrogenase multienzyme complexes and clinical relevancy with chronic lactic acidosis. Ann NY Acad Sci 378:225–235, 1982.
38. Elsas LJ, Danner DJ, Rogers BL: Effect of thiamine on normal and mutant human branched-chain alpha-ketoacid dehydrogenase. In: *Thiamine*, CJ Gubler, M Fujiwara, PM Dreyfus (eds). New York: John Wiley & Sons, 1976, pp 335–351.
39. Elsas LJ, Danner DJ: The role of thiamin in maple syrup urine disease. Ann NY Acad Sci 378:404–421, 1982.
40. Cooper JR, Pincus JH: Subacute necrotizing encephalomyelopathy. In: *Thiamine*, CJ Gubler, M Fujiwara, PM Dreyfus (eds). New York: John Wiley & Sons, 1976, pp 353–360.
41. Clinical Nutrition Case: Thiamine-responsive megaloblastic anemia. Nutr Rev 38:374–375, 1980.
42. Lonsdale D: Recurrent febrile lymphadenopathy treated with large doses of vitamin B-1: Report of two cases. Dev Pharmacol Ther 1:254–264, 1980.
43. Fujiwara M, Itokawa Y, Kimura M: Experimental studies on the relationships between thiamine and divalent cations, calcium and magnesium. In: *Thiamine*, CJ Gubler, M Fujiwara, PM Dreyfus (eds). New York: John Wiley & Sons, 1976, pp 63–82.
44. Lonsdale D: Thiamine metabolism in disease. CRC Crit Rev Lab Sci 5:298–313, 1975.
45. Read DJC: The aetiology of sudden infant death syndrome: Current ideas on breathing and sleep and possible links to deranged thiamine neurochemistry. Aust NZ J Med 8:322–336, 1978.
46. Lonsdale D, Mercer RD: Primary hypoventilation syndrome. Lancet 2:487, 1972.
47. Lonsdale D, Nodar RH, Orlowski JP: The effect of thiamine on abnormal brainstem auditory evoked potentials. Cleve Clin Q 46:83–88, 1979.
48. Davis RE, Icke GC, Hilton JMN: High thiamine levels in sudden infant death syndrome. N Engl J Med 303:462, 1980.

49. Alvarez OM, Gilbreath RL: Thiamin influence on collagen during the granulation of skin wounds. J Surg Res 32:24–31, 1982.
50. Alvarez OM, Gilbreath RL: Effect of dietary thiamin on intermolecular collagen crosslinking during wound repair: A mechanical and biochemical assessment. J Trauma 22:20–24, 1982.
51. Botez MI, Joyal C, Maag U, Bachevalier J: Low blood thiamine levels in phenytoin-treated epileptics. Nutr Rep Int 24:415–423, 1981.

6
Riboflavin (Vitamin B-2)

Introduction	117
Absorption, Metabolism, and Excretion	119
Dosage Forms	120
Clinical Studies	120
The Preterm Infant	120
Phototherapy of Hyperbilirubinemia	120
Other Pediatric Deficiency	121
Pregnancy and Lactation	121
The Elderly	122
Diabetes Mellitus	122
Cataract	122
Alcoholism	123
Sickle Cell Anemia	123
Exercise and Athletic Training	123
Hyperthyroidism	124
Hemodialysis	124
Cogenital Methemoglobinemia	124
Organic Acidurias	124
Toxicity and Side Effects	125
Interactions	125
References	125

INTRODUCTION

Chemically, riboflavin is the tricyclic compound, 7,8-dimethyl-10-(1'-D-ribityl) isoalloxazine (see Fig. 1 for structure). Isolated in pure form, it is an orange powder, sparingly soluble in water. Its solutions have a greenish-yellow fluorescence that can be seen in urine after large doses of the vitamin. Riboflavin is quite stable to heat but sensitive to decomposition by light [1].

In the body, riboflavin is converted to its coenzyme forms: riboflavin 5'-phosphate (FMN, flavin mononucleotide) and flavin adenine dinucleotide

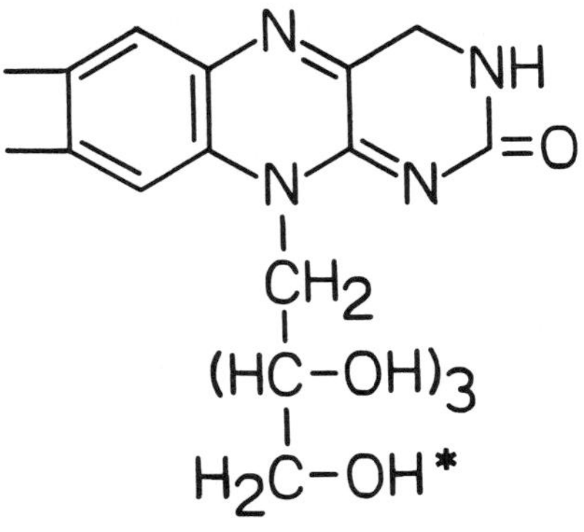

Fig. 1. Structural formula of riboflavin. * Replacement of this hydroxyl group by phosphate gives flavin mononucleotide (FMN).

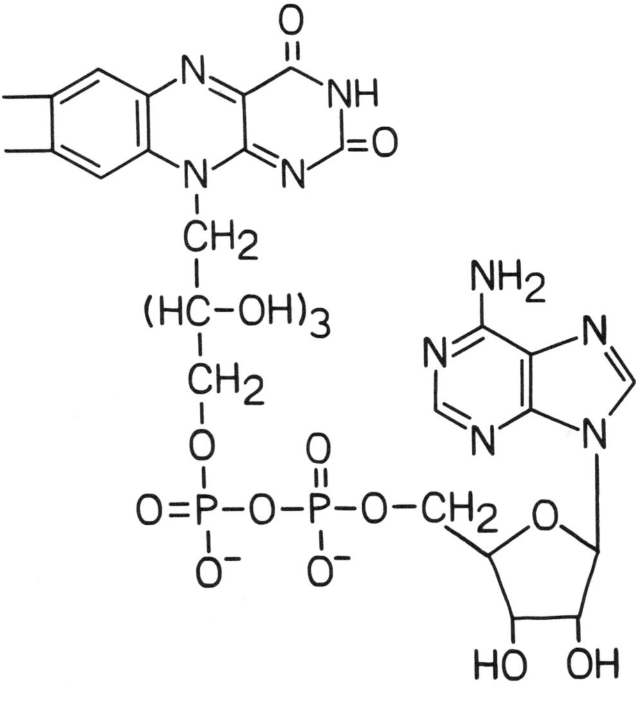

Fig. 2. Structural formula of flavin adenine dinucleotide (FAD).

(FAD) (See Fig. 2). These coenzymes are essential for the function of numerous enzymes involved in the transfer of electrons from a substrate either to oxygen or to another reducible acceptor. Examples are the amino acid oxidases, xanthine oxidase, succinic dehydrogenase complex, glutathione reductase, cytochrome reductase, and lactic acid dehydrogenase. FMN and FAD are essential coenzymes for the generation of energy (ATP) by catabolism of glucose, fatty acids, and amino acids. Flavoenzymes are also essential for the conversion of pyridoxine and folic acid to their coenzyme forms [1].

Overt clinical signs of riboflavin deficiency, such as cheilosis, angular stomatitis, and glossitis, are rarely seen in the United States and other developed countries, but biochemical deficiency, manifested as diminished coenzyme saturation of erythrocyte glutathione reductase, is rather frequently observed in women taking oral contraceptives, children and adolescents from low socioeconomic backgrounds, children with chronic heart disease, and the elderly [1].

The best food sources of riboflavin are milk and other dairy products. Other good sources are meat, eggs, legumes, and green leafy vegetables. The RDA varies from 0.4 mg/d for infants to 1.8 mg/d for pregnant women, with about 1.6 mg/d recommended for most others [1]. Individuals who ingest little milk or cheese are likely to be in a marginal status of riboflavin nutriture unless they pay special attention to obtaining other good dietary sources of the vitamin or unless they take a vitamin supplement regularly.

ABSORPTION, METABOLISM, AND EXCRETION

Riboflavin occurs in foods as such and as FMN and FAD. The latter two are hydrolyzed by gut enzymes to yield free riboflavin, the form in which the vitamin is absorbed. Absorption occurs in the proximal small intestine by an active saturable process. When megadoses are given on an empty stomach, the limit to absorption is reached at about 25 mg [2]. On an empty stomach, only 15% of a 30-mg dose was absorbed [2]. In the intestinal mucosa, absorbed riboflavin is phosphorylated to FMN and liberated to the circulation. Some FMN in blood is hydrolyzed back to riboflavin. Both FMN and riboflavin are largely bound to plasma albumin. Conversion to FAD takes place in the liver and other tissues, and FAD then becomes bound, in some cases covalently, with apoenzymes to form flavoproteins. The kidney excretes both riboflavin and FMN, the latter being dephosphorylated in the bladder. Negligible amounts of riboflavin are excreted in bile, feces, and sweat [1,2]. Although riboflavin is the primary flavin excreted in human urine, other flavin compounds are also present, comprising 30–40% of total urinary flavins. These include 7-alpha- and 8-alpha-hydroxyriboflavin, small amounts of 10-hydroxyethyl-flavin (probably from intestinal microbial action on the vitamin), and traces of lumiflavin (arising from photodecomposition) [3].

DOSAGE FORMS

Riboflavin is available over the counter as 25-, 50-, and 100-mg tablets. By far the greatest part of commercially produced riboflavin is used in combination with other vitamins in various multivitamin pharmaceutical formulations. Such vitamin combinations are often prescribed by physicians to ensure adequate nutriture of riboflavin even if the nutritional status with respect to other vitamins may not require improvement.

CLINICAL STUDIES

General comment. No uses for pharmacologic doses of riboflavin have been reported, except for a few rare congenital disorders. In most cases, supplementation is needed only to correct marginal deficiencies. For this purpose, doses of a few mg a day, i.e., one or two times the RDA, are generally satisfactory. Biochemical deficiency of riboflavin is conveniently determined by measuring the degree of FAD cofactor-unsaturation of erythrocyte glutathione reductase (EGR); the degree of enzyme stimulation by FAD added in vitro is referred to as the activation coefficient [4].

The Preterm Infant

Rönnholm [5] has recently studied the riboflavin status of 39 very low-birth-weight Finnish infants, 19 of whom received 0.3 mg/d of supplemental riboflavin and 20 no supplemental riboflavin, and all of whom were fed human milk. At age 6 weeks, 47% of the infants without supplementation had EGR activation coefficients indicative of riboflavin deficiency. The riboflavin status of the infants receiving the 0.3 mg/d supplement was significantly better, although still marginal in some cases. Breast milk supplied only 20–40% of this amount. The author suggests that a supplement of 0.4–0.5 mg/d would be advisable in the breast-fed infant.

Phototherapy of Hyperbilirubinemia

Riboflavin is sensitive to destruction by the phototherapy used in treatment of hyperbilirubinemia, and breast-fed infants may show biochemical deficiency of the vitamin if phototherapy is prolonged to 2–3 days' total exposure [6]. On infant formulas providing 0.3 mg/d of riboflavin, such deficiency did not develop [7]. Since photodynamic activation of riboflavin may cause alteration in intracellular DNA, it is recommended that riboflavin supplementation of infants be postponed until after phototherapy [5].

Other Pediatric Deficiency

Biliary atresia in infants and children may be associated with impaired absorption of riboflavin and may require a daily supplement to correct biochemical deficiency [2]. Steier et al. [8] found 8 of 27 children with congenital heart disease and 3 of 4 with rheumatic heart disease to show biochemical riboflavin deficiency; a supplement of 2.4 mg/d of riboflavin returned their EGR activation coefficients to normal. Lopez et al. [9] surveyed an adolescent population aged 13–19 years and of low socioeconomic status in New York City, finding a 26.6% prevalence of deficiency among those not taking vitamin supplements.

Pregnancy and Lactation

Vir et al. [10], in a study of riboflavin status in 60 pregnant and 20 nonpregnant women in Northern Ireland, found biochemical deficiency to be absent in the nonpregnant subjects and present in about 21% of the pregnant women during the second and third trimesters; in the early postpartum period, this increased to 30%. Mean dietary intake was higher than the recommended daily allowance.

Clarke [11] examined mean corpuscular riboflavin concentrations (MCRC) in the blood of 22 Canadian women at term, who, by dietary history, took at least one glass of milk a day, and also in the blood of 12 mothers with milk aversion. In the latter group, blood values of eight were suboptimum (MCRC < 31 mcg/dl), and three were deficient (MCRC < 25 mcg/dl) and had angular stomatitis. In the milk-drinking group, only two were suboptimum, and none were deficient or had angular stomatitis.

In pregnancy anemia, Decker et al. [12] found a better hemoglobin response to combination therapy with 3 mg riboflavin and 100 mg ferrous sulfate t.i.d. than to ferrous sulfate alone. These workers had previously reported, based on a survey of 200 pregnant Swiss women, an increased incidence of deficient riboflavin status among those with low hemoglobin levels [13]. A similar interaction of riboflavin with iron nutriture has been noted by Buzina et al. [14] in anemic Yugoslavian school children.

Bates et al. [15] studied riboflavin status in 60 lactating Gambian women who were receiving only about 0.5 mg/d of the vitamin in their diet. All were markedly deficient by the EGR stimulation test. A 2 mg/d supplement restored normal or near-normal activation coefficients in these subjects and maintained them throughout lactation, indicating that a total intake of about 2.5 mg/d of riboflavin was adequate for lactation in this group.

A riboflavin supplement supplement of 2–5 mg/d may be advisable during pregnancy and lactation if consumption of dairy products is low or if the quality of the diet is otherwise suspect.

The Elderly

In healthy elderly individuals consuming good diets, incidence of biochemical riboflavin deficiency is small [16,17]. Garry et al. found 45% of 270 elderly Americans surveyed to be taking some supplemental riboflavin in multivitamin products [16]. However, Lopez et al. [18] found a 23.6% incidence of riboflavin deficiency in 157 patients, 60 years of age or older, who were hospitalized long-term for chronic illness; even among 51 receiving vitamin supplements, 9 were riboflavin-deficient. Chen and Fan-Chiang [19] found biochemical riboflavin deficiency in 34% of institutionalized elderly and 28% of noninstitutionalized elderly subjects aged 60–95 years in Kentucky. Low appetite, chronic illness, and poor dietary habits may induce deficiencies of many essential nutrients in the aged, including riboflavin.

Diabetes Mellitus

Riboflavin deficiency among diabetics is reported to be more frequent than among nondiabetics at the same socioeconomic level, possibly in part due to the increased urinary excretion of riboflavin observed in this disease [1].

Cataract

Riboflavin deficiency by itself does not seem to lead to cataract in laboratory rats, but if the animals are fed diets high in galactose (present in lactose, i.e., milk sugar, combined with glucose), riboflavin-deficient rats show a higher incidence of galactose-induced cataracts than riboflavin-sufficient animals [20]. Riboflavin deficiency in itself reduces the levels of glutathione reductase, glutathione peroxidase, and reduced glutathione in the lens, while the level of lipid peroxides increases, changes believed adverse to resistance of the lens to cataract-inducing agents such as galactose [20,21]. Bhat and Gopalan [22] reported a higher incidence of riboflavin deficiency in Asiatic Indian cataract patients than in controls, and 7 of 15 cataract patients showed abnormal galactose tolerance, as evidenced by higher levels of blood galactose at 90 and 120 minutes after galactose ingestion compared with controls. Riboflavin supplementation (10 mg/d) did not correct the galactose intolerance.

Galactose intolerance may be caused by inherited deficiencies in galactokinase or galactose-1-phosphate uridyl transferase. Prchal et al. [23] detected, in 22 patients under age 50 with presenile cataracts, 5 patients who were heterozygous for galactokinase deficiency and 2 for galactose uridyl transferase deficiency. Riboflavin deficiency was present in 8 of the 22, but not particularly related to galactose intolerance. In a subsequent study of the riboflavin status of presenile and senile cataract patients, compared with young and old subjects with clear lenses, these investigators found no evidence of an association between riboflavin deficiency and presenile cataract formation [24]. Older cataract patients had more riboflavin deficiency, but there was an absence of riboflavin deficiency in older patients with clear lenses. An epidemiologic study by Bhat [25] in India, comparing riboflavin, thiamin, and vitamin B-6 status of cataract patients with controls, showed biochemical deficiency of riboflavin in the patients relative to the controls, but not of thiamin or vitamin B-6.

The evidence in total suggests that adequate riboflavin nurture is important for optimal lens resistance to noxious agents such as galactose. Further clinical studies are desirable in this area.

Alcoholism

Riboflavin deficiency is common in chronic alcoholism and may contribute to neurologic disturbances [26]. The dosages used to treat riboflavin deficiency in alcoholics, in contrast to those employed for other vitamins, are modest. For example, in one multivitamin preparation used intravenously in hospitalized English alcoholics, only 4 mg of riboflavin was present in a mixture supplying 250 mg thiamin·HCl, 50 mg pyridoxine·HCl, 160 mg nicotinamide, and 500 mg ascorbic acid per day [27].

Sickle Cell Anemia

Varma et al. [28] observed elevated EGR activity coefficients in 13 of 27 sickle cell anemia patients of various ages, suggestive of riboflavin deficiency. Abnormal coefficients were present in some despite dietary intakes meeting the RDA. Effects of riboflavin supplementation are being investigated by this group.

Exercise and Athletic Training

Haralambie [29] examined the riboflavin status of German male athletes in training, finding biochemical inadequacy in 8 of 18. Oral administration of a single dose of 10 mg riboflavin on an empty stomach produced a moderate lowering of neuromuscular irritability (musculi vastus medialis and lateralis quadricipitis), particularly for 0.1-msec rectangular stimuli.

Roe and her coworkers have carried out a series of studies in young women to evaluate the effect of jogging and other aerobic exercise on their riboflavin requirement [30–32]. Evidence was obtained that exercise increased the requirement slightly but significantly both in young women of normal weight [30] and in moderately overweight women undertaking aerobic exercise programs to assist in weight reduction [31,32]. A similar increase in apparent riboflavin requirement has been observed in older women (50–67 years of age) during a regular aerobic exercise program [33]. Indications were obtained that even during nonexercise periods, the present RDA for young women is not quite adequate [30].

It appears that a modest riboflavin supplement may be desirable during athletic training or other regularly undertaken exercise programs.

Hyperthyroidism

Excess thyroid activity appears to elevate the riboflavin requirement [34].

Hemodialysis

Since there may be some loss of riboflavin (as well as other water-soluble vitamins) into dialysis fluid, hemodialysis patients should receive vitamin supplements containing riboflavin [1].

Congenital Methemoglobinemia

Kaplan and Chirouze [35] reported successful treatment of recessive congenital methemoglobinemia in a 33-year-old man and an infant with high-dose riboflavin. Both had previously been kept under satisfactory control by administration of ascorbic acid, 1,000 mg/d in the man and 200 mg/d in the infant. Methemoglobin level was kept below 5% in the man by riboflavin at 30 mg/d and in the infant by administration of 20 mg/d, following withdrawal of ascorbic acid.

Organic Acidurias

Multiple acyl-CoA dehydrogenase deficiency (glutaric aciduria type II) usually results in death in early infancy. However, a similar organic aciduria reported in a 32-month-old child was successfully treated with high-dose riboflavin [36]. Harpey et al. [37] record a case of multiple acyl-CoA dehydrogenase deficiency occurring in a pregnant woman who had lost seven infants by stillbirth or death in early infancy. Although apparently healthy between pregnancies, the mother developed malaise, fatigability, mild jaundice, and pruritus in the third trimester of the seven pregnancies with poor outcome. Organic aciduria was found in a workup during the next succeeding pregnancy, which was controlled by administration of 20 mg/d of riboflavin orally; a normal full-term girl was born who was maintained on cow's milk formula and 10 mg/d of riboflavin orally for several weeks until glutaric aciduria no longer occurred. Later, the mother's disorder worsened after a common cold infection, although she was not pregnant. This time 50 mg/d of riboflavin was necessary to control the organic aciduria, and she was subsequently maintained on this dose. The authors consider the acyl-CoA dehydrogenase deficiency in this case to have been secondary to a disturbance in riboflavin metabolism, e.g., a defect in synthesis or transport of FAD or in binding of FAD to its apoenzymes.

TOXICITY AND SIDE EFFECTS

No toxicity or side effects have been reported in the clinical use of riboflavin. In animal studies, it has proved to have remarkably low toxicity [1].

INTERACTIONS

Oral contraceptive (estrogen-progestin) use does not appear to affect riboflavin status adversely if the diet is good [38], but Newman et al. [39] found, in oral contraceptive (OC) users in a low socioeconomic neighborhood in New York City, a 43% prevalence of biochemical riboflavin deficiency versus 11% in nonuser controls. About three-quarters of those who had been on OCs for 3 years or more showed evidence of deficiency. A supplement of riboflavin is advisable during OC use if there is any doubt about the quality of the diet [34].

Chlorpromazine and other phenothiazine derivatives, as well as imipramine and other tricyclic compounds, inhibit the incorporation of riboflavin into FAD and flavoproteins [40]. Riboflavin supplements of 2–5 mg/d are recommended. Boric acid intoxication causes massive excretion of riboflavin, and high-dose riboflavin administration may be therapeutic in such cases [34].

REFERENCES

1. Cooperman JM, Lopez R: Riboflavin. In: *Handbook of Vitamins*, LJ Machlin (ed). New York: Marcel Dekker, 1984, pp 299–327.
2. Jusko WJ, Levy G: Absorption, protein binding, and elimination of riboflavin. In: *Riboflavin*, RS Rivlin (ed). New York: Plenum Press, 1975, pp 99–152.
3. Chastain JL, Mc Cormick DB: Flavin catabolites: Identification and quantitation in human urine. Am J Clin Nutr 46:830–834, 1987.
4. Prentice AM, Bates CJ: A biochemical evaluation of the erythrocyte glutathione reductase

(EC 1.6.4.2) test for riboflavin status. 2. Dose-response relationships in chronic marginal deficiency. Br J Nutr 45:53–65, 1981.
5. Rönnholm KAR: Need for riboflavin supplementation in small prematures fed with human milk. Am J Clin Nutr 43:1–6, 1986.
6. Gromisch DS, Lopez R, Cole HS, Cooperman JM: Light (phototherapy)-induced riboflavin deficiency in the neonate. J Pediatr 90:118–122, 1977.
7. Tan KL, Chow MT, Karim SMM: Effect of phototherapy on neonatal riboflavin status. J Pediatr 93:494–497, 1978.
8. Steier M, Lopez R, Cooperman JM: Riboflavin deficiency in infants and children with heart disease. Am Heart J 92:139–142, 1976.
9. Lopez R, Schwartz JV, Cooperman JM: Riboflavin deficiency in an adolescent population in New York City. Am J Clin Nutr 33:1283–1286, 1980.
10. Vir SC, Love AHG, Thompson W: Riboflavin status during pregnancy. Am J Clin Nutr 34:2699–2705, 1981.
11. Clarke HC: Milk, cellular equilibrium, and nutritional evolution. Int J Vitam Nutr Res 51:293–296, 1981.
12. Decker K, Dotis B, Glatzle D, Hinselmann M: Riboflavin status and anaemia in pregnant women. Nutr Metab 21(suppl 1):17–19, 1971.
13. Decker K, Hinselmann M, Glatzle D: Anämie und Schwangerschaft. In: *Brubacher und Ritzel Erster schweizerischer Ernährungsbericht*. Bern: Huber, 1975, pp 221–232.
14. Buzina R, Jusic M, Milanovic N, et al.: The effects of riboflavin administration on iron metabolism parameters in a school-going population. Int J Vitam Nutr Res 49:136–143, 1979.
15. Bates CJ, Prentice AM, Watkinson M, et al.: Riboflavin requirements of lactating Gambian women: A controlled supplementation trial. Am J Clin Nutr 35:701–709, 1982.
16. Garry PJ, Goodwin JS, Hunt WC: Nutritional status in a healthy elderly population: Riboflavin. Am J Clin Nutr 36:902–909, 1982.
17. Alexander M, Emanuel G, Golin T, et al.: Relation of riboflavin nutriture in healthy elderly to intake of calcium and vitamin supplements: Evidence against riboflavin supplementation. Am J Clin Nutr 39:540–546, 1984.
18. Lopez R, Fisher LV, Cooperman JM: Riboflavin deficiency in an aged population. Fed Proc 38:451, 1979.
19. Chen LH, Fan-Chiang WL: Biochemical evaluation of riboflavin and vitamin B-6 status of institutionalized and non-institutionalized elderly in Central Kentucky. Int J Vitam Nutr Res 51:232–238, 1981.
20. Review: Riboflavin deficiency, galactose metabolism and cataract. Nutr Rev 34:72–79, 1976.
21. Hirano H, Hamajima S, Horiuchi S, et al.: Effects of B-2-deficiency on lipoperoxide and its scavenging system in the rat lens. Int J Vitam Nutr Res 53:377–382, 1983.
22. Bhat KS, Gopalan C: Human cataract and galactose metabolism. Nutr Metab 17:1–8, 1974.
23. Prchal JT, Conrad ME, Skalka HW: Association of presenile cataracts with heterozygosity for galactosaemic states and with riboflavin deficiency. Lancet 1:12–13, 1978.
24. Skalka HW, Prchal JT: Cataracts and riboflavin deficiency. Am J Clin Nutr 34:861–863, 1981.
25. Bhat KS: Nutritional status of thiamine, riboflavin and pyridoxine in cataract patients. Nutr Rep Int 36:685–692, 1987.
26. Langohr HD, Petruch F, Schroth G: Vitamin B-1, B-2, and B-6 deficiency in neurological disorders. J Neurol 225:95–108, 1981.
27. Majumdar SD, Patel S, Shaw GK, et al.: Vitamin C utilization status in chronic alcoholic patients after short-term intravenous therapy. Int J Vitam Nutr Res 51:274–278, 1981.

28. Varma RN, Mankad VN, Phelps DD, et al.: Depressed erythrocyte glutathione reductase activity in sickle cell disease. Am J Clin Nutr 38:884–887, 1983.
29. Haralambie G: Vitamin B-2 status in athletes and the influence of riboflavin administration on neuromuscular irritability. Nutr Metab 20:1–8, 1976.
30. Belko AZ, Obarzanek E, Kalkwarf HJ, et al.: Effects of exercise on riboflavin requirements of young women. Am J Clin Nutr 37:509–517, 1983.
31. Belko AZ, Obarzanek E, Roach R, et al.: Effects of aerobic exercise and weight loss on riboflavin requirements of moderately obese, marginally deficient young women. Am J Clin Nutr 40:553–561, 1984.
32. Belko AZ, Meredith MP, Kalkwarf HJ, et al.: Effects of exercise on riboflavin requirements: Biological validation in weight reducing women. Am J Clin Nutr 41:270–277, 1985.
33. Yoon J-S, Trebler L, Roe DA: Effect of exercise on the riboflavin requirements of older women. Fed Proc 46:1166A, 1987.
34. Rivlin RS: Hormones, drugs and riboflavin. Nutr Rev 37:241–245, 1979.
35. Kaplan JC, Chirouze M: Therapy of recessive congenital methaemoglobinaemia by oral riboflavine. Lancet 2:1043–1044, 1978.
36. Gregerson N, Wintzensen H, Kolvraa S, et al.: C-6-C-10-Dicarboxylic aciduria: Investigations of a patient with riboflavin responsive multiple acyl-CoA dehydrogenation defects. Pediatr Res 16:861, 1982.
37. Harpey J-P, Charpentier C, Goodman SI, et al.: Multiple acyl-CoA dehydrogenase deficiency occurring in pregnancy and caused by a defect in riboflavin metabolism in the mother. J Pediatr 103:394–398, 1983.

7
Niacin and Niacinamide

Introduction	129
Absorption, Metabolism, and Excretion	131
Dosage Forms	131
Clinical Studies	132
Hyperlipidemias	132
Alcoholism	134
Diabetes Mellitus	134
Athletic Training and Exercise	135
Genetic Disease	135
Other Possible Uses	135
Toxicity and Side Effects	136
Interactions	136
References	137

INTRODUCTION

Niacin (nicotinic acid) has the chemical name pyridine-3-carboxylic acid, and niacinamide (nicotinamide) is pyridine-3-carboxylic acid amide (see Fig. 1 for structures). In the body, either form of the vitamin (sometimes referred to as vitamin B-3) may be converted to the coenzymes nicotinamide adenine dinucleotide (NAD) and nicotinamide adenine dinucleotide phosphate (NADP) (see Fig. 2) [1]. Both nicotinic acid and nicotinamide are white crystalline solids, the former being moderately soluble in water (1 g in 60 ml at 25°C), the latter very soluble (1 g in 1 ml at 25°C) [2]. Nicotinic acid has a tart taste, nicotinamide a bitter one. Both are quite stable to heat and oxygen [1].

NAD and NADP are coenzymes for many dehydrogenases that catalyze the removal of hydrogen from oxidizable substrates [1]. They are essential for energy (ATP) generation via glycolysis and the Krebs cycle [1,2]. Reduction of NADP in the pentose phosphate pathway yields NADPH, which is used in the biosynthesis of fatty acids and other products [2]. NAD is also important in DNA metabolism as a high-energy donor for ADP-ribosylation [3]. Nicotin-

130 / Pharmacology of Micronutrients

Fig. 1. Structural formulas of niacin (nicotinic acid—left) and niacinamide (nicotinamide—right).

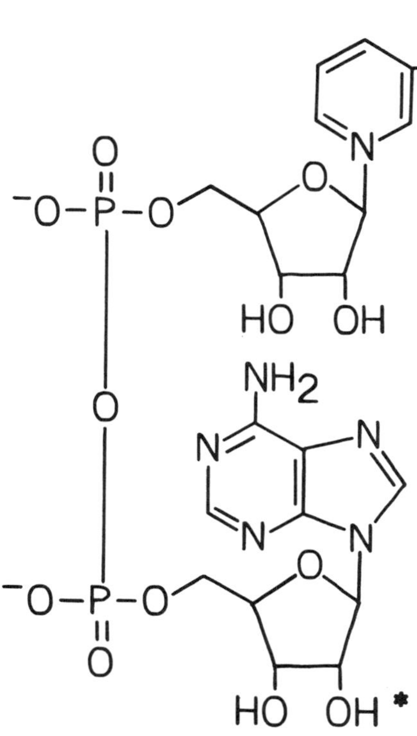

Fig. 2. Structural formula of nicotinamide adenine dinucleotide (NAD). *This hydroxyl group is replaced by phosphate in nicotinamide adenine dimicleotide phosphaste (NADP).

amide in the form of its coenzymes is largely protein (enzyme)-bound and widely distributed in all metabolically active tissues [1].

Niacin deficiency leads initially to nonspecific symptoms such as lassitude, anorexia, weakness, indigestion, and irritability, progressing eventually to the classical deficiency disease, pellagra, characterized by dermatitis, diarrhea, and dementia [4]. In the past, this disease was seen mostly in populations subsisting largely on corn, with inadequate supplementation of protein foods. The available niacin content of corn is low, and it is also a poor source of the essential amino acid, tryptophan. This amino acid allows endogenous synthesis of niacin by a minor catabolic pathway such that approximately 60 mg of dietary tryptophan yields 1 mg niacin [5,6]. In the developed countries, niacin and tryptophan deficiency sufficient to lead to pellagra-like symptoms has only rarely been seen, and then primarily in alcoholics [7].

Food sources of niacin per se include animal protein foods, beans, nuts, whole grains, enriched bread and cereals, coffee, and tea [1]. Because of their relatively high content of tryptophan, the animal protein foods are good sources of what is termed "niacin equivalents." The average American diet is calculated to supply 16–34 mg/d of niacin equivalents, of which 8–17 mg are actual niacin while the balance comes from 500–1000 mg/d of dietary tryptophan [8]. The RDA for adults is 15–20 mg/d of niacin equivalents [8].

ABSORPTION, METABOLISM, AND EXCRETION

Animal and in vitro studies indicate that both niacin and niacinamide are absorbed in the small intestine by passive diffusion, or at least by processes not readily saturable, and that the amide is absorbed more rapidly than the acid [9–11]. Niacin in foods occurs mostly in its coenzyme forms, which are hydrolyzed during digestion, yielding nicotinamide, which seems to be absorbed as such without further hydrolysis in the GI tract [11]. Nicotinamide is the primary circulating form of the vitamin and is converted into its coenzyme forms in the tissues [1]. Nicotinamide, but not nicotinic acid, passes readily from plasma into the cerebrospinal fluid and is taken up into brain cells by a high-affinity accumulation system [1]. Nicotinamide is released by turnover of NAD and NADP in the tissues. Part is recycled to coenzyme synthesis, and part is converted to metabolites (principally N^1-methylnicotinamide and N^1-methyl-2-pyridone-5-carboxylamide), which are excreted in the urine [12].

When pharmacologic doses of nicotinic acid and nicotinamide are administered, urinary excretion of N^1-methyl-nicotinamide and 2-pyridone derivatives increases, along with that of unaltered nicotinic acid and nicotinamide. Nicotinuric acid comprises a moderate proportion of the niacin metabolites in urine when nicotinic acid is administered in either large or small doses, and small amounts of nicotinuric acid may also be found after large doses of nicotinamide, probably formed by way of nicotinic acid released from nicotinamide by nicotinamide deaminase [1,13]. Nicotinic acid and nicotinamide in pharmacologic dosage are rapidly and completely absorbed, peak plasma levels being reached in 1/2 to 2 hours after dosage [13].

DOSAGE FORMS

Both nicotinic acid and nicotinamide are available over the counter as tablets in 50-, 100-, 200-, and 500-mg strengths. Timed-release tablets (500 mg) are also available.

CLINICAL STUDIES
Hyperlipidemias

The principal clinical use of nicotinic acid is for the reduction of elevated blood lipid levels [14]. *Nicotinamide is not useful for this purpose.* Nicotinic acid therapy, introduced by Altschul et al. [15] in 1955, is the oldest of the pharmacologic modalities for treatment of the hyperlipidemias. It decreases total and LDL-cholesterol levels, raises HDL-cholesterol levels, and lowers triglyceride levels [16]. Nicotinic acid is probably the most effective of the lipid-lowering drugs in elevating HDL-cholesterol levels, many studies reporting rises of 10–15 mg/dl [16]. In the Coronary Drug Project [17], nicotinic acid therapy was shown to decrease recurrent nonfatal myocardial infarction by 40% without a concurrent mortality increase from nonatherosclerotic causes, as observed with clofibrate.

Nicotinic acid therapy is begun, after appropriate dietary modifications, with 500 mg or smaller doses orally t.i.d. Flushing of the face and trunk is initially experienced, due to prostaglandin-mediated dilation of cutaneous capillaries, but tolerance develops rather quickly [16]. One aspirin tablet taken about 1/2 hour before the niacin dose is helpful when flushing is persistent. Flushing can also be minimized if nicotinic acid is taken with meals, but this may increase incidence of upper gastrointestinal disturbances, in which case administration of nicotinic acid before meals may be tried [16]. The dosage may be increased every 2–4 weeks until the patient is taking 3–6 g/d in divided doses. At this dosage level, most studies have reported 20–30% reductions in LDL-cholesterol levels [16].

Abnormal liver function test results are occasionally seen with nicotinic acid therapy, and Hunninghake [16] considers elevations of serum glutamic oxaloacetic transaminase (SGOT) and alkaline phosphatase levels to be the best indicators of this problem. If liver function returns to normal following reduction of nicotinic acid dosage, it is often possible to increase the dosage later without incurring the same effect. Infrequently observed abnormalities requiring dosage reduction are hyperuricemia and hyperglycemia [16].

Nicotinic acid may be used together with other cholesterol-lowering drugs and is particularly well suited for combination therapy with agents that sequester bile acids in the gut. For example, Illingworth et al. [18] treated 13 patients with heterozygous familial hypercholesterolemia sequentially with a low-cholesterol, fat-restricted diet (2–4 months), followed by diet plus colestipol (10 g b.i.d. with meals for 1 month, 15 g b.i.d. for 3–9 months), followed by diet plus colestipol plus nicotinic acid (250 mg t.i.d. with meals initially, increased every 2–4 weeks to a total dose of 3–8 g/d). Aspirin was given for the first 4–6 weeks along with the nicotinic acid to reduce flushing, but was not needed thereafter. Mean plasma cholesterol levels fell from 415 mg/dl on diet alone to 327 mg/dl on diet plus colestipol and to 246 mg/dl on diet plus colestipol plus nicotinic acid. HDL-cholesterol levels were 40 mg/dl on diet alone, 43 mg/dl on diet plus colestipol, and 53 mg/dl on diet plus both drugs. Corresponding ratios of LDL:HDL were 8.4, 5.9, and 3.3. The beneficial synergistic action of nicotinic acid is attributed to its ability to inhibit low-density lipoprotein synthesis [19]. Routine liver function tests did not reveal any abnormalities at the nicotinic acid dosages given (mean 5.7 ± 1.7 g/d) [18].

Similar favorable results have been reported by Kane et al. [20] in long-term studies of patients with heterozygous familial hypercholesterolemia who showed little response to dietary therapy. With colestipol alone, mean serum cholesterol levels decreased 16–25%. Addition of clofibrate to the colestipol regimen produced a total 28% mean decrease from the levels obtained with dietary treatment alone. In contrast, serum cholesterol levels fell a total of 45% when colestipol was combined with niacin; mean LDL-cholesterol and triglyceride levels decreased by about one-half, while mean HDL-cholesterol levels increased from 57 to 76 mg/dl. In the combined colestipol-niacin regimen, dosages were 10 g colestipol t.i.d., taken after meals. The dose of niacin was increased progressively as tolerated over a period of 3 months to a total of 6–7.5 g/d (average 6.8 g/d in 22 patients) [20].

An aspirin tablet taken 1/2 hour before niacin dosage minimized flushing. Side effects of colestipol were mainly constipation, controlled with bran. No abnormalities of liver function were noted if niacin dosage was increased no faster than 2.5 g/month; more rapid increase produced transient moderate elevations of plasma aspartate aminotransferase or alkaline phosphatase activity. While taking colestipol and niacin, 4 of the 22 patients had slight elevations of fasting serum glucose levels from 115 to 120 mg/dl, and 6 patients showed uric acid levels above 8.0 mg/dl without development of gout. Eight patients taking niacin experienced occasional mild gastric irritation, controlled in seven of them by taking niacin with meals and using antacids [20].

A remarkable result of the niacin-colestipol treatment was a significant reduction in tendinous xanthomas, measured by quantitative xeroradiography, over a mean 8-month treatment period in each of seven patients with this disorder, indicating that this regimen mobilized cholesterol from tissue pools with slow turnover [20].

In contrast to a minimal additive effect of niacin with clofibrate [20], Morgan and Cohen [21] have reported preliminary favorable results in 19 ambulatory patients with hypercholesterolemia or mixed hyperlipidemia when treated with a combination of niacin and probucol, a drug that does not significantly sequester cholesterol in the GI tract but is believed to inhibit one or more of the early steps in hepatic cholesterol synthesis. Patients were followed for an average of 5 years. Niacin was added slowly, over months, to tolerance or up to 3.5 g/d. Probucol dosage was 250 mg 2 to 4 times daily. Combination therapy was effective in reducing mean serum cholesterol levels from approximately 300 mg/dl with dietary treatment alone to 200 mg/dl with niacin, probucol, and diet [21]. Further reports are awaited.

Mixed results have been obtained with respect to the efficacy of slow-release or timed-release forms of nicotinic acid; both favorable [14] and relatively poor [22] results have been described. Nicotinyl alcohol (Roniacol®, Roche Laboratories), approved for use as a vasodilator (see current PDR®), has been reported also to be an effective cholesterol-lowering agent [14], presumably because of conversion to nicotinic acid in the body.

Alcoholism

The chronic alcoholic whose diet is low in good sources of niacin or niacin equivalents, particularly animal protein foods, is likely to develop niacin deficiency, sometimes progressing to a pellagra-like state with pigmented scaling dermatitis in areas of skin exposed to sunlight [7,22]. There is generally multinutrient deficiency, so that niacin is not administered singly but as part of a regimen including other vitamins, minerals, and amino acids or protein. Majumdar et al. [23], treating hospitalized alcoholics in poor clinical condition, used an intravenous vitamin preparation for the first 5 days that provided daily 160 mg nicotinamide, 250 mg thiamin·HCl, 4 mg riboflavin, 50 mg pyridoxine·HCl, and 500 mg sodium ascorbate. For oral dosage in partially rehabilitated patients or in alcoholics in better condition, they employ a multivitamin tablet providing per day 200 mg nicotinamide, 50 mg thiamin·HCl, 5 mg riboflavin, 5 mg pyridoxine·HCl, and 100 mg vitamin C [24].

Diabetes Mellitus

A preliminary study by Vague et al. [25] from France in seven newly diagnosed, ketotic type I (insulin-dependent) diabetics indicates that high-dose nicotinamide may slow down the destruction of pancreatic beta cells and prolong the period of remission that is frequently observed early in the progression of this disease. There were 16 patients, of whom 7 were randomly assigned nicotinamide dosage (3 g/d), while 9 received placebo. Intensive insulin therapy was started upon diagnosis, and patients were instructed to reduce gradually their insulin dosage, provided that blood glucose remained normal in 3–6 capillary blood tests per day. Three of the seven patients in the nicotinamide group achieved remission for more than 2 years, while none of the nine placebo patients remained in remission for more than 9 months. These observations call for follow-up at other diabetes clinics.

Athletic Training and Exercise

A study of Russian athletes in training showed considerable declines in niacin status by the criterion of N^1-methylnicotinamide excretion, although their diets provided more than the U.S. RDA in niacin equivalents [26]. The increased metabolic rate and sweating induced by heavy exercise appears to increase the requirement for the water-soluble vitamins [27], suggesting that athletes would be benefited in nutritional status by a therapeutic multivitamin supplement. Confirmatory clinical studies are needed.

Genetic Disease

Hartnup's disease is a disorder of amino acid transport in which impaired intestinal absorption of tryptophan leads to a shortage of endogenously synthesized niacin and manifestations of pellagra in the skin and nervous system. It is characterized biochemically by aminoaciduria and indicanuria. Administration of nicotinamide at dosages of 40–250 mg/d has improved the dermatitis and neurologic symptoms [28].

A case of genetic niacinamide dependency without defective tryptophan absorption and without aminoaciduria was reported in a 14-year-old Israeli boy who had a pellagra-like rash and severe neurologic symptoms, including ataxia, dysarthria, nystagmus, and finally coma [29]. There was a striking and rapid improvement upon oral administration of 100 mg/d of nicotinamide. The investigators interpret the results of their study of this patient as compatible with a block in the tryptophan pathway to niacin [29].

Other Possible Uses

Intravenously injected, but not oral, nicotinic acid induces a short-lived (less than 1 hour) fibrinolytic state. The mechanism is not understood, but is exhaustible, as a second dose has no fibrinolytic effect unless administered days to weeks later. However, some patients with liver disease, especially those with cirrhosis and bilirubinemia, may show a response to oral nicotinic acid and may also respond to additional doses on the same day. So far, nicotinic acid has not gained recognized clinical status as a fibrinolytic agent [30]. Studies in vitro [31] and with oral administration to animals in vivo [32] have shown that nicotinic acid reduces synthesis of thromboxane in thrombocytes without affecting prostacyclin synthesis, and Vincent and Zijlstra [32] therefore proposed that nicotinic acid might be a useful antithrombotic agent. Whether this is related to the observed reduction in recurrence of nonfatal myocardial infarctions under nicotinic acid therapy in the Coronary Drug Project [17] is a subject for further clinical investigation.

Administration of large doses of nicotinamide, up to 3 g/d or more together with other vitamins, has been proposed as a treatment for schizophrenia, largely on the basis of a presumed partial enzymatic block in this disease between tryptophan and NAD [1,33]. The consensus of informed opinion, however, after careful review of controlled studies, is that nicotinic acid and nicotinamide are not effective antipsychotic agents [34].

Chouinard et al. [35] conducted clinical trials in depressed patients to evaluate the possibility that the antidepressive action of tryptophan, the precursor of brain serotonin, might be increased by concomitant administration of nicotinamide. They hypothesized that the activity of tryptophan oxidase would be inhibited by increased NAD, and thus more tryptophan be made available for conversion to serotonin. Eleven patients were started on 2 g tryptophan and 0.5 g nicotinamide per day, increasing stepwise to 6 g and 1.5 g respectively by the start of the third week. After a total of 4 weeks' treatment, there was a statistically significant improvement in the mean total scores on both the Hamilton and Beck depression rating scales. Further clinical studies of this treatment would be desirable.

TOXICITY AND SIDE EFFECTS

In animal tests of acute toxicity, nicotinamide is more toxic than nicotinic acid, but the toxicity of both compounds is very low. The LD_{50} values for rats and mice are 3.5–5.0 g/kg of body weight subcutaneously and 5–7 g/kg orally [1].

In humans a lethal dose has never been determined for either nicotinic acid or nicotinamide. Nicotinamide is generally well tolerated. Side reactions of nicotinic acid appearing early in the course of therapy include flushing, dizziness, pruritus, and nausea; these usually disappear after a few days. Other side effects of nicotinic acid that have occasionally been reported include possible activation of peptic ulcer, gastrointestinal discomfort, dryness of skin, keratosis nigricans, hypotension, and transient headaches. Some degree of increased uric acid level and decreased glucose tolerance is occasionally seen. Liver dysfunction, manifest as SGOT and alkaline phosphatase elevation, is sometimes found. Most side reactions are dose-related and may be ameliorated by reduction of dosage. Some patients remain intolerant of nicotinic acid, however, or may develop intolerance during long-term therapy [1,2,14,16–18].

INTERACTIONS

Long-term treatment of tuberculosis with isoniazid, a niacin antagonist, may lead to a pellagra-like psychosis and encephalopathy [36]; a therapeutic niacin supplement should be given.

Laboratory rats poisoned by intraperitoneal injections of paraquat showed longer survival and reduced mortality if injected daily with 500 mg nicotinic acid per kg body weight for 5 days [37].

REFERENCES

1. Hankes LV: Nicotinic acid and nicotinamide. In: *Handbook of Vitamins*, LJ Machlin (ed). New York: Marcel Dekker, 1984, pp 329–377.
2. Horwitt MK: Niacin. In: *Modern Nutrition in Health and Disease*, 6th ed., RS Goodhart, ME Shils (eds). Philadelphia: Lea & Febiger, 1980, pp 204–208.
3. Weiner M, van Eys J: Nicotinamide-adenine-dinucleotides as high-energy donors for transribosylation reactions. In: *Nicotinic Acid: Nutrient-Cofactor-Drug*, M Weiner, J van Eys (eds). New York: Marcel Dekker, 1983, pp 193–207.
4. Weiner M, van Eys J: The discovery of nicotinic acid as a nutrient. In: *Nicotinic Acid: Nutrient-Cofactor-Drug*, M Weiner, J van Eys (eds). New York: Marcel Dekker, 1983, pp 3–16.
5. Patterson JI, Brown RR, Linkswiler H, Harper AE: Excretion of tryptophan-niacin metabolites by young men: Effects of tryptophan, leucine, and vitamin B-6 intakes. Am J Clin Nutr 33:2157–3167, 1980.
6. Horwitt MK, Harper AE, Henderson LM: Niacin-tryptophan relationships for evaluating niacin equivalents. Am J Clin Nutr 34:423–427, 1981.
7. Bonjour JP: Vitamins and alcoholism. VI. Niacin. Int J Vitam Nutr Res 50:430–435, 1980.
8. Food and Nutrition Board: *Recommended Dietary Allowances*, 9th ed. Washington, DC: National Academy of Sciences, 1980.
9. Turner JB, Hughes DE: The absorption of some B-group vitamins by surviving rat intestine preparations. Q J Exp Physiol 47:107–123, 1962.
10. Henderson LM, Gross CJ: Transport of niacin and niacinamide in perfused rat intestine. J Nutr 109:646–653, 1979.
11. Henderson LM: Digestion and absorption of NAD and CoA. Fed Proc 43:2427–2428, 1984.
12. Sauberlich HE, Skala JH, Dowdy RP: *Laboratory Tests for the Assessment of Nutritional Status*. Boca Raton, FL: CRC Press, 1974, pp 70–74.
13. Weiner M, van Eys J: Pharmacokinetics. In: *Nicotinic Acid: Nutrient-Cofactor-Drug*, M Weiner, J van Eys (eds). New York: Marcel Dekker, 1983, pp 229–241.
14. Weiner M, van Eys J: Treatment of the hyperlipidemias. In: *Nicotinic Acid: Nutrient-Cofactor-Drug*, M Weiner, J van Eys (eds). New York: Marcel Dekker, 1983, pp 253–271.
15. Altschul R, Hoffer A, Stephen JD: Influence of nicotinic acid on serum cholesterol in man. Arch Biochem Biophys 54:558–559, 1955.
16. Hunninghake DB: Pharmacologic therapy for the hyperlipidemic patient. Am J Med 74:19–22, 1983.
17. Coronary Drug Project Research Group: Clofibrate and niacin in coronary heart disease. JAMA 231:360–381, 1975.
18. Illingworth DR, Phillipson BE, Rapp JH, Connor WE: Colestipol plus nicotinic acid in treatment of heterozygous familial hypercholesterolaemia. Lancet 1:296–297, 1981.
19. Levy RI, Langer T: Hypolipidemic drugs and lipoprotein metabolism. Adv Exp Med Biol 10:155–162, 1972.
20. Kane JP, Malloy MJ, Tun P, et al.: Normalization of low-density lipoprotein levels in het-

erozygous familial hypercholesterolemia with a combined drug regimen. N Engl J Med 304:251–258, 1981.
21. Morgan J, Cohen L: Hyperlipidemia: Further observations on individualized therapy with the combined use of niacin and probucol. Fed Proc 46:1470A, 1987.
22. Hodges RE: Nutrition and the nervous system. In: *Nutrition in Medical Practice*, RE Hodges (ed). Philadelphia: WB Saunders, 1980, pp 136–163.
23. Majumdar SK, Shaw GK, Thomson AD: Vitamin utilization status in chronic alcoholics. Int J Vitam Res 51:54–58, 1981.
24. Majumdar SK, Shaw GK, Thomson AD: Blood vitamin status in chronic alcoholics after a single dose of polyvitamin. A preliminary report. Postgrad Med J 57:164–166, 1981.
25. Vague P, Vialettes B, Lassman-Vague V, Vallo JJ: Nicotinamide may extend remission phase in insulin-dependent diabetes. Lancet 1:619–620, 1987.
26. Borisov IM: Niacin (nicotinamide) allowances in students of a sports college. Vopr Pitan 6:43–46, 1977.
27. Haralambie G: Vitamin B-2 status in athletes and the influence of riboflavin administration on neuromuscular irritability. Nutr Metab 20:1–8, 1976.
28. Halvorsen K, Halvorsen S: Hartnup disease. Pediatrics 31:29–38, 1963.
29. Freundlich E, Stalter M, Yatziv S: Familial pellagra-like skin rash with neurological manifestations. Arch Dis Child 56:146–148, 1981.
30. Weiner M, van Eys J: Induction of fibrinolysis by nicotinic Acid. In: *Nicotinic Acid: Nutrient-Cofactor-Drug*. M Weiner, J van Eys (eds). New York: Marcel Dekker, 1983, pp 281–291.
31. Deitemeyer D, Yunker R, Subbiah MTR: Effect of in vivo nicotinic acid administration on prostaglandin synthesis in pigeon aorta and thrombocytes. Nutr Rep Int 23:1089–1093, 1981.
32. Vincent SE, Zijlstra FJ: Nicotinic acid inhibits thromboxane synthesis in platelets. Prostaglandins 15:629, 1978.
33. Hawkins D, Pauling L: *Orthomolecular Psychiatry: Treatment of Schizophrenia*, San Francisco: W.H. Freeman, 1973.
34. Weiner M, van Eys J: Miscellaneous pharmacological and clinical actions. In: *Nicotinic Acid: Nutrient-Cofactor-Drug*, M Weiner, J van Eys (eds). New York: Marcel Dekker, 1983, pp 293–300.
35. Chouinard G, Young SN, Annable L, Sourkes TL: Tryptophan-nicotinamide combination in depression. Lancet 1:249, 1977.
36. Devadetta S: Isoniazid-induced encephalopathy (letter). Lancet 2:440, 1965.
37. Brown OR, Heitkamp M, Song C-S: Niacin reduces paraquat toxicity in rats. Science 212: 1510–1512, 1981.

8
Vitamin B-6

Introduction	139
Absorption, Metabolism, and Excretion	141
Dosage Forms	141
Clinical Studies	141
Pregnancy	141
Lactation	143
Hyperemesis Gravidarum (Morning Sickness)	143
Toxemia of Pregnancy	144
Premenstrual Syndrome (PMS)	144
Oral Contraceptive Use	144
The Elderly	145
Diabetes Mellitus	146
Oxalosis, Hyperoxaluria, and Oxalate Urolithiasis	146
Renal Failure	148
Carpal Tunnel Syndrome	149
Alcoholism	150
Asthma	150
Sickle Cell Anemia	151
Cancer	151
Other Clinical Uses and Proposed Uses	152
Toxicity and Side Effects	153
Interactions	154
References	155

INTRODUCTION

Vitamin B-6 is the generic term for 3-hydroxy-2-methyl-pyridine derivatives having the biological activity of pyridoxine (pyridoxol). It occurs naturally in three forms, or vitamers—pyridoxine, pyridoxal, and pyridoxamine—in which alcohol, aldehyde, and amine groups, respectively, are located at the 4-position of the pyridine ring (see Fig 1). The vitamin functions as a coenzyme in the form of pyridoxal 5'-phosphate (PLP) and sometimes as pyridoxamine 5'-phosphate [1].

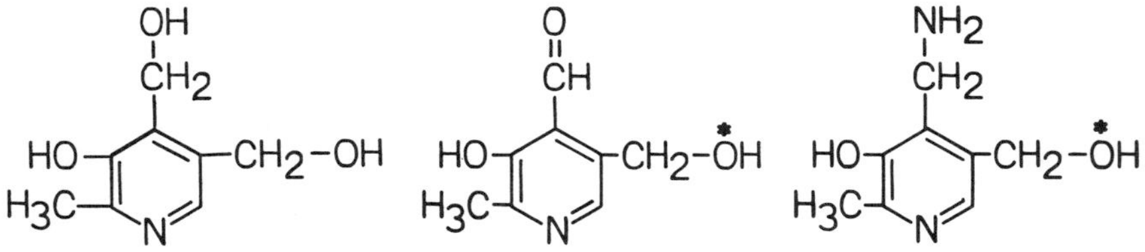

Fig. 1. Structural formulas of pyridoxine (left), pyridoxal (center), and pyridoxamine (right). * Phosphorylation at this point gives the coenzyme forms pyridoxal phosphate (PLP) and pyridoxamine phosphate (PMP).

PLP is an essential coenzyme in nearly all aspects of amino acid metabolism, including transamination, nonoxidative deamination, decarboxylation, and desulfhydration. In decarboxylation reactions, it functions in the synthesis of various neurotransmitters, including gamma-aminobutyric acid (GABA), serotonin, dopamine, and norepinephrine. Biosynthesis of epinephrine (adrenalin) and histamine is also dependent on PLP-containing enzymes. Other reactions involving PLP as a coenzyme include desulfhydration of cysteine and homocysteine, the biosynthesis of porphyrins and heme, and certain aspects of carbohydrate and lipid metabolism [1]. Vitamin B-6 is essential for both synthesis and degradation of polyamines [2] and is involved in essential fatty acid metabolism [3].

The body is capable of considerable storage of vitamin B-6 in the form of PLP bound to skeletal muscle phosphorylase, but this stored vitamin is not readily available if dietary vitamin B-6 is restricted. It only becomes fully available, along with liberated amino acids, when muscle protein breaks down as a result of starvation or inanition [4,5]. Strenuous exercise also causes release of stored PLP, producing temporary increases of plasma PLP and total vitamin B-6 levels, which rapidly subside to or below pre-exercise levels upon completion of exercise [6,7]. A regular if not daily intake is necessary to maintain a satisfactory vitamin B-6 status.

Good dietary sources of vitamin B-6 include meat and fish, whole grains, beans, and cruciferous vegetables. The RDA is 0.3 mg/day for infants, increasing to about 2.2 mg/day for adults and 2.6 mg/day for pregnant women. The American diet does not always meet the need for this vitamin, deficiency having been observed most frequently in pregnant women, women taking oral contraceptives, and the elderly [1].

Vitamin B-6 supplements (usually pyridoxine·HCl) in larger than nutritional amounts have been used for decades for certain conditions such as the nausea and vomiting of early pregnancy, but often without firm evidence gained by controlled double-blind studies and usually without a real understanding of its mechanism of action. Basic and clinical research on vitamin B-6 is currently very active, so we may expect to see considerable advances in knowledge of its proper therapeutic applications during the coming years [8].

ABSORPTION, METABOLISM, AND EXCRETION

Vitamin B-6 is absorbed principally in the jejunum, but also in the ileum, by passive diffusion. When vitamin B-6 is provided pharmaceutically as pyridoxine·HCl, it is converted by the liver to pyridoxine phosphate and then oxidized by pyridine phosphate oxidase to PLP; the liver is believed to be the chief source of circulating PLP. Plasma pyridoxal appears to derive, in part at least, from the action of plasma alkaline phosphatase on PLP. The vitamers in the circulation are bound to proteins, mainly albumin and hemoglobin. PLP is very tightly bound to proteins, while pyridoxal is more loosely bound and thought, therefore, to be the circulating form that is primarily available to body tissues; pyridoxal enters the body cells by passive diffusion and is retained there by phosphorylation to PLP. PLP is widely distributed in the lean body tissues, with highest concentration in the liver [9,10].

The major catabolic product of vitamin B-6 metabolism is 4-pyridoxic acid, which is excreted in the urine together with some unidentified catabolic products and small amounts of pyridoxine, pyridoxal, and pyridoxamine and their phosphorylated derivatives [1,10,11].

DOSAGE FORMS

Pyridoxine hydrochloride is available over the counter as tablets in 10-, 25-, 50-, 100-, 250-, and 500-mg strengths. The OTC marketing of tablet strengths over 50 mg is highly questionable in view of recent reports of chronic neurotoxicity problems (see Toxicity and Side Effects section, pp. 153–154). The hydrochloride and nitrate are present in numerous multivitamin preparations of various pharmaceutical forms, including tablets, capsules, pediatric drops, and injectable solutions.

CLINICAL STUDIES
Pregnancy

Laboratory animal studies by Kirksey and her associates clearly showed the adverse effects of marginal vitamin B-6 nutriture during pregnancy and lactation on central nervous system (CNS) cytoarchitecture [12], myelination [13], and dendritic arborization [14] in the offspring. There is therefore concern that inadequate maternal intake of vitamin B-6 during pregnancy and lactation may be detrimental to CNS development in the fetus and infant [15]. The rising incidence of teenage pregnancy in the United States and the poor dietary habits of many teenagers make this a matter of special concern for their infants. Driskell and Moak [16] observed biochemical deficiency of vitamin B-6 in approximately one-fifth of both white (n = 96) and black (n = 90) adolescent girls in Virginia.

On usual dietary intakes of vitamin B-6, maternal plasma PLP levels tend to fall as pregnancy progresses, especially during the third trimester, possibly because of sequestration of PLP or pyridoxal by the fetus [15]. This is accompanied by biochemical but normally not clinical signs of B-6 deficiency in the mother. Cord plasma PLP is significantly increased by maternal oral supplements of 2-2.6 mg pyridoxine·HCl/d and appears to be saturated at intakes of 7.5–10 mg/d. Intakes of 7.5–10 mg/d also prevent the third-trimester decline in maternal plasma PLP [15]. Bernard et al. [17] have proposed, however, that such maintenance of plasma PLP levels may be unphysiologic, based on their studies in human pregnancy, which show that an increase in unphosphorylated pyridoxal levels compensates for the decline in PLP concentrations; pyridoxal may be more readily taken up by maternal and fetal tissues, they suggest, than PLP.

Schuster et al. [18] reported higher 1-min Apgar scores for infants born to mothers supplemented with 7.5 mg or more of pyridoxine·HCl/d than for infants whose mothers received 5 mg/d or less. The Apgar scores at 5 min do not show this difference [15], and Vir et al. [19] found, in a group of unsupplemented pregnant women ingesting 1–2 mg/d of dietary vitamin B-6, no differences in birth weight or anthropometric measurements related to the B-6 nutriture of the mother. Temesvari et al. [20] concluded from their studies that pyridoxine supplementation may decrease significantly the blood oxygen affinity of the newborn and thus provide the advantage of enhanced tissue oxygenation during the immediate postnatal period.

A subset of pregnant women seem to be at elevated risk of vitamin B-6 deficiency, namely those who have been on long-term oral contraceptive (estrogen-progestin) therapy before pregnancy. Estrogens and estrogen-containing oral contraceptives modify certain metabolic pathways requiring PLP coenzyme, causing, for example, elevated xanthurenic acid excretion after a test load of tryptophan [21]. Plasma PLP, plasma total vitamin B-6, and red cell B-6 levels tend to be lower in oral contraceptive users than in nonusers [21]. Roepke and Kirksey [22] found that long-term (> 30 months) use of oral contraceptives prior to pregnancy resulted in lower B-6 levels in maternal plasma at 5 months' gestation and at delivery, compared to levels in women on similar B-6 intakes who had used oral contraceptives for 1–30 months before becoming pregnant. Levels of vitamin B-6 were lower in cord blood in long-term users than in nonusers. During lactation, long-term users also had depressed

B-6 levels in their milk. Women terminating use of oral contraceptives in order to become pregnant should therefore take an adequate pyridoxine·HCl supplement (e.g. 10 mg/d) before and during pregnancy.

Lactation

The RDA for the infant is 0.3 mg/d, an intake met in commercial infant formulas by supplementation with pyridoxine·HCl. Without maternal supplementation, breast milk supplies only about one-fifth this amount [22,23], but clinical signs of vitamin B-6 deficiency do not appear, and at 6 months of age, plasma PLP levels are similar in formula-fed and breast-fed infants. It is thought that the RDA for the infant may be higher than needed or that vitamin B-6 in breast milk may be more bioavailable than pyridoxine·HCl. The concentration of vitamin B-6 in breast milk increases with maternal supplementation, approaching with supplements of 10–20 mg/d a level that would meet the RDA [22].

Early work suggested that massive doses of pyridoxine·HCl (600 mg/d) might lower prolactin levels and suppress lactation, but this was not confirmed by later megadose studies. Studies in Kirksey's laboratory showed that usual supplementation doses of 4–20 mg/d of pyridoxine·HCl had no significant effect on the milk volume of nursing mothers [22].

Preterm infants given total parenteral nutrition may show extreme elevations of plasma B-6 vitamin levels, which drop back to normal when oral feeding is begun, suggesting that commercial parenteral alimentation formulas are too high in pyridoxine·HCl for premature infants [23]. Preterm infants may also show a moderate temporary elevation of plasma B-6 vitamers on oral formulas, which declines after several days, probably reflecting maturation of the enzymes involved in vitamin B-6 degradation [23]. No adverse effects were noted from these temporary elevations of plasma B-6 vitamers.

Hyperemesis Gravidarum (Morning Sickness)

The first use of vitamin B-6 in pharmacologic doses (50 mg/d of pyridoxine·HCl) was in treatment of the nausea and vomiting of early pregnancy [24,25]. Despite absence of data obtained by double-blind, placebo-controlled trials, use has continued for more than 4 decades on the basis of clinical impression. Since dosages of 10–20 mg/d of pyridoxine·HCl seem to correct most if not all abnormalities of B-6 metabolism during pregnancy, there is probably no reason to use larger doses for treatment of this disorder. The mechanism of therapeutic action is obscure, but may involve regulation of polyamine synthesis and/or catabolism; plasma diamines and other polyamine levels are high at 10–14 weeks of gestation, when incidence of nausea and vomiting is usually greatest, and fall subsequently until another rise occurs in late pregnancy, peaking at about the 36th week [2]. One regulator of polyamine concentrations is the enzyme diamine oxidase (DAO), for which PLP is a cofactor, and which is produced in the placenta. Low DAO activity in plasma is associated with increased incidence of spontaneous abortion and stillbirth [26–28]. Martner-Hewes et al. [29] found, in a study of 122 pregnant Hispanic teenagers, that plasma DAO activity in early pregnancy was positively associated with dietary vitamin B-6 intake and was lowest in those with biochemical deficiency (EGPT index above 1.25).

Toxemia of Pregnancy

Although an early trial with a daily supplement of 10 mg/d of pyridoxine·HCl appeared to indicate a reduction in incidence of pregnancy toxemia, later studies with 20 mg/d failed to confirm this clinical benefit, and administration of pyridoxine during established toxemia also failed to confer clinical benefit [2,30]. However, very low cord blood levels of PLP have been recorded for infants born of toxemic mothers, so that pyridoxine·HCl supplementation (5–10 mg/d) of pregnant women at risk of pre-eclampsia is recommended to protect the neurologic development of the fetus [31].

Premenstrual Syndrome (PMS)

Oral pyridoxine·HCl has been used for decades to treat PMS, although the mechanism of its action remains obscure. London et al. [32] have reviewed the studies carried out on this therapy, some of which have had well-controlled placebo designs. Significant relief of symptoms has been reported in many patients in these trials, using 50 or 100 mg/d dosages. Barr employed 100 mg pyridoxine·HCl/d on cycle days 10 through 3 of the next cycle, obtaining a highly significant response compared with placebo in a 48-patient double-blind study [33].

Oral Contraceptive Use

Oral contraceptives (estrogen-progestin) cause disturbance of some aspects of vitamin B-6 metabolism, e.g., elevating the excretion of xanthurenic acid after a tryptophan load, while not affecting other aspects, e.g., erythrocyte aminotransferase activities and excretion of 4-pyridoxic acid, in women receiving dietary intakes meeting the RDA [21]. However, since, at comparable dietary intakes of B-6, oral contraceptive (OC) users tend to show lower plasma and erythrocyte levels of PLP and total vitamin B-6 than nonusers [21], and since adverse effects on vitamin B-6 nutritional status are observed during pregnancy following long-term OC use [22], it appears that some deep-seated vitamin B-6 deficiency state is induced by chronic use of these contraceptive drugs and that a degree of supplementation with the vitamin is desirable [34]. Studies by various investigators using dosages of 20–100 mg/d of pyridoxine·HCl show that 20 mg/d will normalize tryptophan metabolism in the majority of OC users, with a few requiring larger dosages [35].

Supplementation may also improve the mental state of OC users, probably by enhancing the PLP-dependent conversion of tryptophan to the neurotransmitter, serotonin. Baumblatt and Winston [36] reported that oral administration of 50 mg pyridoxine·HCl once daily, starting at the first sign of depression, completely or partially relieved depressive symptoms in about three-quarters of 58 patients. In double-blind trials involving administration of 20 mg pyridoxine·HCl or placebo twice daily, followed by reversal of drug and placebo after 2 months, Adams et al. found that depression was relieved in about one-half of a total of 39 depressed OC users [37,38]. The subjects who responded were those who initially showed marked biochemical deficiency; those not initially deficient showed no response.

The Elderly

An age-related change occurs in vitamin B-6 metabolism, manifesting in reduced plasma levels of PLP and total vitamin B-6 in the elderly compared with those of young adults on the same dietary intake [39,40]. Driskell and coworkers [41,42] found 25–30% biochemical B-6 deficiency, as measured by the erythrocyte alanine aminotransferase stimulation test, in Americans over age 60 and living at home. Chen and Fan-Chiang [43] observed biochemical B-6 deficiency in 56.6% of institutionalized elderly and 43.5% of noninstitutionalized elderly in central Kentucky. In Northern Ireland, Vir and Love [44] found a similar prevalence of inadequate vitamin B-6 nutriture in institutionalized elderly. Supplements of 2–2.5 mg/d of pyridoxine·HCl do not always correct this inadequacy [41–44]. Lee and Leklem [39] found that on the same dietary intake of vitamin B-6 (2.3–2.4 mg/d), women aged 55.3 ± 4.0 years had plasma PLP and total B-6 levels about two-thirds as high as those of women aged 24.4 ± 3.2 years. Administration of 8 mg/d of pyridoxine (about 10 mg pyridoxine·HCl) elevated plasma levels in both groups; there was still a difference in favor of the younger group but not a statistically significant one.

Talbott et al. [45], in a preliminary study with 11 subjects aged 65–81 years, of whom 5 had below-normal plasma PLP levels, found that indices of cell-mediated and humoral immunity were improved by pyridoxine·HCl supplementation, especially in those with initially low PLP levels. The dosage regimen was 100 mg/d for 2 months.

The physician should always consider the possibility of subclinical vitamin B-6 deficiency in the elderly patient, particularly if appetite or dietary habits are poor. Under such conditions, a 10–20 mg/d supplement of pyridoxine·HCl can restore biochemical indices of vitamin B-6 nutriture in most elderly patients to youthful levels [40].

Diabetes Mellitus

Several studies have shown the frequent occurrence of a low vitamin B-6 status in both children and adults with diabetes mellitus [46–50]; the cause is unknown. Hollenbeck et al. [49] observed deficient whole blood B-6, plasma B-6, and plasma PLP levels in four of six young women with insulin-dependent (type I) diabetes mellitus; the other two patients had values just above the lower limit of normal. One subject had displayed normal blood and plasma B-6 levels while taking 10 mg/d of pyridoxine·HCl before being brought into the study. The study diet supplied the RDA level of vitamin B-6.

McCann and Davis [50] compared serum pyridoxal levels in 50 patients suffering from diabetic neuropathy with the pyridoxal levels of randomly selected diabetic patients matched for age and sex who did not show evidence of neuropathy; duration of diabetes was not significantly different between the two groups. The mean serum pyridoxal concentration was significantly (about 30%) lower in the group with neuropathy. While it is not known whether B-6 deficiency is primarily responsible for the development of diabetic neuropathy, it is of interest that vitamin B-6 deficiency induced by isoniazid therapy for tuberculosis is associated with onset of peripheral neuropathy [51].

Based on the present extent of knowledge, a pyridoxine·HCl supplement of 10 mg/d should provide insurance against diabetic complications that might be related to vitamin B-6 deficiency.

Oxalosis, Hyperoxaluria, and Oxalate Urolithiasis

Type I primary hyperoxaluria is a rare autosomal recessive genetic disease characterized by defective metabolism of glyoxylate, increased endogenous production of oxalate, increased urinary excretion of oxalate, glycolate, and glyoxylate, and recurrent calcium oxalate urolithiasis. A very rare type II primary hyperoxaluria has been described in which excretion of oxalate and L-glycerate is increased, but not that of glycolate. The disease leads to widespread deposition of calcium oxalate crystals in the tissues (oxalosis), nephrocalcinosis, stone formation, and progressive kidney failure [52]. Secondary oxalosis, on the other hand, may result from inhibition of oxalate excretion due to kidney failure from any cause and is occasionally seen as a result of ethylene glycol poisoning.

If primary hyperoxaluria develops early in infancy, it is very difficult to diagnose in time to prevent irreversible kidney damage [53]. Rose et al. [54] describe successful treatment of one infant with high-dose pyridoxine (400 mg/d). Initially normal at birth, plasma oxalate and glycolate levels were grossly elevated at 5–6 weeks of age. Pyridoxine treatment rapidly lowered glycolate levels, but oxalate levels declined to normal only gradually over an 8-month period, possibly aided by addition of thiamin and magnesium hydroxide and chloride to the therapeutic regimen. Early diagnosis in this case was facilitated by knowledge of the previous death of a sibling due to oxalosis. Another successfully treated case of type I hyperoxaluria is described by Alinei et al. [55]. They obtained a sustained normalization of renal function in an infant (a 3-month-old female) by oral administration of 300 mg/d of pyridoxine·HCl. Within 7 months, oxalate excretion was reduced nearly to normal and remained so at age 2 years. These investigators recommend that an ultrasonographic examination of the kidneys and a urinary oxalate determination should be carried out in any infant with unexplained chronic renal failure [55].

When children of older ages manifest symptoms of oxalosis and hyperoxaluria, the disease may be milder and easier to control. Yendt and Cohanim [56] describe their treatment of two children with type I and two with type II hyperoxaluria. In the type I cases, supplements of only 2 mg pyridoxine per day, administered in a multivitamin tablet, substantially reduced oxalate excretion, and 25 mg/d of pyridoxine (presumably the hydrochloride) reduced oxalate excretion to normal. The type II cases were more refractory, one showing moderate reductions of urinary oxalate excretion with pyridoxine supplements of 25 and 50 mg/d, the other requiring 200 mg/d. In treating childhood primary hyperoxaluria, the authors suggest starting with large doses of pyridoxine to reduce urinary oxalate levels as quickly as possible and then reducing the dosage in stepwise fashion at 3-month intervals to establish the minimal dose needed to maintain the initial effect [56].

Both high and low doses of pyridoxine have been used in treatment of recurrent oxalate stones in adults. Smith and Williams [57] reported good results in most but not all patients with administration of 100–150 mg pyridoxine·HCl daily. Will and Bijvoet [58] obtained reductions of 60–70% in oxalate excretion in two hyperoxaluric women with dosages of 250 mg q.i.d., noting no ill effects clinically during follow-ups of 30 months for one patient and 20 months for the other. Such high dosages, however, raise a concern for the possible eventual development of neurotoxic effects [59,60].

Harrison et al. [61] describe case histories of two patients (a 42-year-old man and a 26-year-old woman) who had hyperoxaluria and had formed numerous kidney stones. The man was treated with 400 mg pyridoxine (hydrochloride?) daily for 4 months, the woman with 100 mg/d for 15 months. Excretion of urinary glycolate and oxalate was reduced to normal levels and remained there after cessation of pyridoxine dosage. The patients did not relapse even after long periods of further observation, indicating that they did not have primary hyperoxaluria. The causes of abnormal oxalate metabolism in these two patients are unclear.

Prien and Gershoff [62] treated 149 adults with recurrent kidney stones with 10 mg of pyridoxine once daily plus 100 mg magnesium oxide t.i.d. over a period of 5 years. Average number of stones per patient per year dropped from 1.3/year in the 4.5-year period preceding medication to 0.1 while on medication. Furthermore, all the stones were formed in only 17 of the 149 patients, the remainder being symptom-free. The investigators favor the hypothesis that magnesium supplementation improved the solvent characteristics of urine for oxalate. It appears that quite moderate dosages of pyridoxine, at least when supplemental magnesium is taken, are effective in reducing stone formation in adult patients subject to recurrent oxalate stones.

Renal Failure

Biochemical abnormalities of the uremic state include decline of plasma PLP levels, reduced erythrocyte aminotransferase activity and cofactor saturation, and elevated plasma oxalate concentrations. For example, Dobbelstein et al. [63] found a marked biochemical B-6 deficiency in 48 of 69 uremic patients (37 on dialysis, 32 treated conservatively by diet); EGOT activation coefficients were restored to normal by administration of 300 mg/d of pyridoxine·HCl for 2 weeks. Kopple et al. [64] and Teehan et al. [65] also found 60–70% of patients with advanced renal failure to be deficient in vitamin B-6 on ordinary dietary intakes of the vitamin. Plasma PLP levels were normal in patients with stable motor nerve conduction velocity and low transfusion requirements (to correct anemia), while they were low in patients with declining motor nerve velocity or high transfusion requirements [65]. Oral pyridoxine·HCl at 100 or 200 mg/d restored plasma PLP levels to normal in most patients within 2 weeks. Kleiner et al. [66] also observed restoration of erythrocyte aminotransferase activity to normal within two weeks when 300 mg/d of pyridoxine·HCl was administered to dialysis patients.

Other desirable effects observed with pyridoxine supplementation in renal failure include improvement of the plasma amino acid profile [66], lowering of plasma oxalate levels [67], an increase in HDL cholesterol without change in total cholesterol or triglyceride levels [66], and improvement in immune function [63,68,69]. Improvements have not been noted in anemia or peripheral neuropathy.

Kopple et al. [70] carried out a careful investigation of minimum pyridoxine supplementation needs of patients in renal failure. Deficiency was correctible within 3 weeks with 10 or 50 mg/d of pyridoxine·HCl. For long-term maintenance, 5 mg/d was sufficient for nondialyzed patients and for patients receiving intermittent peritoneal dialysis, while 10 mg/d was required for those on maintenance hemodialysis. Larger supplements may be required for patients who are septic or who are taking drugs antagonistic to vitamin B-6, e.g., isoniazid or hydralazine.

Carpal Tunnel Syndrome

Many studies have demonstrated dramatic amelioration of the pain and stiffness of carpal tunnel syndrome by high-dose supplementation (100–300 mg/d) with pyridoxine·HCl [71–75]. The original findings of the benefits of vitamin B-6 supplementation in this syndrome were made by Folkers and his associates and are graphically described by Folkers in his recent address accepting the Priestley Medal, the highest award of the American Chemical Society [76]. Most of the patients studied were biochemically deficient in vitamin B-6, and in one patient who was surgically explored, it appeared that the deficiency was associated with edematous and fibrotic changes in the subendothelial layers of synovia, causing compression of the median nerve [72]. These patients appear to have a higher than average requirement for vitamin B-6. Smith et al. [77] found that carpal tunnel patients who are not biochemically deficient in B-6 may not respond to pyridoxine therapy or may respond only minimally. On the other hand, Driskell et al. [78] observed substantial improvement in the neurophysiologic symptoms of 27 of 28 carpal tunnel syndrome patients treated with 100 mg/d of pyridoxine·HCl for 5–28 weeks, even though initially only one-fourth of the patients were biochemically deficient in B-6 as judged by PLP stimulation of erythrocyte alanine aminotransferase activity. Folkers [76] reports that many patients with the syndrome are also deficient in riboflavin, which is needed to convert pyridoxine to PLP, so that combination therapy with pyridoxine·HCl and riboflavin is needed in these patients.

Resolution of the syndrome is slow. For example, Salkeld and Stotz [75], treating a group of 23 patients with severe carpal tunnel syndrome, found that their initial deficient or marginal B-6 status (erythrocyte aminotransferase activation test) was corrected in most after 6 weeks on 100 mg pyridoxine·HCl daily, but symptom score continued to improve substantially during another 6 weeks of supplementation. In another group of patients whose erythrocyte transaminase tests showed adequate or optimal B-6 status and who were treated surgically, postoperative administration of pyridoxine to a few of them appeared to accelerate recovery [75]. Once maximal improvement in the syndrome has been established, it may be possible to reduce the dosage of pyridoxine·HCl for long-term maintenance.

Carpal tunnel syndrome has been noted in long-term hemodialysis patients [79], but clinical trials to explore specifically the effect of pyridoxine on this syndrome in such cases have not been reported; accumulation of 4-pyridoxic acid and other B-6 metabolites could be a problem in such patients.

Alcoholism

Chronic alcoholism leads to deficiencies of several water-soluble vitamins, including B-6 [80]. In large part, poor dietary intake of essential nutrients is responsible. Factors specifically affecting vitamin B-6 nutriture have been reviewed by Bonjour [81] and Li and Lumeng [82]. As compared with normal controls, alcoholics show depressed levels of total vitamin B-6 in serum and cerebrospinal fluid and of PLP in serum and plasma. However, the erythrocyte aspartate aminotransferase stimulation test may be normal, possibly because of blockage of the PLP site on the apoenzyme by ethanol or its metabolite, acetaldehyde, giving the impression of a good saturation of the apoenzyme with PLP [81]. Fatty or cirrhotic changes in the liver reduce vitamin B-6 storage and may decrease rate of synthesis of PLP. Chronic alcoholics have a markedly subnormal response in plasma PLP elevation when injected IV with 20 mg pyridoxine; response becomes normal or near-normal after some weeks of withdrawal from alcohol and nutritional rehabilitation [81]. Clearance of PLP from the circulation is enhanced by acetaldehyde, which displaces PLP from protein binding, enhancing its degradation by plasma alkaline phosphatase [82].

High-dose pyridoxine (100 mg) decreases incidence of convulsions during withdrawal from alcohol and may have beneficial effects on the neuropathy and sideroblastic anemia of alcoholism [81,82]. Intravenous PLP seems to be more rapid and effective in raising plasma PLP levels than intravenous pyridoxine. The proper dosage and route of administration of pyridoxine, whether oral, IV, or IM, is highly variable and dependent on the clinical condition of the individual patient, so that general guidelines for supplementation are not available. The course followed by plasma PLP values during therapy provides the best indication of the effect of supplementation.

Asthma

An intriguing preliminary report by Reynolds and Natta [83] describes dramatic reductions in the frequency and severity of wheezing or asthmatic attacks in 15 adult patients with bronchial asthma when they were given 50 mg pyridoxine (60 mg pyridoxine·HCl) twice daily for varying periods (3–10 months). All patients had markedly low plasma and erythrocyte PLP levels initially. Their response to the pyridoxine supplement was compared with that of healthy controls. Whereas the control subjects exhibited steep rises in plasma and erythrocyte PLP, the PLP response in the asthmatics was variable, inconsistent, and generally low. In spite of this, the frequency, duration, and severity of asthmatic attacks was markedly reduced.

An earlier double-blind study by Collipp et al. [84] with 76 children found a significant improvement in asthma-related clinical symptoms and a reduction in the dosage of bronchodilators and cortisone needed to relieve symptoms when they were supplemented with 200 mg oral pyridoxine daily.

The reasons for low PLP status and poor PLP response in the adult asthmatics studied by Reynolds and Natta are obscure, but either impaired synthesis or increased degradation of PLP could be responsible. These investigators suggest as one possible explanation for the beneficial effect of high-dose vitamin B-6 its ability to decrease hemoglobin oxygen affinity [20,85]. Further and larger-scale studies are being carried out.

Sickle Cell Anemia

Reynolds and Natta [86], in a small study with five homozygous sickle cell anemia patients, obtained a striking reduction in the number, severity, and duration of painful crises when they administered 100 mg pyridoxine (125 mg pyridoxine·HCl) daily for 2 months. Plasma PLP levels, usually very low in these patients, were markedly elevated by this therapy, and there were slight but not significant elevations in hemoglobin concentration and hematocrit, suggesting increased erythrocyte lifespan. The mechanism of the pyridoxine effect is obscure, but may involve an antisickling action of plasma PLP, which has been demonstrated in vitro [87]. Natta and Reynolds [88] have extended their observations to a total of 16 sickle cell anemia patients, noting increases in plasma and erythrocyte PLP levels and small but not statistically significant rises in hemoglobin and hematocrit on a dosage regimen of 50 mg pyridoxine·HCl twice daily. Concurrent riboflavin supplementation may be needed, as the hepatic enzyme responsible for converting pyridoxine to pyridoxal, pyridine phosphate oxidase, is FMN-dependent [89].

Cancer

Animal studies indicate that certain tryptophan metabolites are carcinogenic [90], and increased urinary excretion of these metabolites has been noted in nearly 50% of bladder cancer patients [91]. These abnormalities are correctible by administration of 25 mg pyridoxine daily [92]. Byar and Blackard [93] therefore undertook a prospective 45-month clinical trial of 121 patients with stage 1 bladder cancer randomized to placebo, 25 mg pyridoxine once daily, or thiotepa instilled intravesically once weekly. Over the first 10 months of the study, no differences between placebo and pyridoxine were apparent, but in patients followed for longer periods, pyridoxine was slightly but significantly better than placebo, although not as effective in preventing recurrences as thiotepa.

Oral pyridoxine (25–200 mg/d) has been reported to be useful in ameliorating symptoms of radiation sickness after radiotherapy of cancer [94,95].

A recent report by Gridley et al. [96] describes inhibition of tumor growth and enhancement of immune status by a high level of dietary vitamin B-6 in mice inoculated with herpes simplex virus type 2-transformed (H238) cells. The highest B-6 concentration tested, which reduced tumor volume 32% compared with that of the requirement level of the vitamin, was 74.3 mg pyridoxine per kg of diet, equivalent to less than 40 mg pyridoxine in a day's human food intake (dry weight) of about 500 g.

Other Clinical Uses and Proposed Uses

There are several rare genetic disorders that are responsive to high-dose vitamin B-6. These include pyridoxine-responsive infant seizures [97] and irritability and hyperkinesis [98]; dosages of pyridoxine required for control vary from 25 to 400 mg/d, depending on the individual case. Other congenital B-6-dependencies are vitamin B-6-responsive anemia, cystathioninuria, xanthurenic aciduria, and homocystinuria [1,99]. Patients with hyperornithinemia and gyrate atrophy of the choroid and retina have shown declines in plasma ornithine levels and encouraging stabilization of visual function when given oral pyridoxine [100]. Most, but not all, patients with sideroblastic anemia due to inadequate delta-aminolevulinate synthase activity show a response to high-dose pyridoxine [101]. Plasma PLP levels are low in children and adults with celiac disease, due to malabsorption of the vitamin [102]; abnormal tryptophan metabolism in adult celiacs is corrected by pyridoxine supplementation [103].

Folkers and coworkers have found that individuals who experience the so-called Chinese restaurant syndrome when they eat foods heavily seasoned with monosodium glutamate (MSG) may be helped by a supplement of 50 mg of pyridoxine daily [104]. The syndrome may include sensations of warmth, stiffness, weakness, tingling, headache, lightheadedness, heartburn, or gastric discomfort. Many, but not all, susceptible subjects tested showed evidence of biochemical B-6 deficiency. Some individuals who had their B-6 deficiency corrected by supplementation still showed the syndrome when challenged with MSG, indicating that vitamin B-6 deficiency is not the only cause of the disorder.

Folkers [76] believes that the basic biochemical defects of carpal tunnel syndrome may be present in a number of disorders for which vitamin B-6 supplementation should be tried, including nocturnal paralysis of the arm and hand, rheumatism, arthritis, painful shoulders, pain or stiffness in the knees, painful elbows, edema of the feet and ankles, tennis elbow, trigger fingers, bursitis, periarthritis of the shoulder, Dupuytren's contracture, and De Quervain's disease. He and his associates have also obtained some success in treating psoriasis with high-dose supplements of vitamin B-6 and riboflavin, although considerable time is required to bring results [76].

Coburn [97] has carefully reviewed available published investigations of the possible utility of vitamin B-6 supplements in various mental disorders, including mental retardation, schizophrenia, depression, autism, and hyperkinesis. Although biochemical B-6 deficiency is frequently noted in these disorders, especially in depression, he concludes that there is little clear-cut evidence that correction of the deficiency with pyridoxine supplements improves the mental condition (except for depression associated with oral contraceptive use). Nevertheless, in the management of these patients, general health benefits would be expected to follow correction of deficiencies of vitamin B-6 as well as of other essential nutrients.

PLP has been demonstrated in vitro to be an effective inhibitor of platelet aggregation, particularly second-wave aggregation in platelet-rich plasma containing adenosine diphosphate, thrombin, or adrenaline, apparently by forming Schiff-base bonds with platelet surface proteins through its active aldehyde group [105,106]. In healthy human volunteers, intravenous PLP not only inhibited platelet aggregation but prolonged whole-blood and thrombin clotting time [107]. However, a 6-week clinical trial of oral dosage at 100 mg/d failed to show a significant change in platelet aggregation [108]. In addition, it has been reported that middle-aged individuals classified as being at high risk for coronary heart disease, when compared with a similar age-sex group classified as low risk, exhibit higher plasma levels of homocysteine [109], a methionine metabolite that has been hypothesized to contribute to atherogenesis [110] and that requires vitamin B-6, folic acid, and vitamin B-12 for reconversion to methionine [111,112]. Clinical studies are needed to determine the relationship of plasma homocysteine level to nutritional status with respect to vitamin B-6 and other vitamins.

TOXICITY AND SIDE EFFECTS

Until recently it had been thought that the usual form in which supplemental vitamin B-6 is given, pyridoxine·HCl, was remarkably safe, and as a matter of fact many individuals have taken doses of as much as 300–400 mg/d for months or years without apparent ill effect. However, recently Schaumburg et al. [59] reported seven cases of sensory neuropathy in individuals who had been chronically self-medicating with 2–5 g/d of pyridoxine·HCl. Symptoms of unsteady gait, numb feet, and numbness and clumsiness of the hands developed within 2–4 months. Dramatic clinical improvement occurred upon withdrawal of pyridoxine, but there was some residual neuropathy. Subsequently, Berger and Schaumburg [60] reported a case of a woman who showed peripheral nerve dysfunction after prolonged ingestion of 500 mg/d of pyridoxine·HCl, and Parry and Bredesen [113] reported neurotoxic symptoms in one woman who had taken 200 mg/d for 3 years or more. Differences in individual sensitivity to pyridoxine at high doses appear to be an important factor.

Doses of 50–200 mg of pyridoxine appeared to aggravate seizures in one patient with encephalitis and to contribute to deterioration of the electroencephalographic pattern in another [1]. One study reported indications of impaired memorization after as little as 10–15 days' ingestion of 100–1,000 mg/d of pyridoxine [114]. Symptoms of vitamin B-6 dependency were noted in normal adults after 33 days on 200 mg/d of pyridoxine·HCl upon withdrawal of the supplement [1].

Since doses of 25–50 mg/d of pyridoxine·HCl have been repeatedly shown to produce nearly the same plasma PLP concentrations in adults as a dose of 500 mg/d [8,10], there seems little reason to use larger dosages for chronic administration except possibly for some of the very rare B-6-dependent genetic disorders that are unresponsive to moderate B-6 doses; high-dose treatment of such disorders should be closely monitored for appearance of neurologic symptoms. Doses larger than 50 mg/d may be needed as cotherapy with certain drugs (see below).

INTERACTIONS

Bhagavan [115] has reviewed the interactions of vitamin B-6 with a variety of other therapeutic agents. For the most part, these drugs contain amino groups capable of reacting with the aldehyde groups of pyridoxal and PLP to form Schiff bases. Various hydrazine-derived drugs are vitamin B-6 antagonists and may in chronic administration induce such acute B-6 deficiency signs and symptoms as irritability, peripheral neuropathy, and convulsions. Examples are the antidepressants iproniazid, nialamide, and isocarboxazid; the MAO inhibitor, phenalzine; the antituberculosis drug, isonicotinylhydrazide (INH); the antihypertensive, hydralazine; and the anticancer drug, procarbazine. Pyridoxine at 100 mg/d is recommended during chronic administration of INH [115].

The broad-spectrum antibiotic cycloserine also reacts with vitamin B-6, so that symptoms of vitamin B-6 deficiency appear usually within the first 2 weeks of therapy. These side effects of cycloserine can usually be reversed by coadministration of pyridoxine supplements (> 50 mg/d) [115].

In the case of patients receiving L-DOPA for Parkinson's disease, a minimum amount of vitamin B-6 must be present to provide PLP cofactor for the decarboxylase enzyme that converts L-DOPA to dopamine, but excess vitamin B-6 can reduce the clinical efficacy of L-DOPA by Schiff base formation. Coadministration with L-DOPA of a peripheral decarboxylase inhibitor such as carbidopa minimizes the effect of vitamin B-6 status on the action of the anti-Parkinson drug [115].

Chronic administration of penicillamine, as in treatment of Wilson's disease, may result in adverse neurologic effects due to vitamin B-6 inactivation. Coadministration of pyridoxine (> 50 mg/d) is recommended [115].

Interactions have also been reported between vitamin B-6 and amphetamines, amitryptyline, amiodarone, and 6-azauridine. Overdosage of isonicotinylhydrazide has been successfully treated with high doses of pyridoxine (gram for gram with the amount of INH); intravenous administration of pyridoxine·HCl may be necessary in emergency cases, IV doses of 70–357 mg/kg body weight having been used in such cases without adverse effects [115].

REFERENCES

1. Driskell JA: Vitamin B-6. In: *Handbook of Vitamins*, LJ Machlin (ed). New York: Marcel Dekker, 1984, pp 379–401.
2. Campbell RA, Keniston RC: Hyperpolyaminemia and vitamin B-6. In: *Vitamin B-6: Its Role in Health and Disease*, RD Reynolds, JE Leklem (eds). New York: Alan R. Liss, 1985, pp 347–386.
3. Cunnane SC, Manku MS, Horrobin DF: Accumulation of linoleic and gamma-linolenic acids in tissue lipids of pyridoxine-deficient rats. J Nutr 114:1754–1761, 1984.
4. Black AL, Guirard BM, Snell EE: Increased muscle phosphorylase in rats fed high levels of vitamin B-6. J Nutr 107:1962–1968, 1977.
5. Black AL, Guirard BM, Snell EE: The behavior of muscle phosphorylase as a reservoir for vitamin B-6 in the rat. J Nutr 108:670–677, 1978.
6. Dreon DM, Butterfield GE: Vitamin B-6 utilization in active and inactive young men. Am J Clin Nutr 43:816–824, 1986.
7. Manore MM, Leklem JE, Walter MC: Vitamin B-6 metabolism as affected by exercise in trained and untrained women fed diets differing in carbohydrate and vitamin B-6 content. Am J Clin Nutr 46:995–1004, 1987.
8. Reynolds RD, Leklem JE: Implications on the role of vitamin B-6 in health and disease—a summary. In: *Vitamin B-6: Its Role in Health and Disease*, RD Reynolds, JE Leklem (eds). New York: Alan R. Liss, 1985, pp 481–489.
9. Henderson LM: Intestinal absorption of B-6 vitamers. In: *Vitamin B-6: Its Role in Health and Disease*, RD Reynolds, JE Leklem (eds). New York: Alan R. Liss, 1985, pp 25–33.
10. Ubbink JB, Serfontein WJ, Becker PJ, DeVilliers LS: Effect of different levels of oral pyridoxine supplementation on plasma pyridoxal-5'-phosphate and pyridoxal levels and urinary vitamin B-6 excretion. Am J Clin Nutr 46:78–85, 1987.
11. Sauberlich HE, Canham JE: Vitamin B-6. In: *Modern Nutrition in Health and Disease*, RS Goodhart, ME Shils (eds). Philadelphia: Lea & Febiger, 1980, pp 216–229.
12. Morré DM, Kirksey A, Das GD: Effects of vitamin B-6 deficiency on the developing central nervous system of the rat. Gross measurements and cytoarchitectural alterations. J Nutr 108:1250–1259, 1978.
13. Morré DM, Kirksey A, Das GD: Effects of vitamin B-6 deficiency on the developing central nervous system of the rat. Myelination. J Nutr 108:1260–1265, 1978.
14. Morré DM, Kirksey A: The effect of deficiency of vitamin B-6 on selected neurons of the developing rat brain. Nutr Rep Int 21:301–312, 1980.
15. Kirksey A, Udipi SA: Vitamin B-6 in pregnancy and lactation. In: *Vitamin B-6: Its Role in Health and Disease*, RE Reynolds, JE Leklem (eds). New York: Alan R. Liss, 1985, pp 57–77.

16. Driskell JA, Moak SW: Plasma pyridoxal phosphate concentrations and coenzyme stimulation of erythrocyte alanine aminotransferase activities of white and black adolescent girls. Am J Clin Nutr 43:599–603, 1986.
17. Barnard HC, DeKock JJ, Vermaak WJH, Potgieter GM: A new perspective in the assessment of vitamin B-6 nutritional status during pregnancy in humans. J Nutr 117:1303–1306, 1987.
18. Schuster K, Bailey LB, Mahan CS: Effect of maternal pyridoxine.HCl supplementation on the vitamin B-6 status of mother and infant and on pregnancy outcome. J Nutr 114:977–988, 1984.
19. Vir SC, Love AHG, Thompson W: Vitamin B-6 status during pregnancy. Int J Vitam Nutr Res 50:403–411, 1980.
20. Temesvari P, Szilaggi T, Eck E, Boda D: Effects of an antenatal load of pyridoxine (vitamin B-6) on the blood oxygen affinity and prolactin levels in newborn infants and their mothers. Acta Paediatr Scand 72:525–529, 1983.
21. Miller LT: Oral contraceptives and vitamin B-6 metabolism. In: *Vitamin B-6: Its Role in Health and Disease*, RD Reynolds, JE Leklem (eds). New York: Alan R. Liss, 1985, pp 243–255.
22. Roepke JLB, Kirksey A: Vitamin B-6 nutriture during pregnancy and lactation. II. The effect of long-term use of oral contraceptives. Am J Clin Nutr 32:2257–2264, 1979.
23. McCoy E, Strynadka K, Brunet K: Vitamin B-6 intake and whole blood levels of breast and formula fed infants: Serial whole blood vitamin B-6 levels in premature infants. In: *Vitamin B-6: Its Role in Health and Disease*. RD Reynolds, JE Leklem (eds). New York: Alan R. Liss, 1985, pp 79–86.
24. Willis RS, Winn WW, Morris AT, et al.: Clinical observations in treatment of nausea and vomiting in pregnancy with vitamin B-1 and vitamin B-6. Am J Obstet Gynecol 44:265–271, 1942.
25. Weinstein BB, Wohl Z, Mitchell MD, Sustendal GF: Oral administration of pyridoxine hydrochloride in the treatment of nausea and vomiting of pregnancy. Am J Obstet Gynecol 47:389–394, 1942.
26. Weingold AB, Southren AL, Lee BO: A fetal salvage program: The use of plasma DAO as a fetal monitor. Int J Fertil 16:24–35, 1971.
27. Southren AL, Kobayashi Y, Weingold AB, Carmody NC: Serial plasma diamine oxidase (DAO) assays in first- and second-trimester complications of pregnancy. Am J Obstet Gynecol 96:502–510, 1968.
28. Gahl WA, Raubertas F, Vale AM, Golubjatnikov R: Maternal serum diamine oxidase in fetal death and low birth-weight infants. Br J Obstet Gynecol 89:202–207, 1982.
29. Martner-Hewes PM, Hunt IF, Murphy NJ, et al.: Vitamin B-6 nutriture and plasma diamine oxidase in pregnant Hispanic teenagers. Am J Clin Nutr 44:907–913, 1986.
30. Rose DP: The interaction between vitamin B-6 and hormones. Vitam Horm 36:53–99, 1978.
31. Brophy MH, Siiteri PK: Pyridoxal phosphate and hypertensive disorders of pregnancy. Am J Obstet Gynecol 121:1075–1079, 1975.
32. London RS, Murphy L, Kitlowski KE: Treatment of premenstrual syndrome with vitamin B-6: Physicians' attitudes and perceptions. In: *Vitamin B-6: Its Role in Health and Disease*, RD Reynolds, JE Leklem (eds). New York: Alan R. Liss, 1985, pp 469–477.
33. Barr W: Pyridoxine supplements in the premenstrual syndrome. Practitioner 228:425–427, 1984.
34. Miller LT: Do oral contraceptive agents affect nutrient requirements—vitamin B-6? J Nutr 116:1344–1345, 1986.

35. Brin M: Abnormal tryptophan metabolism in pregnancy and with the oral contraceptive pill. II. Relative levels of vitamin B-6 vitamers in cord and maternal blood. Am J Clin Nutr 24:704–708, 1971.
36. Baumblatt MJ, Winston F: Pyridoxine and the pill. Lancet 1:832–833, 1970.
37. Adams PW, Rose DP, Folkard J, et al.: Effect of pyridoxine hydrochloride upon depression associated with oral contraception. Lancet 1:897–904, 1973.
38. Adams PW, Wynn V, Seed M: Vitamin B-6, depression and oral contraception. Lancet 2:516–517, 1974.
39. Lee CM, Leklem, JE: Differences in vitamin B-6 status indicator responses between young and middle-aged women fed constant diets with two levels of vitamin B-6. Am J Clin Nutr 42:226–234, 1985.
40. Kirsch A, Bidlack WR: Nutrition and the elderly: Vitamin status and efficacy of supplementation. Nutrition 3:305–314, 1987.
41. Hampton DJ, Chrisley BM, Driskell JA: Vitamin B-6 status of the elderly in Montgomery County, Va. Nutr Rep Int 16:743–750, 1977.
42. Chrisley BM, Driskell JA: Vitamin B-6 status of adults in Virginia. Nutr Rep Int 19:553–560, 1979.
43. Chen LH, Fan-Chiang WL: Biochemical evaluation of riboflavin and vitamin B-6 status of institutionalized and non-institutionalized elderly in central Kentucky. Int J Vitam Nutr Res 51:232–238, 1981.
44. Vir SC, Love AHG: Vitamin B-6 status of hospitalized aged. Am J Clin Nutr 31:1383–1391, 1978.
45. Talbott MC, Miller LT, Kerkvliet NI Pyridoxine supplementation: Effect on lymphocyte responses in elderly people. Am J Clin Nutr 46:659–664, 1987.
46. Mooradian AD, Morley JE: Micronutrient status in diabetes mellitus. Am J Clin Nutr 45:877–895, 1987.
47. Davis RE, Calder JS, Curnow DH: Serum pyridoxal and folate concentrations in diabetics. Pathology 8:151–159, 1976.
48. Wilson RG, Davis RE: Serum pyridoxal concentrations in children with diabetes mellitus. Pathology 9:95–98, 1977.
49. Hollenbeck CB, Leklem JE, Riddle MC, Connor WE: The composition and nutritional adequacy of subject-selected high carbohydrate, low fat diets in insulin-dependent diabetes mellitus. Am J Clin Nutr 38:41–51, 1983.
50. McCann VJ, Davis RE: Serum pyridoxal concentrations in patients with diabetic neuropathy. Aust NZ J Med 8:254–261, 1978.
51. Jones WA, Jones GP: Peripheral neuropathy due to isoniazide. Lancet 1:1073, 1953.
52. William HE, Smith LH JR: Primary hyperoxaluria. In: *The Metabolic Basis of Inherited Disease*, 5th ed., JB Stanbury, et al. (eds). New York: McGraw-Hill, 1983, pp 204–228.
53. Morris MC, Chambers TL, Evans PWG, et al.: Oxalosis in infancy. Arch Dis Child 57:224–228, 1982.
54. Rose GA, Arthur LJH, Chambers TL et al.: Successful treatment of primary hyperoxaluria in a neonate. Lancet 1:1298–1299, 1982.
55. Alinei P, Grignard JP, Jaeger P: Pyridoxine treatment of type I hyperoxaluria. N Engl J Med 311:798–799, 1984.
56. Yendt ER, Cohanim M: Response to a physiologic dose of pyridoxine in type I primary hyperoxaluria. N Engl J Med 312:953–957, 1985.
57. Smith LH, Williams HE: Treatment of primary oxaluria. Mod Treatm 4:522–530, 1967.
58. Will EJ, Bijvoet OLM: Primary oxalosis: Clinical and biochemical response to high-dose pyridoxine therapy. Metabolism 28:542–548, 1979.

59. Schaumburg H, Kaplan J, Windebank A, et al.: Sensory neuropathy from pyridoxine abuse: A new megavitamin syndrome. N Engl J Med 309:445–448, 1983.
60. Berger A, Schaumburg HH: More on neuropathy from pyridoxine abuse. N Engl J Med 311:986–987, 1984.
61. Harrison AR, Kasidas GP, Rose GA: Hyperoxaluria and recurrent stone formation apparently cured by short courses of pyridoxine. Br Med J 282:2097–2098, 1981.
62. Prien El Sr, Gershoff SF: Magnesium oxide-pyridoxine therapy for recurrent calcium oxalate calculi. J Urol 112:509–512, 1974.
63. Dobbelstein H, Koener WF, Mempel W, et al.: Vitamin B-6 deficiency in uremia and its implications for the depression of immune responses. Kidney Int 5:233–239, 1974.
64. Kopple JD, Jones M, Fukuda S, Swendseid ME: Amino acid and protein metabolism in renal failure. Am J Clin Nutr 31:1532–1540, 1978.
65. Teehan BP, Smith LJ, Sigler MH, et al.: Plasma pyridoxal 5'-phosphate levels and clinical correlations in chronic hemodialysis patients. Am J Clin Nutr 31:1932–1936, 1978.
66. Kleiner MJ, Tate SS, Sullivan JF, Chami J: Vitamin B-6 deficiency in maintenance dialysis patients: Metabolic effects of repletion. Am J Clin Nutr 33:1612–1619, 1980.
67. Balcke P, Schmidt P, Zazgornik J, et al.: Effect of vitamin B-6 administration on elevated plasma oxalic acid levels in haemodialysed patients. Eur J Clin Invt 12:481–483, 1982.
68. Sjögren U, Thysell H, Lindholm T: The influence of vitamin B-6 supplementation on the bone marrow morphology in patients in regular haemodialysis treatment. Scand J Urol Nephrol 13:101–103, 1979.
69. Casciato DA, McAdam LP, Kopple JD, et al.: Immunologic abnormalities in hemodialysis patients: Improvement after pyridoxine therapy. Nephron 38:9–16, 1984.
70. Kopple JD, Mercurio K, Blumenkrantz J, et al.: Daily requirement for pyridoxine supplements in chronic renal failure. Kidney Int 19:694–704, 1981.
71. Ellis JM, Kishi T, Azuma J, Folkers K: Therapy of the carpal tunnel syndrome with vitamin B-6. IRCS Med Sci 4:193, 1976.
72. Ellis JM, Azuma J, Watanabe T, et al.: Survey and new data on treatment with pyridoxine of patients having a clinical syndrome including the carpal tunnel and other defects. Res Commun Chem Pathol Pharmacol 13:744, 1976.
73. Folkers K, Ellis JM, Saji S, Kaji M: Biochemical evidence for a deficiency of vitamin B-6 in the carpal tunnel syndrome based on a cross-over clinical study. Proc Natl Acad Sci USA 75:3410, 1978.
74. Del Tredici AM, Bernstein AL, Chinn K: Carpal tunnel syndrome and vitamin B-6 therapy. In: *Vitamin B-6: Its Role in Health and Disease*, RD Reynolds, JE Leklem (eds). New York: Alan R. Liss, 1985, pp 459–462.
75. Salkeld RM, Stotz R: Vitamin B-6 deficiency: An etiological factor of the carpal tunnel syndrome. In: *Vitamin B-6: Its Role in Health and Disease*, RD Reynolds, JE Leklem (eds). New York: Alan R. Liss, 1985, pp 463–467.
76. Folkers K: Priestley Medal Address: Contemporary therapy with vitamin B-6, vitamin B-2, and coenzyme Q-10. Chem Eng News, April 21, 1986:27–30, 55–56.
77. Smith GP, Rudge PJ, Peters TJ: Biochemical studies of pyridoxal and pyridoxal phosphate status and therapeutic trial of pyridoxine in patients with carpal tunnel syndrome. Ann Neurol 15:104–107, 1984.
78. Driskell JA, Wesley RL, Hess IE: Effectiveness of pyridoxine hydrochloride treatment on carpal tunnel syndrome patients. Nutr Rep Int 34:1031–1040, 1986.
79. Jain, VK, Cestero RVM, Baum J: Carpal tunnel syndrome in patients undergoing maintenance hemodialysis. JAMA 242:2868–2869, 1979.

80. Mezey E: Alcoholic liver disease: Roles of alcohol and malnutrition. Am J Clin Nutr 33:2709–2718, 1980.
81. Bonjour JP: Vitamins and alcoholism. III. Vitamin B-6. Int J Vitam Nutr Res 50:215–230, 1980.
82. Li T-K, Lumeng L: Vitamin B-6 metabolism in alcoholism and chronic liver disease. In: *Vitamin B-6: Its Role in Health and Disease*, RD Reynolds, JE Leklem (eds). New York: Alan R. Liss, 1985, pp 257–269.
83. Reynolds RD, Natta CL: Depressed plasma pyridoxal phosphate concentrations in adult asthmatics. Am J Clin Nutr 41:684–688, 1985.
84. Collipp PJ, Goldzier S, Weiss N, et al.: Pyridoxine treatment for childhood bronchial asthma. Ann Allergy 35:93–97, 1975.
85. Kark JA, Tarasoff PG, Bongiovanni R, et al.: Pyridoxal phosphate as an antisickling agent in vitro. J Clin Invest 71:1224–1229, 1983.
86. Reynolds RD, Natta CL: Vitamin B-6 and sickle cell anemia. In: *Vitamin B-6: Its Role in Health and Disease*, RD Reynolds, JE Leklem (eds). New York: Alan R. Liss, 1985, pp 301–316.
87. Kark JA, Tarassoff PG, Bongiovanni R, et al.: Pyridoxal phosphate as an antisickling agent in vitro. J Clin Invest 71:1224–1229, 1983.
88. Natta CL, Reynolds RD: Apparent vitamin B-6 deficiency in sickle cell anemia. Am J Clin Nutr 40:235–239, 1984.
89. Adelekan DA, Adekile AD, Thurnham DI: Dependence of pyridoxine metabolism on riboflavin status in sickle cell patients. Am J Clin Nutr 46:86–90, 1987.
90. Bryan GT, Brown RR, Price JM: Mouse bladder carcinogenicity of certain tryptophan metabolites and other aromatic nitrogen compounds suspended in cholesterol. Cancer Res 24:569, 1964.
91. Price JM, Brown RR: Studies on the etiology of carcinoma of the urinary bladder. Acta Un Int Cancer 18:684, 1962.
92. Brown RR, Price JM, Satler EJ, Wear JB: The metabolism of tryptophan in patients with bladder cancer. Acta Un Int Cancer 16:299, 1960.
93. Byar D, Blackard C: Comparisons of placebo, pyridoxine, and topical thiotepa in preventing recurrence of stage 1 bladder cancer. Urology 10:556–561, 1977.
94. Maxfield JR, McIlwain AJ, Robertson JE: Treatment of radiation sickness with vitamin B-6 (pyridoxine hydrochloride). Radiology 41:383–388, 1943.
95. Mattie H, Emery EW, Hill ID, Laurence DR: Treatment of radiation sickness with pyridoxine hydrochloride in outpatients of a radiotherapy unit. Br Med J 3:215–216, 1967.
96. Gridley DS, Stickney DR, Nutter RL, et al.: Suppression of tumor growth and enhancement of immune status with high levels of dietary vitamin B-6 in BALB/c mice. J Natl Cancer Inst 78:951–959, 1987.
97. Coburn SP: Metabolic and clinical studies of vitamin B-6 in mental disorders. In: *Vitamin B-6: Its Role in Health and Disease*, RD Reynolds, JE Leklem (eds). New York: Alan R. Liss, 1985, pp 123–159.
98. Brenner A, Wapnir RA: A pyridoxiase-dependent behavior disorder. Am J Dis Child 132:773–776, 1978.
99. Mudd SH: Pyridoxine-responsive genetic disease. Fed Proc 30:970–976, 1971.
100. Berson EL, Hanson AH, Rosner B, Shih VE: A two year trial of low protein, low arginine diets or vitamin B-6 for patients with gyrate atrophy. Birth Defects 18:209–218, 1982.
101. Sauberlich HE: Interaction of vitamin B-6 with other nutrients. In: *Vitamin B-6: Its Role*

in Health and Disease, RD Reynolds, JE Leklem (eds). New York: Alan R. Liss, 1985, pp 193–217.
102. Reinken L, Zieglauer H: Vitamin B-6 absorption in children with acute celiac disease and in control subjects. Am J Clin Nutr 108:1562–1565, 1978.
103. Knowlessar OD, Haeffner LJ, Benson GD: Abnormal tryptophan metabolism in patients with adult celiac disease with evidence for deficiency of vitamin B-6. J Clin Invest 43:894–903, 1964.
104. Folkers K, Shizukuishi S, Scudder SL, et al.: Biochemical evidence for a deficiency of vitamin B-6 in subjects reacting to monosodium L-glutamate by the Chinese restaurant syndrome. Biochem Biophys Res Commun 100:972–977, 1981.
105. Editorial: Is vitamin B-6 an antithrombotic agent? Lancet 1:1299–1300, 1981.
106. Experimental Nutrition Report: Inhibition of platelet aggregation and clotting by pyridoxal-5'-phosphate. Nutr Rev 40:55–57, 1982.
107. Subbarao K, Kuchibhotia J, Kakkar VV: Pyridoxal 5'-phosphate—a new physiological inhibitor of blood coagulation and platelet function. Biochem Pharmacol 28:531–534, 1979.
108. Schoene NW, Chanmugam P, Reynolds RD: Effect of oral vitamin B-6 supplementation on in vitro platelet aggregation. Am J Clin Nutr 43:825–830, 1986.
109. Swift ME, Schultz TD: Relationship of vitamins B-6 and B-12 to homocysteine levels: Risk for coronary heart disease. Nutr Rep Int 34:1–14, 1986.
110. McCully KS: Homocysteine theory of arteriosclerosis development and current status. Atherosclerosis Rev 11:157, 1983.
111. Freeman JM, Finkelstein JD, Mudd SH: Folate-responsive homocystinuria and "schizophrenia". A defect in methylation due to deficient 5,10-methylene-tetrahydrofolate reductase activity. N Engl J Med 292:491, 1975.
112. Schuh S, Rosenblatt DS, Cooper BA, et al.: Homocystinuria and megaloblastic anemia responsive to vitamin B-12 therapy. An inborn error of metabolism due to a defect in cobalamin metabolism. N Engl J Med 310:686, 1984.
113. Parry GJ, Bredesen DE: Sensory neuropathy with low-dose pyridoxine. Neurology 35:1466–1468, 1985.
114. Molimard R, Marillaud A, Paille A, et al.: Impairment of memorization by high doses of pyridoxine in man. Biomedicine 32:88–92, 1980.
115. Bhagavan HN: Interaction between vitamin B-6 and drugs. In: *Vitamin B-6: Its Role in Health and Disease*, RD Reynolds, JE Leklem (eds). New York: Alan R. Liss, 1985, pp 401–415.

9
Folic Acid

Introduction	161
Absorption, Metabolism, and Excretion	164
Dosage Forms	164
Clinical Studies	164
Pregnancy	164
Infertility	166
The Infant	166
The Elderly	166
Alcoholism	167
Neurologic and Neuromuscular Disorders	167
Gastrointestinal Disease	168
Postmenopausal Homocysteinemia	168
Genetic Folate Disorders	169
Hemolytic Anemias	169
Cancer	169
Toxicity and Side Effects	171
Interactions	171
References	172

INTRODUCTION

As supplied commercially, folic acid (pteroyl glutamate) is in an oxidized state relative to its usual occurrence in foods and in the body. In mammals, the most common circulating folate is a methylated and reduced derivative, 5-methyl-tetrahydrofolate, in which the ring atoms numbered 5 through 8 have been reduced (see Fig. 1). Folic acid is a yellow solid that is only slightly soluble in water. Its sodium and potassium salts are readily soluble [1,2].

As found naturally, folate compounds are not only in a reduced pteroyl form, but they may contain any of the following one-carbon moieties attached at the N-5, N-10, or 5,10-positions: methyl, formyl, formimino, methenyl, or methylene. Also, as folate occurs within plant and animal cells, from two to eight additional glutamic acid molecules are linked to the glutamate end of the mol-

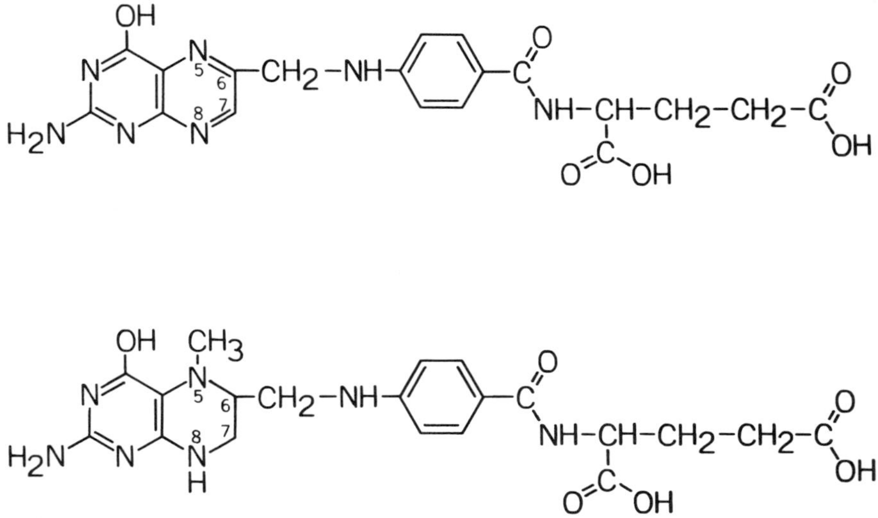

Fig. 1. Structural formulas of folic acid (upper) and 5-methyltetrahydrofolic acid.

ecule through successive gamma-carboxylamide linkages. The folyl polyglutamates are not transported as such across cell walls, but are formed within the cell. The generic term for the various folic acid derivatives is folate or folacin [1,2].

Tetrahydrofolyl compounds serve as cofactors or cosubstrates for enzymes involved in catalysis of a large variety of metabolic processes, all of which involve transfer of one-carbon moieties from one substrate to another. In RNA and DNA synthesis, folate is fundamentally involved in that two of the carbon atoms of purines are derived from the N-5, N-10-methenyl and the N-10-formyl tetrahydrofolates. In addition, deoxythymidylate, needed for DNA synthesis, is synthesized from deoxyuridylate by the action of thymidylate synthetase with N-5,N-10-methylene tetrahydrofolate as the coenzyme. In amino acid metabolism, tetrahydrofolic acid acts as an acceptor of the formimino group from N-formiminoglutamic acid (FIGLU) derived from catabolism of histidine. Increased urinary excretion of FIGLU is one indicator of folate deficiency [1,2].

In mammalian metabolism, methyl tetrahydrofolate acts as a methyl donor in only one reaction, namely the methylation of homocysteine to methionine, with simultaneous regeneration of tetrahydrofolate for further functioning in transfer of one-carbon moieties. The homocysteine methylation reaction is vitamin-B-12-dependent, so that vitamin B-12 deficiency can induce a secondary folate deficiency by blocking release of active tetrahydrofolate from methyl tetrahydrofolate (the so-called "methyl trap" phenomenon) [1,2].

Because folate is essential for DNA synthesis, one of the most notable results of its deficiency is the occurrence of abnormalities in rapidly dividing cell populations, e.g., in cells lining the GI tract and in bone marrow. Hypersegmentation of neutrophils occurs, and the bone marrow becomes megaloblastic, resulting in macrocytic anemia, leukopenia, and thrombopenia [2]. Very early or mild folate deficiency may manifest simply as an increase in neutrophil hypersegmentation, e.g., shift to predominance of five-lobed over three-lobed neutrophils, without abnormality in other hematologic indices [3].

Serum folate concentrations below 6 ng/ml are considered unsatisfactory and are likely to place the individual at risk of mild deficiency if long maintained, while values below 3 ng/ml are definitely indicative of deficiency [4]. Red blood cell folate is considered to be a better index of the long-term folate status; values below 140 ng/ml RBC are indicative of deficiency, and levels of 140–159 ng/ml are suspect [4]. Typical RBC folate levels that have been observed in large groups of healthy individuals range from 160 to about 500 ng/ml, with mean levels of 316 ng/ml in one study [4] and 347 ng/ml in another [5]. The normal mean serum folate concentration is about 8 ng/ml [5]. A high incidence of folate deficiency, by criteria of low serum and RBC folate levels, has been observed in low-income pregnant women [2,6], chronic alcoholics [7], adolescent girls [8], and in hospitalized patients [9–11]. The Second National Health and Nutrition Examination Survey (NHANES II) found a 9–10% incidence of serum folate levels <3.0 ng/ml in men and women aged 45–74 years [12]. By the criterion of low RBC folate levels, women aged 20–44 years were the population group at greatest risk of folate deficiency [12].

Good food sources of folate include vegetables (e.g., broccoli, spinach, and cabbage), some fruits (oranges, avocados), egg yolk, and whole wheat bread. Milk and meat, except organ meat, are rather poor sources. Losses occur with heat during food preparation, and folate may be lost in discarded cooking water [1]. The RDA for adults and children above age 4 is 400 mcg/d, and for pregnant women 800 mcg/d [13]. Several experts have recently proposed a downward revision of recommended daily intakes for normal healthy individuals to 3 mcg folate/kg/d [14,15], 200–300 mcg/d for nonpregnant women [16], and 500 mcg/d for pregnant women [14], but these revised values have not been accepted by official bodies. If the diet is well balanced with respect to raw or properly cooked vegetables, it should meet the folate requirement for most individuals. However, pregnant women will usually require a vitamin supplement to obtain a folate intake of 800 mcg/d [1,2], or even 500 mcg/d [14].

ABSORPTION, METABOLISM, AND EXCRETION

Folyl polyglutamates occurring in food are hydrolyzed to the monoglutamate forms by brush border and intracellular intestinal folate conjugase [17] and absorbed in the jejunum as monoglutamates into the portal blood [18,19]. Besides the conjugase of the intestinal mucosa, there is also some conjugase activity in bile and pancreatic juice, although it is uncertain how much it contributes to total conjugase activity in the GI tract [18,20]. Absorption of folyl monoglutamates at physiologic concentrations occurs by a specific saturable system, while at higher (pharmacologic) concentrations, there is also absorption by simple diffusion [19]. Folates are converted in the liver to methyl tetrahydrofolate [1], which is the circulating form of the vitamin [21]. Folate-binding proteins are present in plasma [22] and in cell membranes [23], as well as within the cell where they bind the folyl polyglutamates [24]. Folyl monoglutamates transported into the cell are converted to polyglutamates and remain trapped intracellularly in that form except for trace amounts released by hydrolysis as monoglutamate; the latter is produced in the course of normal intracellular turnover of folyl polyglutamates [1,2]. The folyl polyglutamates are widely distributed in body tissues. Total body folate has been estimated at 5–10 mg, with one-half of that amount being present in the liver [1]. The turnover of the folate pool is very low, less than 1% per day [1].

Urinary excretion of folate is only a small fraction of the daily intake. Bile contains considerable folate, and there is enterohepatic circulation of the vitamin. The feces contain much more folate than the amount ingested, due to bacterial synthesis in the bowel [1,2].

DOSAGE FORMS

Folic acid is available over the counter as tablets containing 400 micrograms and 800 micrograms. Tablets containing 1 mg of folic acid are available on prescription. Folic Acid Injection is an aqueous solution of the sodium salt of pteroyl glutamate. Leucovorin Calcium Injection is also available. Leucovorin is folinic acid, 5-$CHOH_4PteGlu$, also known as citrovorum factor. This injectable form is used to provide reduced folate to cancer patients who are under treatment with methotrexate, which inhibits the folate-reducing enzyme, dihydrofolate reductase [25].

CLINICAL STUDIES
Pregnancy

Assessment of the folate status of 269 low-income pregnant women (70 adolescents and 199 mature women ≥18 years of age) at their first prenatal visit to a maternity clinic in Gainesville, Florida, showed that 29% were at "high risk," with RBC folate levels <140 ng/ml and 11% at "medium risk," with RBC folate at 140–160 ng/ml [6]. Similar degrees of folate deficiency have been reported by Clark et al. [8] for adolescent girls in Alabama, a real concern in view of the high rate of teenage pregnancy. Folate deficiency was much more prevalent than iron deficiency in the low-income Florida group; the finding is somewhat surprising in view of the availability of low-cost vegetables and fruit in this region. Maternal folate deficiency is associated with congenital defects in the offspring in animal studies [26] and with obstetric complications in humans [27]. Laurence et al. [28] demonstrated a marked, significant protective effect of folic acid supplementation (4 mg/d) against recurrence of neural tube defects (anencephaly, encephalocele, spina bifida) in a randomized, controlled, double-blind trial with pregnant women who had already had one child with such a developmental defect. Instructions given the participants were that they should begin taking folic acid or placebo when contraceptive precautions were suspended and conception intended, and should continue through 6–9 weeks after the first missed period. There were no neural tube defects in infants born to folic acid-supplemented women or to placebo-supplemented women whose diets were scored as "excellent" with regard to folate content, suggesting that the 4-mg/d dosage was higher than needed.

Tolarova [29], in a study of similar design but oriented toward the incidence of harelip and cleft palate in the offspring of women whose previous newborns had had these defects, observed only one recurrence of the defect in the infants born to 85 mothers receiving a daily folic acid supplement (10 mg), while there were 15 recurrences in a nonsupplemented control group of 212 women. A low maternal folic acid status has been linked with an increased incidence of small-for-date neonates [30]. However, in spite of the abundant evidence that a low folate status is detrimental in pregnancy, there are limited observations, thus far unexplained and unconfirmed by repeated studies, that high plasma folate levels in pregnancy may be associated with an increased incidence of infection [31]. Further clinical trials must resolve this troubling question.

The folate RDA for pregnant women is 800 mcg/d, an amount not available in most American diets. Folic acid is usually not prescribed for the pregnant woman as a single supplement but as part of pregnancy multivitamin formulations that provide the RDA amount. Because of the essentiality of this vitamin for DNA synthesis in the rapidly dividing cells of the embryo, fetus, and placenta, it is important that good folate nutriture be established as early in pregnancy as possible and preferably before conception to diminish the likelihood of fetal malformations resulting from low folate status.

Folic acid is one of a number of micronutrients, including vitamins A, C,

E, and beta-carotene, found to be deficient in abruptio placentae, which appears to be a condition of multiple vitamin deficiency [32].

Infertility

Dawson and Sawers [32] describe folic acid treatment of three cases of infertility that were associated with folate deficiency. The three women (two aged 29 years, one aged 32 years), who had experienced infertility of several years' duration, were found to have erythrocyte macrocytosis together with low serum and RBC folate levels. Two of the patients were gluten-intolerant and, upon jejunal biopsy, showed subtotal villous atrophy. All three were placed on 5 mg folic acid t.i.d., which returned hematologic indices to normal. All became pregnant during the following 3–15 months. The authors recommend that folate status be evaluated in cases of infertility.

The Infant

Late in pregnancy the fetus concentrates folate in its plasma severalfold higher than in maternal plasma; during breast feeding, infant plasma folate levels remain substantially above adult values [34–36]. Preterm infants, however, may not have had sufficient time to establish good folate stores. For example, Strelling et al. [37] reported finding significant folate deficiency in 14 of 37 preterm English infants of birthweight 2,000 grams or less, manifested by megaloblastic erythropoiesis. Administration of 100–200 mcg/d of folic acid, either orally or IM, produced optimum hematologic responses.

Commercial infant formulas are adequately supplemented with folic acid. Homemade formulas made with boiled cow's milk, however, may be seriously deficient in folate [38]; infants who are fed such formulas should receive a separate multivitamin supplement providing 100–200 mcg folic acid daily.

The Elderly

Low folate status is fairly common in the elderly of Great Britain and the United States, according to reports of a number of investigators; causes are not entirely clear, but include poor dietary habits and possibly malabsorption [39–43]. Baker et al. [44] found that subjects over the age of 70 did not absorb natural folyl polyglutamates of yeast as efficiently as subjects aged 24–42 years, suggesting a decline of intestinal conjugase activity with age; synthetic folic acid, however, was absorbed with equal efficiency by old and young. On the other hand, Bailey et al. [45] observed no differences between young and elderly subjects in intestinal conjugase activity or ability to absorb folate from synthetic pteroyl-glutamyl-gamma-hexaglutamate. The most likely causes for low folate status in the elderly include low income and poor dietary habits [44,45], gastrointestinal surgery, and alcohol and prescription drug usage affecting folate absorption and utilization [48–50]. Low plasma and RBC folate levels may be present in the aged without overt anemia [41]. Prescription of a multivitamin supplement providing the RDA of 400 mcg of folic acid should ordinarily take care of folate deficiency in the elderly.

If macrocytic anemia is present, the possibility of vitamin B-12 deficiency must also be considered. Since large doses of folic acid (e.g. 400 mcg/d IM or 5 mg/d orally) may produce hematologic responses in vitamin B-12 deficiency without preventing the neurologic sequelae of B-12 deficiency [1,2], it is desirable to first determine the response to daily administration of 100 mcg folic acid orally (or IM if malabsorption is suspected), an amount too small to mask B-12 deficiency but large enough to induce a reticulocytosis reaching a peak at about 7 days [2, 51]. If folate deficiency is thus confirmed, the patient may be maintained with 400–500 mcg folic acid daily [51].

Alcoholism

Folate deficiency is common among chronic alcoholics, in part because of poor diet and in part because of adverse effects of alcohol on food folate absorption and on metabolism of folate [52–56]. Alcohol ingestion increases urinary folate excretion and prolongs turnover time [57], as well as inhibiting the hemopoietic activity of folic acid [58]. After cessation of drinking, a normal folate status can be maintained by the RDA intake of 400 mcg, but if drinking is continued, the alcoholic may need 2 mg or more of folic acid daily to maintain normoblastic erythropoiesis [52].

Neurologic and Neuromuscular Disorders

Folate deficiency is frequently seen in psychiatric and psychogeriatric patients, and there is a particular association of this deficiency with depression [59–62]. Reynolds et al. suggest that the relation between folate deficiency and mood derives from the fact that methyl tetrahydrofolate is necessary for regeneration from brain homocysteine of methionine and S-adenosyl-methionine, the major methyl donor in the brain [62]. Folic acid derivatives are also coenzymes in the synthesis of neurotransmitters from tryptophan [63] and tyrosine [64]. It has been reported that hospitalized endogenous depressives and schizophrenics whose folate deficiency is treated with folic acid supplements spend less time in the hospital and show better symptomatic and social recoveries than patients whose folate deficiency is not corrected [65].

Botez and his coworkers have reported numerous cases of folate-responsive nervous disorders [66–68]. Conditions that were highly or partially responsive to folic acid therapy in folate-deficient patients included depression, muscular and intellectual fatigue, restless legs syndrome, depressed ankle jerks, diminution of vibration sense in the knees, stocking-type tactile hypoesthesia, confusional syndrome, and nonsenile dementia [67,69]. The usual oral dosage of folic acid was 10 mg/d, together with 15 mg IM once a week in patients with malabsorption. Recovery, estimated by clinical, psychological, and electrophysiologic tests, takes place in most patients within 3 months, although some require 12 months [66–69]. After convalescence, some continued to require 5–30 mg/d of folic acid orally to prevent recurrence of neurologic symptoms, although most could be maintained on 5 mg twice weekly [67].

Botez and Lambert [70] noted a significant correlation between low folate status and incidence of the restless legs syndrome (cramps, numbness, creeping sensations) in pregnant women, though not all those with low folate status experienced the syndrome. Administration of 500 mcg folic acid daily was effective in preventing the syndrome in most patients.

Folate deficiency has been reported to cause hypotonia in infants, reversible by folic acid supplementation [71]. Botez [71] describes two siblings, a boy of 8 years and a girl of 12, who manifested clumsiness in walking, difficulty in running, and inability to participate in sports. Family diet, altered because of irritable colon syndrome in the parents, had been folate-deficient for years. Oral folic acid therapy (5 mg/d) restored the children to normal.

Gastrointestinal Disease

Absorption of folic acid and folyl polyglutamates is impaired in tropical sprue, in celiac disease (gluten-sensitive enteropathy), and in general when there is jejunal villous atrophy [2, 71–73]. Folate deficiency caused by malabsorption may be treated with pharmacologic doses of folic acid.

Folic acid supplementation has been recommended for optimum cell regeneration in hepatitis [74–76]. Patients with Ménétrier's disease (giant hypertrophic gastritis) may become folate-deficient because of leakage into the gastric lumen [77].

Postmenopausal Homocysteinemia

Brattström et al. [77] found, in a small study, that postmenopausal women had significantly higher plasma homocysteine levels, measured as homocysteine-cysteine mixed disulfide (MDS), than premenopausal women or young men. After a methionine load test, the rise of MDS levels in postmenopausal women was significantly greater than the rise induced by methionine in premenopausal women, young men, and old men. Folic acid therapy (5 mg/d for 4 weeks) in the postmenopausal women resulted in substantial reductions of MDS concentrations both before (−31%) and after (−28%) a methionine load. The authors consider these observations to be of potential clinical importance in view of the fact that congenital homocysteinemia is associated with inductions of juvenile arteriosclerosis, recurrent thromboembolic complications, and osteoporosis. In milder or more slowly developing degree, these are also problems of women in the postmenopausal years.

Genetic Folate Disorders

A number of congenital megaloblastic anemias have been reported involving: defective folate absorption and transport [79]; homocystinuria due to low 5,10-methylene tetrahydrofolate reductase activity, both responsive [80] and unresponsive [81] to high-dose folic acid; and dihydrofolic acid reductase deficiency [82]. Mental retardation and neuropathy are characteristic sequelae. Oral folic acid (20 mg/d) was effective in one case of methylene tetrahydrofolate reductase deficiency [80] and in congenital malabsorption [79], but folinic acid (N^5-formyl tetrahydrofolic acid), 5 mg/d IM, was more effective in elevating folate concentration in cerebrospinal fluid [79] and is essential in cases of dihydrofolate reductase deficiency [82,83]. Folinic acid therapy produced a good clinical response in a 7-month-old infant with low methylene tetrahydrofolate reductase activity [84].

Taunton et al. [85] reported folic acid treatment of a woman and her 19-month-old daughter who manifested hypoglycemia due to low fructose-1,6-diphosphatase (FDPase) activity. For reasons not understood, pharmacologic doses of folic acid stimulate FDPase activity, and therapy with 15 mg/d of folic acid was therefore tried in the two patients. There was good symptomatic relief and improvement in ability to maintain normal blood glucose levels. In this connection, it is also of interest to note that Botez and Bachevalier [71] observed a significant impairment of folic acid absorption, relative to normal controls, in 17 patients with reactive hypoglycemia.

Branda et al. [86] report a case of congenital absorption and transport defects in a young man who required 20 mg/d of folic acid to attain normal plasma and RBC folate concentrations and to fully correct anemia; smaller doses were inadequate.

Hemolytic Anemias

In patients with hemolytic anemia, megaloblastosis may occur due to folate deficiency that is believed to result from elevated need for folate because of increased DNA synthesis by the hyperactive bone marrow [1]. Supplementation over and above the dietary supply is therefore needed.

Cancer

Yunis and his associates [87,88] have presented evidence that chromosome breaks in dividing cells occur predominantly at fragile sites that are rich in thymine, and that the frequency of such breaks in human cell cultures is greatly increased if the medium lacks thymine and folinic acid. In addition, if lymphocytes were taken from human subjects after they had received 5 mg/d of folinic acid orally for 3 days, breakage induced by caffeine in FTD medium was suppressed [87]. Based on this evidence, they have hypothesized that folate deficiency may predispose dividing cells, through chromosome breakage and subsequent rearrangements that may activate oncogenes, to development of abnormalities including cancer [87,88].

Butterworth and coworkers [89] carried out an important study on the relation of folate nutriture to cervical dysplasia, which has a strong tendency to progress to carcinoma in situ [90–93]. The Butterworth study involved the participation of young women who were using oral contraceptives and who had mild or moderate cervical dysplasia. Initially, the mean RBC folate level of oral contraceptive users without dysplasia was compared with that of nonusers and found to be lower (189 versus 269 ng/ml), while in oral contraceptive users with dysplasia mean RBC folate was even lower (161 ng/ml). Forty-seven oral contraceptive users with dysplasia were than given oral supplements of either 10 mg/d of folic acid (5-mg capsule b.i.d.) or 10 mg/d of ascorbic acid (placebo) under double-blind conditions for 3 months. Evaluation of smears and biopsies at the end of the trial showed significantly better mean scores in the folate-supplemented group than in the placebo-supplemented group. Final versus initial scores were also significantly better than initial scores in the folic acid-supplemented group, while they were unchanged in the placebo group. The investigators conclude that the dysplastic process in the cervix may be arrested or in some cases reversed by oral folic acid supplementation. A generous folic acid intake would seem desirable to minimize the possibility of development of cervical dysplasia, particularly in young women taking oral contraceptives.

At the same medical center, Heimburger et al. [93] carried out clinical studies to test a hypothesis advanced by Krumdieck of that group that exposure to cigarette smoke results in folic acid deficiency affecting principally the bronchial epithelium through direct chemical inactivation, rendering it more susceptible to neoplastic transformation by carcinogens in tobacco smoke. Comparing plasma and erythrocyte folic acid levels in 25 nonsmokers, 27 smokers without metaplasia, and 42 smokers with metaplasia, they found that the smokers with metaplasia had significantly lower levels of RBC and plasma folic acid than the nonsmokers and that smokers with metaplasia had significantly lower plasma levels than smokers without metaplasia.

Preliminary reports by the same group indicate a significant reversal of premalignant bronchial metaplasia in cigarette smokers by folic acid (10 mg/d) and vitamin B-12 (500 mcg/d) supplementation for 4 months in comparison with placebo-supplemented controls [95,96].

TOXICITY AND SIDE EFFECTS

Folic acid has a low acute and chronic toxicity for humans, no adverse effects being noted in adults taking 400 mg/d for 5 months or 10 mg/d for 5 years [1]. Herbert et al. [2] consider doses up to 15 mg/d to be without toxic effects; however, others have reported that some individuals may experience sleep disturbances and irritability while taking 15 mg/d [97]. There seems little need to prescribe supplements beyond 5 mg/d except in cases of genetic folate disorders, malabsorption, or use of drugs that have antifolate activity. For the majority of patients, after their folate deficiency has been corrected, RDA amounts should suffice for maintenance.

INTERACTIONS

It must be borne in mind that folic acid administration can mask vitamin B-12 deficiency, correcting megaloblastic anemia without halting the adverse neurologic consequences of B-12 deficiency [1,2]. Deficiency of vitamin B-12 is most likely in the older patient and in those with impaired B-12 absorption due to lack of intrinsic factor, gastric atrophy, gastrectomy, or small intestine disorders affecting the ileum, which is the main site of B-12 absorption [2]. If true folate deficiency is present, there will be a reticulocyte response to a very small (100-mcg) daily dose of folic acid [2,51]. Simultaneous folate and vitamin B-12 deficiency may be present, requiring administration of both vitamins. (Vitamin B-12 is usually administered IM when malabsorption is present).

Megaloblastic anemia occurs in as many as 50% of epileptics receiving anticonvulsant drugs, including diphenyl hydantoin alone or in combination with other drugs; primidone alone or in combination with other drugs; and barbiturates alone or in combination with other drugs [98]. A case of severe gingival hyperplasia induced by anticonvulsant therapy was corrected within 8 days by oral folic acid at 15 mg/d [99]. At least one study has suggested an antagonistic effect of high-dose folic acid on anticonvulsant drug efficacy [1], but doses of 1–5 mg/d have been reported to be sufficient to overcome folate deficiency without impairing the anticonvulsant drug effect [100]. An increase in seizure frequency calls for a reduction in folic acid dosage.

Other drugs that may adversely affect folic acid absorption or metabolism include antacids, cholestyramine, colestipol, triamterene, antimalarials, oral contraceptives, sulfasalazine, trimethoprim, and pyrimethamine [89,98,101–104]. The newer acid-lowering antiulcer drugs, cimetidine and ranitidine, may also lower folic acid absorption somewhat [105]. Supplements of 1–2 mg/d of folic acid are usually adequate to compensate for the antifolate activity of any of these drugs. Although aspirin decreases folic acid absorption in laboratory rats, it appears to increase folate absorption in humans [106].

Human experiments indicate that folic acid supplements impair zinc absorption appreciably, possibly by forming an unabsorbable chelate or complex with zinc that is excreted in the feces [107]. Trials in pregnant women who were receiving supplements of 100 mg/d of ferrous iron and 350 mcg/d of folic acid indicated about a 50% reduction in the absorption of 50 mg of zinc (from $ZnSO_4$), measured over a 4-hour period subsequent to zinc ingestion [108]. Folic acid alone, without iron, reduced apparent zinc absorption about 20%. The mechanism is obscure, as the total number of micromoles of zinc ion used in these experiments far exceeded that of folic acid [108]. The effect of pharmaceutical folic acid and iron supplements on zinc status in pregnancy could be important if the diet is low in zinc [107,108]. In contrast, the extensive human and animal experiments of Keating et al. [109] showed no inhibition of zinc absorption by folic acid. In their rat-feeding studies, the molar ratios of folate to Zn were 1:20 in one trial and 4:1 in the other. The human study (six healthy young men) involved ingestion of 25 mg Zn (as $ZnSO_4$ in 240 ml water) after an overnight fast and sequential measurement of serum Zn levels; in a separate experiment, 10 mg of folic acid in 50 ml of water was given immediately after the ingestion of the $ZnSO_4$ solution. Despite the wide range of molar ratios used in the animal studies and the relatively large amount of folic acid given in the human studies, there was no discernible impairment of Zn absorption or utilization [109].

Such effects, if real, can perhaps be minimized by taking folic acid and iron supplements other than at mealtimes, or by taking them only at one meal of the day. It is not known whether the folyl polyglutamates and organically bound zinc and iron found in food interact in the same way as pharmaceutical folic acid and the zinc and iron salts used experimentally. Obviously, much more clinical research is needed in this area.

REFERENCES

1. Brody T, Shane B, Stokstad ELR: Folic acid. In: *Handbook of Vitamins,* LJ Machlin (ed). New York: Marcel Dekker, 1984, pp 459–496.
2. Herbert V, Colman N, Jacob E: Folic acid and vitamin B-12. In: *Modern Nutrition in Health and Disease,* 6th ed., RS Goodhart, ME Shils (eds). Philadelphia: Lea & Febiger, 1980, pp 229–259.
3. Bills T, Spatz L: Neutrophilic hypersegmentation as an indicator of incipient folic acid deficiency. Am J Clin Pathol 68:263–267, 1977.
4. Sauberlich HE, Dowdy RP, Skala JH: *Laboratory Tests for the Assessment of Nutritional Status.* Boca Raton, FL: CRC Press, 1974, pp 49–60.
5. Milne DB, Johnson LAK, Mahalko JR, Sandstead HH: Folate status of adult males living in a metabolic unit: Possible relationships with iron nutriture. Am J Clin Nutr 37: 768–773, 1983.

6. Bailey LB, Mahan CS, Dimperio D: Folacin and iron status in low-income pregnant adolescents and mature older women. Am J Clin Nutr 33:1997–2001, 1980.
7. Halsted CH: Folate deficiency in alcoholism. Am J Clin Nutr 33: 2736–2740, 1980.
8. Clark AJ, Mossholder S, Gates R: Folacin status in adolescent females. Am J Clin Nutr 46: 302–306, 1987.
9. Weinsier RL, Hunker EM, Krumdieck CL, Butterworth CE Jr: Hospital malnutrition. A prospective evaluation of general medical patients during the course of hospitalization. Am J Clin Nutr 32: 418–426, 1979.
10. Leevy CM, Cardi L, Frank O, et al.: Incidence and significance of hypovitaminemia in a randomly selected municipal hospital population. Am J Clin Nutr 17:259–271, 1965.
11. Lawrence VA: Demographic analysis of serum folate and folate-binding capacity in hospitalized patients. Acta Haematol 69: 289–293, 1983.
12. Senti FR, Pilch SM: Analysis of folate data from the Second National Health and Nutrition Examination Survey (NHANES II). J Nutr 115:1398–1402, 1985.
13. Food and Nutrition Board: *Recommended Dietary Allowances*, 9th ed., Washington, DC: National Academy of Sciences, 1980.
14. Herbert V: Recommended dietary intakes (RDI) of folate in humans. Am J Clin Nutr 45:661–670, 1987.
15. Reisenauer AM, Halsted CH: Human folate requirements. J Nutr 117:600–602, 1987.
16. Sauberlich HE, Kretsch MJ, Skala JH, et al.: Folate requirement and metabolism in nonpregnant women. Am J Clin Nutr 46:1016–1028, 1987.
17. Reisenauer AM, Krumdieck CL, Halsted CH: Folate conjugase: Two separate activities in human jejunum. Science 198:196–197, 1977.
18. Halsted CH: The intestinal absorption of folates. Am J Clin Nutr 32:846–855, 1979.
19. Rosenberg IH: Digestion and absorption of dietary folate. Fed Proc 43:2428–2429, 1984.
20. Horne DW, Krumdieck CL, Wagner C: Properties of folic acid gamma-glutamyl hydrolase (conjugase) in rat bile and plasma. J Nutr 111:442–449, 1981.
21. Hoppner K, Lampi B: Reversed phase high pressure liquid chromatography of folates in human whole blood. Nutr Rep Int 27 911–919, 1983.
22. Eichner ER, McDonald CR, Dickson VL: Elevated serum levels of unsaturated folate binding protein: Clinical correlates in a general hospital population. Am J Clin Nutr 31:1988–1992, 1978.
23. Corrocher R, Pachor ML, Bambara LM, DeSandre G: Evidence for a folic acid binding protein in human cell membrane. Acta Haematol 66:202–209, 1981.
24. Zamierowski MM, Wagner C: Effect of folacin deficiency on folacin-binding proteins in the rat. J Nutr 107:1937–1945, 1977.
25. Hillman RS: Vitamin B-12, folic acid, and the treatment of megaloblastic anemias. In: *Goodman and Gilman's The Pharmacological Basis of Therapeutics,* 7th ed., AG Gilman, LS Goodman, TW Rall, F Murad (eds). New York: Macmillan, 1985, pp 1323–1337.
26. Nelson MM, Asling CW, Evans HM: Production of multiple congenital abnormalities in young by maternal pteroylglutamic acid deficiency during gestation. J Nutr 48:61–79, 1952.
27. Rothman D: Folic acid in pregnancy. Am J Obstet Gynecol 108:149–175, 1970.
28. Laurence KM, James N, Miller MH, et al.: Double-blind randomised controlled trial of folate treatment before conception to prevent recurrence of neural-tube defects. Br Med J 282:1509–1511, 1981.
29. Tolarova M: Periconceptional supplementation with vitamins and folic acid to prevent recurrence of cleft lip. Lancet 2:217, 1982.
30. Gandy G, Jacobson W: Influence of folic acid on birthweight and growth of the erythroblastotic infant. Arch Dis Child 52:1–6, 1977.

31. Sandstead H, Cherry F, Bazzano G, et al.: Folate-Zn interaction in human pregnancy. Fed Proc 46:748A, 1987.
32. Sharma S, et al.: Comparison of blood levels of vitamin A, beta-carotene and vitamin E in abruptio placentae with normal pregnancy. Int J Vitam Nutr Res 56:3–9, 1986.
33. Dawson DW, Sawers AH: Infertility and folate deficiency. Case reports. Br J Obstet Gynaecol 89:678–680, 1982.
34. Ek J, Magnus EM: Plasma and red blood cell folate in breastfed infants. Acta Paediatr Scand 68:239–243, 1979.
35. Ek J: Plasma and red cell folate values in newborn infants and their mothers in relation to gestational age. J Pediatr 97:288–292, 1980.
36. Tamura T, Yoshimura Y, Arakawa T: Human milk folate and folate status in lactating mothers and their infants. Am J Clin Nutr 33:193–197, 1980.
37. Strelling, MK, Blackledge DG, Goodall HB: Diagnosis and management of folate deficiency in low birthweight infants. Arch Dis Child 54:271–277, 1979.
38. Ek J, Magnus E: Plasma and red cell folacin in cow's milk-fed infants and children during the first 2 years of life: The significance of boiling pasteurized milk. Am J Clin Nutr 33:1220–1224, 1980.
39. Elsborg L: Reversible malabsorption of folic acid in the elderly with nutritional folate deficiency, Acta Haematol 55:140, 1976.
40. Krehl WA: The influence of nutritional environment on aging. Geriatrics 29:65, 1974.
41. Bates CJ, Fleming M, Paul AA, et al.: Folate status and its relation to vitamin C in healthy elderly men and women. Age Ageing 9:241–248, 1980.
42. Morgan AG, Kelleher J, Walker E, et al.: A nutritional survey of the elderly: Hematological aspects. Int J Vitam Nutr Res 43:461, 1973.
43. Rosenberg IH, Bowman BB, Cooper BA, et al.: Folate nutrition in the elderly. Am J Clin Nutr 36:1060–1066, 1982.
44. Baker H, Jaslow SP, Frank O: Severe impairment of dietary folate utilization in the elderly. J Am Geriatr Soc 26:218–221, 1978.
45. Bailey LB, Cerda JJ, Bloch BS, et al.: Effect of age on poly- and monoglutamyl folacin absorption in human subjects. J Nutr 114:1770–1776, 1984.
46. Wagner PA, Bailey LB, Krista ML, et al.: Comparison of zinc and folacin status in elderly women from differing socioeconomic backgrounds. Nutr Res 1:565–569, 1981.
47. Bailey LB, Wagner PA, Christakis GJ, et al.: Folacin and iron status and hematological findings in predominantly black elderly persons from urban low-income households. Am J Clin Nutr 32:2346–2356, 1979.
48. Waxman S, Corcino JJ, Herbert V: Drugs, toxins and dietary amino acids affecting vitamin B-12 or folic acid absorption or utilization. Am J Med 48:599–608, 1970.
49. Schuckitt MA, Pastor PA: The elderly as a unique population. Alcoholism Clin Exp Res 2:31–38, 1978.
50. Zimberg A: Alcohol and the elderly. In: *Drugs and the Elderly,* DM Peterson, et al. (ed). Springfield, IL: C.C. Thomas, 1979, p 28.
51. Hodges RE: *Nutrition in Medical Practice*. Philadelphia, W.B. Saunders, 1980, pp 211–213.
52. Bonjour JP: Vitamins and alcoholism. II. Folate and vitamin B-12. Int J Vit Nutr Res 50:96–121, 1980.
53. Lindenbaum J: Folate and vitamin B-12 deficiencies in alcoholism. Semin Hematol 17:119–129, 1980.
54. Halsted CH: Folate deficiency in alcoholism. Am J Clin Nutr 33:2736–2740, 1980.
55. Lindenbaum J, Roman MJ: Nutritional anemia in alcoholism. Am J Clin Nutr 33:2727–2735, 1980.

56. Blocker DE, Thenen SW: Intestinal absorption, liver uptake, and excretion of ^{3}H-folic acid in folic acid-deficient, alcohol-consuming nonhuman primates. Am J Clin Nutr 46:503–510, 1987.
57. Russell RM, Rosenberg IH, Wilson FD, et al.: Increased urinary excretion and prolonged turnover time of folic acid during ethanol ingestion. Am J Clin Nutr 38:64–70, 1983.
58. Sullivan LW, Herbert V: Suppression of hematopoiesis by ethanol. J Clin Invest 43:2048–2062, 1964.
59. Editorial: Folic acid and the nervous system. Lancet 2:836, 1976.
60. Ghadirian AM, Ananth J, Engelsmann F: Folic acid deficiency and depression. Psychosomatics 21:926–929, 1980.
61. Carney MWP: Serum folate values in 423 psychiatric patients. Br Med J iv:512–516, 1967.
62. Reynolds EH, Carney MWP, Toone BK: Methylation and mood. Lancet 2:196–198, 1984.
63. Gal EM, Armstrong JC, Ginsberg B: The nature of in vitro hydroxylation of L-tryptophan by brain tissue. J Neurochem 13:643–654, 1966.
64. Nagatsu T, Levitt M, Udenfriend S: Tyrosine hydroxylase. The initial step in norepinephrine synthesis. J Biol Chem 239:2910–2917, 1964.
65. Carney, MWP, Sheffield BF: Associations of subnormal serum folate and vitamin B-12 and effects of replacement therapy. J Nerv Ment Dis 150:404–412, 1970.
66. Botez MI, Peyronnard JM, Bachevalier J, Charron L: Polyneuropathy and folate deficiency. Arch Neurol 35:581–584, 1978.
67. Botez MI, Fontaine F, Botez T, Bachevalier J: Folate-responsive neurological and mental disorders: Report of 16 cases. Eur Neurol 16:230–246, 1977.
68. Botez MI, Peyronnard JM, Charron L. Polyneuropathies responsive to folic acid therapy. In: *Folic Acid in Neurology, Psychiatry and Internal Medicine*, MI Botez, EH Reynolds (eds). New York: Raven Press, 1979, pp 401–412.
69. Botez MI, Bachevalier J: The blood-brain barrier and folate deficiency. Am J Clin Nutr 34:1725–1730, 1981.
70. Botez MI, Lambert B: A possible correlation between restless legs syndrome and folate deficiency in pregnancy. Nutr Rep Int 18:143–146, 1978.
71. Botez MI, Bachevalier J: Folic acid absorption test in various clinical conditions. Ann Nutr Metab 25:389–395, 1981.
72. Rodriguez MS: A conspectus of research on folacin requirements of man. J Nutr 108:1987–2075, 1978.
73. Halsted CH: Intestinal absorption and malabsorption of folates. Annu Rev Med 31:79–87, 1980.
74. Campbell RE, Pruitt CFW: The effect of vitamin B-12 and folic acid in the treatment of viral hepatitis. Am J Med Sci 229:8, 1955.
75. Leevy CM, TenHove W, Frank O, Baker H: Folic acid deficiency and hepatic DNA synthesis. Proc Soc Exp Biol Med 117:746, 1964.
76. Tamura T, Stokstad ELR: Increased folate excretion in acute hepatitis. Am J Clin Nutr 30:1378–1379, 1977.
77. Zittoun J, Marquet J, Zittoun R: Effect of folate and cobalamin compounds on the deoxyuridine suppression test in vitamin B-12 and folate deficiency. Blood 51:119–128, 1978.
78. Brattström LE, Hultberg BL, Hardebo JE: Folic acid responsive postmenopausal homocysteinemia. Metabolism 34:1073–1077, 1985.
79. Poncz M, Colman N, Herbert V, et al. Therapy of congenital folate malabsorption. J Pediatr 98:76–79, 1981.

80. Freeman JM, Finkelstein JD, Mudd SH: Folate-responsive homocystinuria and "schizophrenia." N Engl J Med 292:491–496, 1975.
81. Wong PWK, Justice P, Hruby M, et al.: Folic acid nonresponsive homocystinuria due to methylenetetrahydrofolate reductase deficiency. Pediatrics 59:749–756, 1977.
82. Tauro GP, Danks DM, Rowe PB, et al.: Dihydrofolate reductase deficiency causing megaloblastic anemia in two families. N Engl J Med 294:466–470, 1976.
83. Walters T: Congenital megaloblastic anemia responsive to N^5-formyl tetrahydrofolic acid administration. J Pediatr 70:686–687, 1967.
84. Harpey J-P, Rosenblatt DS, Cooper BA, et al.: Homocystinuria caused by 5,10-methylenetetrahydrofolate reductase deficiency: A case in an infant responding to methionine, folinic acid, pyridoxine, and vitamin B-12 therapy. J Pediatr 98:275–278, 1981.
85. Taunton OD, Greene HL, Stifel FB, et al.: Fructose-1, 6-diphosphatase deficiency, hypoglycemia, and response to folate therapy in a mother and her daughter. Biochem Med 19:260–276, 1978.
86. Branda RF, Moldow CF, MacArthur JR, et al.: Folate-induced remission in aplastic anemia with familial defect of cellular folate uptake. N Engl J Med 298:469–475, 1978.
87. Yunis JJ, Soreng AL: Constitutive fragile sites and cancer. Science 226:1199–1204, 1984.
88. Yunis JJ, Hoffman WR: Birth of an errant cell. The Sciences, Nov/Dec 1985, pp 28–33.
89. Butterworth CE Jr, Hatch KD, Gore H, et al.: Improvement in cervical dysplasia associated with folic acid therapy in users of oral contraceptives. Am J Clin Nutr 35:73–82, 1982.
90. Richart RM, Barron BA: A follow-up study of patients with cervical dysplasia. Am J Obstet Gynecol 105:386–393, 1969.
91. Stern E, Neely PM: Carcinoma and dysplasia of the cervix: A comparison of rates for new and returning populations. Acta Cytol 7:357–361, 1963.
92. Lerch V, Okagaki T, Austin JH, et al.: Cytologic findings in progression of anaplasia (dysplasia) to carcinoma in situ: Progress report. Acta Cytol 7:183–186, 1963.
93. Koss LG, Stewart FW, Foote FW, et al.: Some histological aspects of behavior of epidermoid carcinoma in situ and related lesions of the uterine cervix. Cancer 9:1160–1211, 1963.
94. Heimburger DC, Krumdieck CL, Alexander CR, et al.: Localized folic acid deficiency and bronchial metaplasia in smokers: Hypothesis and preliminary report. Nutr Int 3:54–60, 1987.
95. Heimburger DC, Krumdieck CL, Butterworth CE Jr: Role of folate in prevention of cancers of the lung and cervix. J Am Coll Nutr 6:425A, 1987.
96. Heimburger DC, Alexander CB, Birch R, et al.: Improvement in bronchial squamous metaplasia in smokers treated with folate and vitamin B-12. Am J Clin Nutr 45:866A, 1987.
97. Hunter R, Barnes J, Oakley HF, Matthews DM: Toxicity of folic acid given in pharmacological doses to healthy volunteers. Lancet 1:61, 1970.
98. Theuer RC, Vitale JJ: Drug and nutrient interactions In: *Nutritional Support of Medical Practice*, HA Schneider, CE Anderson, DB Coursin (eds). New York: Harper & Row, 1977, pp 297–305.
99. Stein GM, Lewis H: Oral changes in a folic acid deficient patient precipitated by anticonvulsant drug therapy. J Periodontol 44:645–650, 1973.
100. Reynolds EH: Mental effects of anticonvulsants and folic acid metabolism. Brain 91:197–214, 1968.
101. Roe DA: Dietary control of human studies related to aging and drug disposition or response. J Am Coll Nutr 1:199–205, 1982.
102. Roe DA: Nutrient and drug interactions. Nutr Rev 42:141–154, 1984.

103. Darcy-Vrillon B, Selhub J, Rosenberg IH: Analysis of sequential events of folypolyglutamate absorption. Fed Proc 46:1003A, 1987.
104. Zimmerman J, Selhub J, Rosenberg IH: Competitive inhibition of folate absorption by dihydrofolate reductase inhibitors, trimethoprim and pyrimethamine. Am J Clin Nutr 46:518–522, 1987.
105. Russell RM, Golner RB, Krasinski SD: Effect of acid lowering agents on folic acid absorption. Fed Proc 46:1159A, 1987.
106. Mason JB, Rosenberg IH: The effect of aspirin on intestinal absorption of folate in humans. Am J Clin Nutr 45:848A, 1987.
107. Milne DB, Canfield WK, Mahalko JR, Sandstead HH: Effect of folic acid supplements on zinc, copper, and iron absorption and excretion. Am J Clin Nutr 39:535–539, 1984.
108. Simmer K, Iles CA, James C, Thompson RPH: Are iron-folate supplements harmful? Am J Clin Nutr 45:122–125, 1987.
109. Keating JN, Wada L, Stokstad ELR, King JC: Folic acid: Effect on zinc absorption in humans and the rat. Am J Clin Nutr 46:835–839, 1987.

10
Vitamin B-12

Introduction 179
Absorption, Metabolism, and Excretion 181
Dosage Forms 182
Clinical Studies 182
    Pernicious Anemia 182
    Other Malabsorptive States 183
    Genetic Disease 184
    Other Uses and Proposed Uses 185
Toxicity and Side Reactions 185
Interactions 185
References 186

INTRODUCTION

The term vitamin B-12 is ordinarily applied to the most common pharmaceutical form of the vitamin, i.e., cyanocobalamin (see Fig. 1 for structural formula). The term may also be used in a generic sense to designate all cobalamins having the biologic activity of cyanocobalamin, for example, hydroxocobalamin (vitamin B-12a), aquacobalamin (vitamin B-12b), and nitrocobalamin (vitamin B-12c), in which the CN group of cyanocobalamin is replaced by OH, H_2O, and NO_2, respectively. Pure synthetic cyanocobalamin occurs as dark red crystals that are slightly soluble in water, forming neutral solutions. Cyanocobalamin is stable in solution even at autoclave temperatures. Vitamin B-12 is the only vitamin containing a metal (cobalt), but cobalt itself does not have B-12 or other nutrient activity in man [1,2].

In the body, vitamin B-12 is converted to two coenzymes. One coenzyme has the 5'-deoxyadenosyl radical in place of the CN group in the above figure. This coenzyme is essential for the conversion of methylmalonyl coenzyme A to succinyl coenzyme A, which enters the Krebs cycle for generation of energy (ATP). It is also essential for the reduction of ribonucleotide triphosphates to deoxyribonucleotide triphosphates, needed for DNA synthesis.

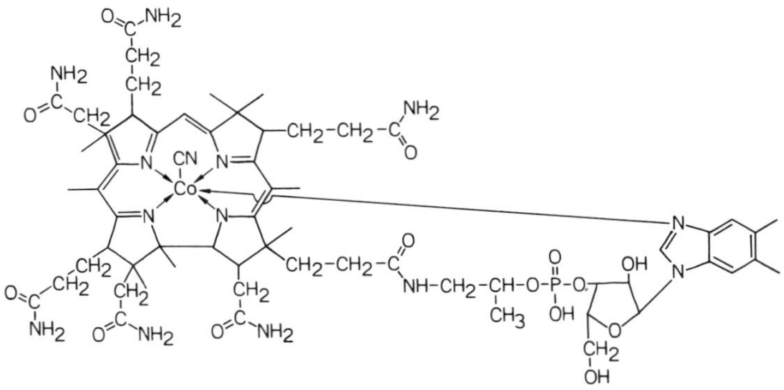

Fig. 1. Structural formula of the most common pharmaceutical form of vitamin B-12, cyanocobalamin.

The other B-12 coenzyme is methylcobalamin, with CH_3 replacing CN in the above figure. This coenzyme functions in the transfer of a methyl group from methyl tetrahydrofolate to homocysteine, forming methionine and active tetrahydrofolate [1,2]. This reaction is essential for regenerating tetrahydrofolate in a form useful for transfer of 1-carbon moieties; therefore, a secondary tetrahydrofolate deficiency is a clinical consequence of vitamin B-12 deficiency [2]. Vitamin B-12 is also essential in the metabolism of valine and isoleucine [1] and of odd-carbon-number fatty acids [1–5], and for the maintenance of myelin [6–8].

Clinical deficiency of vitamin B-12 is usually secondary to abnormality of gastrointestinal function [1,2]. Deficiency due to dietary deprivation is very rare, but is occasionally seen in adults (vegans) who have followed a strict egg- and milk-free vegetarian diet for years. If these individuals previously consumed animal proteins, their accumulated liver stores, which are normally about 1,000-fold more than the daily requirement, may protect them for several years [9]. On the other hand, infants born to mothers who are strict vegetarians may manifest signs of B-12 deficiency during the first few months of life [10,11]. Because of the essentiality of vitamin B-12 for regeneration of active tetrahydrofolate, the manifestations of B-12 deficiency share several of the aspects of folic acid deficiency, e.g., megaloblastic anemia and hypersegmentation of neutrophils. These manifestations can be overcome by high daily doses of folic acid, but unless the underlying B-12 deficiency is rectified, there may be progressive central and peripheral nerve damage. Subacute combined degeneration of the spinal cord and peripheral nerve dysfunction are treatable by cyanocobalamin injections, but long-maintained chronic damage may not be completely reversible [1,2].

Increased segmentation of neutrophils [12,13] and macrocytosis (elevation of mean red cell volume, MCV) [13–15] provide important early clues to possible vitamin B-12 deficiency, even before hemoglobin levels decline below the normal range. Serum levels of cobalamins are normally 200–900 pg/ml; values below 100 pg/ml are usually considered diagnostic of B-12 deficiency [1,2], but Urban et al. [12] noted hypersegmentation of neutrophils when levels fell below 160 pg/ml.

Dietary vitamin B-12 is supplied by animal protein foods, including meat, fish, eggs, and milk. Liver is an especially good source. Fruit and vegetables contain trace amounts at best, possibly of microbial origin [1]. The vitamin B-12 RDA for adults is 3 mcg/d and for pregnant and lactating women, 4 mcg/d [16]. These intakes are easily met by diets providing regular intakes of modest amounts of animal protein foods, including vegetarian diets containing eggs, milk, and cheese [17,18].

ABSORPTION, METABOLISM, AND EXCRETION

Vitamin B-12 in food occurs almost entirely in its coenzyme forms, tightly bound to the enzymes methionine synthetase and methylmalonyl coenzyme-A mutase. The vitamin is freed as hydroxocobalamin during protein digestion in the stomach, after which it binds to R protein originating in saliva. Another protein, specifically required for intestinal absorption of vitamin B-12 and termed "intrinsic factor," is secreted by the parietal cells, but under acid conditions does not form as stable a complex with B-12 as R-protein does. In the small intestine, at higher pH and after the action of pancreatic proteases, intrinsic factor forms a strong complex with vitamin B-12 that remains intact until it reaches the distal intestine, which has specific receptors for absorption of this complex [1,2,19,20]. Administered in pharmacologic dosages, about 1% of cyanocobalamin may be absorbed independently of the intrinsic factor mechanism [2].

After passage through the enterocytes, vitamin B-12 is complexed in the portal blood with plasma proteins called transcobalamins. Transcobalamin I, which carries most of the B-12, is an R protein from which the vitamin is liberated only very slowly and that does not appear to have a key role, if any, in delivering the vitamin to cells. Transcobalamin II is the most important transport protein for B-12, delivering the vitamin to many if not all cells of the body through cell surface receptors specific for this protein. Transcobalamin III, another R-protein, delivers cobalamin rapidly and exclusively to hepatocytes [19,21,22]. The transcobalamin-vitamin B-12 complex is internalized by endocytosis and degraded within lysosomes, with liberation of the vitamin intracellularly [19].

Noncobalamin vitamin B-12 analogs that bind to R-proteins but not to intrinsic factor have been found in human plasma [23], red cells, liver, and brain [24]. Their function, if any, is unknown, but to avoid the possibility that such analogs may mask vitamin B-12 deficiency, radiodilution kits for B-12 assay sold in the United States now contain only purified intrinsic factor as the cobalamin-binding protein rather than a mixture of intrinsic factor with R-proteins [1]. Cobalamin analog concentrations are increased in folate deficiency [25].

Cobalamins, occurring mostly as the coenzymes, adenosylcobalamin and methylcobalamin, are found in all body cells. Total body content in adults is 2–5 mg, most of which is stored in the liver [1].

Some cobalamin is excreted in bile, from which 65–75% or more is reabsorbed in the distal small intestine by the intrinsic factor mechanism, providing efficient conservation of the vitamin. Feces may contain vitamin B-12 that has escaped the enterohepatic cycle, as well as B-12 newly synthesized by gut microflora. Trace amounts of cobalamin are excreted in the urine; if pharmacologic doses are administered, urinary excretion of the vitamin increases markedly [1,2].

DOSAGE FORMS

The most common pharmaceutical form of vitamin B-12 is cyanocobalamin. This is the form used in multivitamin tablets, capsules, and solutions. For parenteral use, it is supplied as clear red solutions in unit dosages of 30, 100, or 1,000 mcg/ml; administration is by intramuscular or deep subcutaneous route [26,27]. Injectable hydroxocobalamin is also available, but no therapeutic advantage over cyanocobalamin has been reported except for treatment of tobacco amblyopia [28].

Vitamin B-12 (cyanocobalamin) is also available over the counter as tablets in 50-, 100-, 250-, 500-, and 1,000-mcg strengths. A 1,000-mcg lozenge for sublingual use is also marketed.

CLINICAL STUDIES
Pernicious Anemia

Megaloblastic anemia caused by failure of intrinsic factor production in the stomach is called pernicious anemia. The tendency to develop this disorder is heritable and usually involves atrophy of the gastric mucosa with achlorhydria. Hepatic stores of vitamin B-12 may prevent appearance of anemia for several years after development of instrinsic factor deficiency. Eventually the plasma vitamin B-12 level falls below 100 pg/ml, and the anemia comes on, accompanied by such clinical signs and symptoms as fatigue, dyspnea, anorexia, weight loss, and pale and smooth tongue. In addition to macrocytic anemia, there is hypersegmentation of neutrophils, leukopenia, and thrombocytopenia [1,2,27].

Definitive diagnosis of vitamin B-12 malabsorption is by the Schilling test [29], which measures 24-hour urinary excretion of a small oral dose of radiolabeled cyanocobalamin after a flushing dose of 1,000 mcg parenteral cyanocobalamin given simultaneously or within 2 hours after the oral dose. Significant urinary radioactivity is found in normal patients, but very little in pernicious anemia patients. Those patients whose anemia is caused by folic acid deficiency have similar hematologic indices, but if their intrinsic factor production is normal, urinary radioactivity excretion will be normal [1,2,27].

It is important to distinguish vitamin B-12 deficiency from folic acid deficiency to prevent the neurologic damage that occurs with prolonged B-12 deficiency. Mistaken diagnosis and treatment as folic acid deficiency can correct the anemia while allowing nerve damage to proceed [1,2,27].

Macrocytosis may precede anemia and is an important sign of potential vitamin B-12 deficiency. A number of clinicians have called attention to the need for follow-up of this finding, occurring together with subnormal (below 200 pg/ml) levels of serum B-12, to prevent unnecessary delays in institution of cyanocobalamin therapy; prompt treatment will prevent development of neurologic symptoms [30–32].

The usual treatment for pernicious anemia is monthly injections, by intramuscular or deep subcutaneous route, of 1,000 mcg parenteral cyanocobalamin. This is considered to give more reliable and consistent results than administering vitamin B-12 orally with pharmaceutically prepared intrinsic factor or giving extremely large oral doses daily to obtain absorption by diffusion. If the patient is in a critically ill state, he should be given both vitamin B-12 (100 mcg) and folic acid (15 mg) intramuscularly, followed by 5 mg oral folic acid and 100 mcg IM cyanocobalamin daily for 1 week; etiologic diagnosis may follow after the hematologic state has improved [2,27].

It must be emphasized to the pernicious anemia patient that he will require vitamin B-12 therapy for the rest of his life. Cessation of therapy will eventually result in relapse, even though it may be delayed for 2 years or more [33].

Other Malabsorptive States

Patients who have had total or partial gastrectomies should have their plasma vitamin B-12 levels measured periodically, as their absorption of the vitamin from food will be affected by loss of intrinsic factor-secreting mucosa. Roos [34,35] noted peripheral neuropathy in such patients, as well as low serum

B-12 values (below 200 pg/ml), at varying periods (average 10 years) after gastrectomy. Treatment is with parenteral cyanocobalamin, as with pernicious anemia.

If there is small intestine bacterial overgrowth, as in surgically produced blind loop, strictures, anastomoses, or diverticula, ingested vitamin B-12 appears to be metabolized by the bacteria to inactive analogs, so that low plasma B-12 levels may result [1,36].

Since the vitamin B-12 complex with intrinsic factor is absorbed mainly in the distal small intestine, there may be loss of B-12 absorption with ileitis or ileal resection, as well as in tropical sprue and celiac disease. Plasma levels of vitamin B-12 should be monitored at intervals in these patients [1,2].

In exocrine pancreatic insufficiency, e.g., in chronic pancreatitis, 40–45% of patients show malabsorption of crystalline cobalamin by the Schilling test, but vitamin B-12 from food seems to be adequately enough absorbed to prevent vitamin B-12 deficiency in most cases. In these patients, the normal degradation of R-protein (haptocorrin, cobalophilin) that takes place in the duodenum is incomplete, interfering with full take-up of cobalamin by intrinsic factor and sometimes with enterohepatic circulation of cobalamin [37].

Russell et al. [38] report that elderly subjects with mild atrophic gastritis and hypochlorhydria absorb protein-bound (food) vitamin B-12 significantly less well than healthy controls. Tetracycline treatment (250 mg q.i.d.) normalized their absorption of protein-bound vitamin B-12, while having no effect on the controls. Both groups absorbed crystalline vitamin B-12 normally before and after antibiotic treatment.

Genetic Disease

Rare genetic disorders of vitamin B-12 metabolism have been reported for infants, including deficiencies in synthesis of adenosylcobalamin and methylcobalamin. Some cases are responsive to high-dose vitamin B-12. For example, Schuh et al. [39] and Rosenblatt et al. [40] describe cases of vitamin B-12 dependency in young infants caused by inadequacy of methylcobalamin synthesis, resulting in homocystinuria and megaloblastic anemia; the condition responded well to 1,000 mcg of hydroxocobalamin intramuscularly twice weekly. Cyanocobalamin was not effective. These investigators were also able to secure the birth of a later sibling in a fully normal state by administering 1,000 mcg hydroxocobalamin twice weekly to the mother from the 25th week of gestation, thereby preventing the mental retardation that had occurred in the earlier sibling. The second child showed normal growth and development at 6 months of age on continued therapy, with twice weekly injections of 1,000 mcg hydroxocobalamin [40].

Infants with hereditary transcobalamin II deficiency develop severe megaloblastic anemia and pancytopenia while having normal plasma levels of vitamin B-12. Excellent clinical responses have been reported to therapy with 1,000 mcg cyanocobalamin given by intramuscular injection weekly [41,42]. Seligman et al. [43] describe a patient with pancytopenia whose transcobalamin II levels were normal but did not bind vitamin B-12; therapy consisted of 1,000 mcg cyanocobalamin IM weekly.

Other Uses and Proposed Uses

There are a number of uses for parenteral vitamin B-12 that are based on clinical impression rather than on firm evidence gained by controlled trials. Probably the most common of these is as a "tonic" for treatment of chronic fatigue and depression, particularly in the elderly. It has also been administered as therapy for trigeminal neuralgia, multiple sclerosis, and other neuropathies, as well as for various psychiatric disorders [27].

Evans et al. [44] describe manifestations of organic psychosis in two women who had low serum B-12 levels but no anemia or spinal cord symptoms. One patient, aged 47 years, had hallucinations and a confusinal state. The other, aged 58 years, showed increasingly bizarre and and irascible behavior. Both responded well to cyanocobalamin injections, given daily for several days, then biweekly, recovering normal mental and emotional function within a few weeks, and are maintained on monthly injections of cyanocobalamin. These investigators recommend consideration of possible vitamin B-12 deficiency in all patients with organic psychiatric symptoms.

TOXICITY AND SIDE REACTIONS

Neither oral nor injectable cyanocobalamin has been found to demonstrate any toxicity when administered in quantities several thousand times larger than the daily requirement. A rare allergic reaction has been reported that may have been due to impurities in the particular injectable preparation of cyanocobalamin. Depot preparations may be painful on injection and have shown no advantage therapeutically over the usual parenteral vitamin B-12 [2].

One manufacturer cautions that patients who have early Leber's disease (hereditary optic nerve atrophy) may suffer severe acute optic atrophy when treated with parenteral vitamin B-12. They also caution that vitamin B-12 may unmask the signs of polycythemia vera.

INTERACTIONS

The metabolism of folic acid is strongly dependent on adequacy of vitamin B-12 nutriture through the required presence of the coenzyme methylcobalamin for regeneration of active tetrahydrofolate from methyl tetrahydrofolate, as discussed earlier in this chapter. Another interrelationship has been demonstrated by Sheppard and Ryrie [25], who observed that folic acid-deficient patients have increased levels of inactive cobalamin analogs in their plasma and that folate supplementation reduced the levels of analogs and increased plasma levels of vitamin B-12.

Absorption of vitamin B-12 from food is impaired by several drugs, including para-aminosalicylic acid, colchicine, neomycin, metformin, phenformin, and potassium chloride [1].

Nitrous oxide anesthesia impairs vitamin B-12 and secondarily folate metabolism by inhibiting synthesis of methylcobalamin. Synthesis of adenosylcobalamin is not affected. Supplementation with both vitamin B-12 and folic acid is advisable after nitrous oxide anesthesia [1].

An early report that vitamin B-12 might be unstable in the presence of high-dose vitamin C proved groundless, an error in analytical procedure having been responsible for this mistaken conclusion [45,46]. Baker et al. [47] analyzed plasma vitamin B-12 levels in ten healthy adults who had been taking 2–10 g/d of vitamin C for at least 2 years, using a microbiologic assay that does not respond to pseudo-vitamin B-12s; in addition, they were taking "therapeutic-strength" multi-vitamin preparations daily. Compared with control subjects who did not take vitamin supplements, plasma vitamin B-12 levels in the vitamin C-supplemented subjects were significantly higher, indicating no adverse effect of high-dose vitamin C on vitamin B-12 status.

Hydroxocobalamin has been used to treat cyanide intoxication, as it combines in the body with cyanide to produce cyanocobalamin, which is excreted in the urine. It may be used as an adjunct to sodium nitroprusside treatment of severe hypertension in order to remove the cyanide released by the metabolic degradation of nitroprusside [1].

REFERENCES

1. Ellenbogen L: Vitamin B-12. In: *Handbook of Vitamins*, LJ Machlin (ed). New York: Marcel Dekker, 1984, pp 497–547.
2. Herbert V, Colman N, Jacob E: Folic acid and vitamin B-12. In: *Modern Nutrition in Health and Disease*, 6th ed., RS Goodhart, ME Shils (eds). Philadelphia: Lea & Febiger, 1980, pp 229–259.
3. Bitman J, Weyant JR, Wood DL, Wrenn TR: Necessity of vitamin B-12 for growth of rats fed on an odd- or even-carbon-number fat. Br J Nutr 39:615–626. 1978.
4. Peifer JJ, Lewis RD: Effects of vitamin B-12 deprivation on phospholipid fatty acid patterns in liver and brain of rats fed high and low levels of linoleate in low methionine diets. J Nutr 109:2160–2172, 1979.
5. Åkesson B, Fehling C, Jägerstad M: Lipid composition and metabolism in liver and brain of vitamin B-12-deficient rat sucklings. Br J Nutr 41:263, 1979.

6. Scott JM, Dinn JJ, Wilson P, Weir DG: Pathogenesis of subacute combined degeneration: A result of methyl group deficiency. Lancet 2:334–337, 1981.
7. Scott JM, Weir DG: The methyl folate trap. Lancet 2:337–340, 1981.
8. Van der Westhuyzen J, Cantrill RC, Fernandes-Costa F, Metz J: Effect of a vitamin B-12-deficient diet on lipid and fatty acid composition of spinal cord myelin in the fruit bat. J Nutr 113:531–537, 1983.
9. Murphy MF: Vitamin B-12 deficiency due to a low-cholesterol diet in a vegetarian. Ann Intern Med 94:57–58, 1981.
10. Davis JR, Goldenring J, Lubin BH: Nutritional vitamin B-12 deficiency in infants. Am J Dis Child 135:566–567, 1981.
11. Clinical nutrition case: Vitamin B-12 deficiency in the breast-fed infant of a strict vegetarian. Nutr Rev 37:142–144, 1979.
12. Urban G, Pietrzik K, Hötzel D: Radiological investigations of cobalamin supply with regard to folate status. Int J Vitam Nutr Res 51:124–131, 1981.
13. Clinical nutrition case: Macrocytosis, mild anemia and delay in the diagnosis of pernicious anemia. Nutr Rev 37:47–48, 1979.
14. Hall CA: Vitamin B-12 deficiency and early rise in mean corpuscular volume. JAMA 245:1144–1146, 1981.
15. Savage D, Lindenbaum J: Relapses after interruption of cyanocobalamin therapy in patients with pernicious anemia. Am J Med 74:765–772, 1983.
16. Food and Nutrition Board: *Recommended Dietary Allowances,* 9th ed. Washington, DC: National Academy of Sciences, 1980.
17. Lewis NM, Kies C, Fox HM: Vitamin B-12 status of lacto-ovo-vegetarian and omnivore subjects fed controlled lacto-vegetarian, vegan and omnivore diets. Nutr Rep Int 34:197–206, 1986.
18. Helman AD, Darnton-Hill I: Vitamin and iron status in new vegetarians. Am J Clin Nutr 45:785–789, 1987.
19. Allen RH: Intestinal absorption and plasma transport of cobalamin (vitamin B-12). Fed Proc 43:2424–2425, 1984.
20. Schilling RF: The role of the pancreas in vitamin B-12 absorption. Am J Hematol 14:197–199, 1983.
21. Carmel R: Cobalamin-binding proteins of man. In: *Contemporary Hematology/Oncology,* R Silber, AS Gordon, J LoBue, FM Muggia (eds). New York: Plenum Press, 1981, pp 79–129.
22. Carmel R: The distribution of endogenous cobalamin among cobalamin-binding proteins in the blood in normal and abnormal states. Am J Clin Nutr 41:713–719, 1985.
23. Kolhouse JF, Kondo H, Allen NC, et al.: Cobalamin analogues are present in human plasma and can mask cobalamin deficiency because current radioisotope dilution assays are not specific for true cobalamin. N Engl J Med 299:785–792, 1978.
24. Kanazawa S, Herbert V: Noncobalamin vitamin B-12 analogues in human red cells, liver, and brain. Am J Clin Nutr 37:774–777, 1983.
25. Sheppard K, Ryrie D: Changes in serum levels of cobalamin and cobalamin analogues in folate deficiency. Scand J Haematol 25:401–406, 1980.
26. *U.S. Pharmacopeia,* 21st revision, Rockville, MD: U.S. Pharmacopeial Convention, 1985.
27. Hillman RS: Vitamin B-12, folic acid, and the treatment of megaloblastic anemias. In: *Goodman and Gilman's The Pharmacological Basis of Therapeutics,* 7th ed., AG Gilman, LS Goodman, TW Rall, F Murad (eds). New York: Macmillan, 1985, pp 1323–1337.
28. Chisholm JA, Bronte-Stewart J, Foulds WS: Hydroxocobalamin versus cyanocobalamin in treatment of amblyopia induced by tobacco. Lancet 2:450, 1967.

29. Schilling RF: Intrinsic factor studies. II. The effect of gestric juice on the urinary excretion of radioactivity after the oral administration of radioactive vitamin B-12. J Lab Clin Med 42:860–866, 1953.
30. Carmel R: Macrocytosis, mild anemia, and delay in the diagnosis of pernicious anemia. Arch Intern Med 139:47–50, 1979.
31. Shojania AM: Physicians' management of suspected vitamin B-12 deficiency. Can Med Assoc J 123:1127–1130, 1980.
32. Hall CA: Vitamin B-12 deficiency and early rise in mean corpuscular volume. JAMA 245:1144–1146, 1981.
33. Savage D, Lindenbaum J: Relapses after interruption of cobalamin therapy in patients with pernicious anemia. Am J Med 74:765–772, 1983.
34. Roos D: Electrophysiological findings in gastrectomized patients with low serum B-12. Acta Neurol Scand 56:247–255, 1977.
35. Roos D: The vibration perception threshold in gastrectomized patients with low serum B-12. Acta Neurol Scand 56:551–562, 1977.
36. Anon.: Vitamin B-12 analogues and intestinal bacteria. Nutr Rev 37:45–46, 1979.
37. Guéant JL, Djalali M, Aouadi R, et al.: In vitro and in vivo evidences that the malabsorption of cobalamin is related to its binding on haptocorrin (R binder) in chronic pancreatitis. Am J Clin Nutr 44:265–277, 1986.
38. Russell RM, Suter PM, Golner B, Samloff IM: Decreased bioavailability of protein bound vitamin B-12 in elderly with mild atrophic gastritis: Reversal by antibiotics. J Am Coll Nutr 6:434A, 1987.
39. Schuh S, Rosenblatt DS, Cooper BA, et al.: Homocystinuria and megaloblastic anemia responsive to vitamin B-12 therapy. N Engl J Med 310:686–690, 1984.
40. Rosenblatt DS, Cooper BA, Schmutz SM, et al.: Prenatal vitamin B-12 therapy of a fetus with methylcobalamin deficiency (cobalamin E disease). Lancet 1:1127–1129, 1985.
41. Rana SR, Colman N, Kong-Oo G, et al.: Transcobalamin II deficiency associated with unusual bone marrow findings and chromosomal abnormalities. Am J Hematol 14:89–96, 1983.
42. Niebrugge DJ, Benjamin DR, Christie D, Scott CR: Hereditary transcobalamin II deficiency presenting as red cell hypoplasia. J Pediatr 101:732–735, 1982.
43. Seligman PA, Steiner LL, Allen RH: Studies of a patient with megaloblastic anemia and an abnormal transcobalamin II. N Engl J Med 303:1209–1212, 1980.
44. Evans DL, Edelsohn GA, Golden RN: Organic psychosis without anemia or spinal cord symptoms in patients with vitamin B-12 deficiency. Am J Psychiatry 140:218–221, 1983.
45. Hogenkamp HPC: The interaction between vitamin B-12 and vitamin C. Am J Clin Nutr 33:1–3, 1980.
46. Marcus M, Prabhudesai M, Wassef S: Stability of vitamin B-12 in the presence of ascorbic acid in food and serum: Restoration by cyanide of apparent loss. Am J Clin Nutr 33:137–143, 1980.
47. Baker H, Pauling L, Frank O: Mega-ascorbate taken with other vitamins permits elevation of circulating vitamins including B-12 in humans. Nutr Rep Int 23:669–677, 1981.

11
Pantothenic Acid

Introduction 189
Absorption, Metabolism, and Excretion 190
Dosage Forms... 190
Clinical Studies 191
Toxicity and Side Effects 191
Interactions... 191
References 191

INTRODUCTION

D-Pantothenic acid is a condensation product of beta-alanine and pantoic acid (see Fig. 1). When isolated in pure form, it is a slightly yellow oily liquid that is soluble in water. The amide bond is fairly stable in neutral solution, but may be hydrolyzed in acid or alkaline solution. There is no enzyme system in mammals either for forming the amide bond between beta-alanine and pantoic acid or for hydrolyzing it [1,2].

In the body, pantothenic acid is used as a building block in the formation of coenzyme A and 4'-phosphopantetheine, the cofactor for acyl carrier protein. Coenzyme A is involved in the transfer of methyl groups to the citric acid cycle for complete oxidation with generation of energy and generally as an acyl group carrier, e.g., in enzymatic reactions involved in the synthesis of fatty acids, cholesterol, and other sterols; in oxidation of fatty acids, pyruvate, and alpha-keto-glutarate; and in biologic acetylations. Acyl carrier protein is required for biosynthesis of fatty acids [1,2].

Pantothenic acid, usually as coenzyme A, is found in all body tissues and is at especially high levels in the liver. It occurs in free form in plasma and in both free and coenzyme A form in red blood cells, with the latter predominating [1].

The vitamin is widely distributed in foods in both free and bound forms, but mostly the latter. Good sources include eggs, organ meats, beef, chicken, yeast, peas, beans, nuts, whole grains, tomatoes, broccoli, and avocados [1,3].

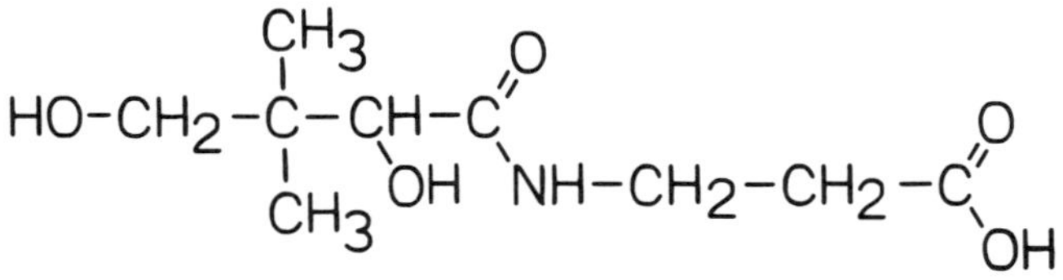

$$HO-CH_2-\underset{\underset{CH_3}{|}}{\overset{\overset{CH_3}{|}}{C}}-\underset{OH}{\overset{|}{CH}}-C\overset{\nearrow O}{\underset{\searrow NH-CH_2-CH_2-C\overset{\nearrow O}{\searrow OH}}{}}$$

Fig. 1. Structural formula of pantothenic acid.

Food pantothenic acid is about 50% bioavailable. Highly processed foods are low in pantothenic acid, and ordinary methods of preparation may cause 20–40% losses.

Clinical deficiency is not observed in the developed countries, and usual Western diets are believed to meet the requirement. Only provisional recommendations have been made for daily intake, ranging from 2 mg/d for the infant to 4–7 mg/d for the adult [4].

ABSORPTION, METABOLISM, AND EXCRETION

In laboratory rats, coenzyme A present in foods is hydrolyzed in the gut to pantotheine and pantothenic acid, which are transported into the intestinal mucosa by simple diffusion [5]. Intestinal tissue, which contains high levels of pantetheinase, degrades pantetheine to pantothenic acid. Pantothenic acid is transported to the tissues in the circulation, and 5 hours after administration of radiolabeled pantothenic acid, 40% of the label was found in muscle and 10% in liver [5]. Conversion to coenzyme A and acyl carrier protein takes place in the tissues. There is evidently a constant synthesis and degradation (turnover) of these bound forms of pantothenic acid, as about the same amount of pantothenic acid is excreted in the urine daily as is ingested [1].

DOSAGE FORMS

Pantothenic acid is available over the counter as calcium pantothenate tablets in 50-, 100-, 250-, and 500-mg strengths. The calcium salt is widely used in multivitamin tablets and capsules. For use in liquid vitamin preparations, oral or parenteral, the reduced form, D-pantothenyl alcohol (panthenol, dexpanthenol) is commercially available; it is oxidized to pantothenic acid in the body.

CLINICAL STUDIES

No well-substantiated clinical applications have been reported for pantothenic acid or its salts. Dexpanthenol has been given in large doses intramuscularly (adult dose, 250–500 mg at a time, repeated every 6 hours for 48–72 hours) after abdominal surgery to minimize paralytic ileus; the mechanism is reported to be stimulation of acetylation of choline to acetylcholine in the intestinal wall [1]. Further clinical verification is needed for this use.

In studies with rabbits, high-dose parenteral pantothenic acid pre- and postoperatively was beneficial in accelerating wound healing, markedly increasing aponeurosis strength, and slightly but not significantly increasing skin strength [6]. No clinical follow-up has been reported as of this writing.

Elevated urinary excretion of pantothenic acid has been observed in diabetes mellitus, and low coenzyme A activity is found in intestinal mucosa of patients with chronic ulcerative or granulomatous colitis, suggesting that these groups might benefit from supplemental pantothenic acid. Chronic alcoholics may also be in a low pantothenic acid status [1]. The "burning feet" syndrome has been treated with calcium pantothenate supplements, based on development of this symptom during pantothenic acid deficiency [1].

TOXICITY AND SIDE EFFECTS

Pantothenic acid toxicity has not been reported in humans. The only side effect reported from daily intakes of 10–20 mg pantothenic acid as the calcium salt has been an occasional diarrhea [1]. Coburn et al. [7], in testing the effects of megadose multivitamins on mental performance of mentally retarded young adults (with negative results), administered 600 mg/d of calcium pantothenate for 20 weeks without adverse effect. Dexpanthenol solutions administered intramuscularly in doses of 250 or 500 mg (1 or 2 ml) should be used in accordance with the instructions of the manufacturer (see current PDR®).

INTERACTIONS

Oral contraceptives (estrogen-progestin) appear to increase the pantothenic acid requirement [1].

REFERENCES

1. Fox HM: Pantothenic acid. In: *Handbook of Vitamins*, LJ Machlin (ed). New York: Marcel Dekker, 1984, pp 437–457.

2. Sauberlich HE: Pantothenic acid. In: *Modern Nutrition in Health and Disease*, 6th ed., RS Goodhart, ME Shils (eds). Philadelphia: Lea & Febiger, 1980, pp 209–216.
3. Walsh JH, Wyse BW, Hansen RG: Pantothenic acid content of 75 processed and cooked foods. J Am Diet Assoc 78:140–144, 1981.
4. Food and Nutrition Board: *Recommended Dietary Allowances*, 9th ed. Washington, DC: National Academy of Sciences, 1980.
5. Shibata K, Gross CJ, Henderson LM: Hydrolysis and absorption of pantothenate and its coenzymes in the rat small intestine. J Nutr 113:2207–2215, 1983.
6. Aprahamian M, Dentinger A, Stock-Damagé C, et al.: Effects of supplemental pantothenic acid on wound healing: Experimental study in rabbit. Am J Clin Nutr 41:578–589, 1985.
7. Coburn SP, Schaltenbrand WE, Mahuren JD, et al.: Effect of megavitamin treatment on mental performance and plasma vitamin B-6 concentrations in mentally retarded young adults. Am J Clin Nutr 38:352–355, 1983.

12
Biotin

Introduction	193
Absorption, Metabolism, and Excretion	194
Dosage Forms	195
Clinical Studies	195
Avidin-Induced Deficiency	195
Total Parenteral Nutrition	195
Seborrheic Dermatitis and Leiner's Disease	196
Biotin-Responsive Multiple Carboxylase Deficiency	196
Methylcrotonyl-CoA Carboxylase Deficiency	196
Biotinidase Deficiency	197
Hyperlipidemia	197
Diabetes Mellitus	198
Sudden Infant Death Syndrome (SIDS)	198
Toxicity and Side Effects	198
Interactions	199
References	199

INTRODUCTION

Biotin is a dicyclic compound in which imidazolidone and tetrahydrothiophene rings are fused (see Fig. 1). There is a valeric acid side chain attached to the sulfur ring, and it is through condensation of the carboxyl group of this side chain with epsilon-amino groups of lysine apoenzymes that the active biotin-dependent holoenzymes are formed. Biotin has three asymmetric carbon atoms, so that eight stereoisomers are possible, but only one, D-biotin, is ordinarily found in nature, and this is the only isomer that is biologically active. Commercially available synthetic biotin is the D-isomer [1,2].

Four biotin-dependent enzymes are found in human and animal tissues, namely pyruvate carboxylase, which catalyzes conversion of pyruvate to oxaloacetate and is essential for gluconeogenesis and lipogenesis; acetyl-CoA carboxylase, which catalyzes conversion of acetyl-CoA to malonyl-CoA, the first committed step in fatty acid synthesis; propionyl-CoA carboxylase, which

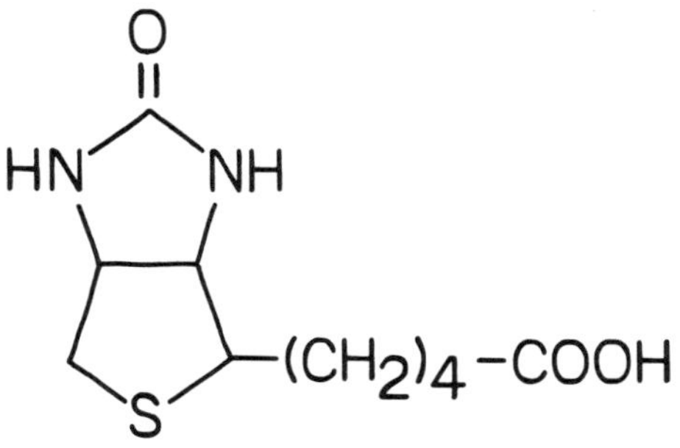

Fig. 1. Structural formula of biotin.

converts propionyl-CoA to methyl-malonyl-CoA and is essential for catabolism of propionate and higher odd-carbon-numbered fatty acids, as well as of branched-chain amino acids, methionine, and threonine; and 3-methycrotonyl-CoA carboxylase, essential for the catabolism of leucine [1,2].

Biotin deficiency of nutritional origin has rarely been observed in man except after chronic consumption of large quantities of raw egg white, which contains the biotin-binding protein avidin [1,2], or after prolonged total parenteral nutrition by alimentation fluids lacking biotin [3]. (Avidin's ability to deactivate biotin is destroyed by cooking.) An erythematous exfoliative dermatitis is the most common early clinical sign of deficiency, with alopecia developing after long-term severe deficiency.

Biotin is widely distributed in most foods, although in low concentrations. Liver and yeast are the best sources, with eggs, beans, peas, nuts, and whole grains following; meat, fruit, and refined cereal and bread products are poor sources [2]. It is thought that bacterial synthesis in the intestine can fill part of the biotin requirement. The recommended dietary intake is 35 mcg/d for infants, increasing through childhood up to the adult recommended intake of 100–200 mcg/d [4]. Usual Western diets meet these recommendations.

ABSORPTION, METABOLISM, AND EXCRETION

Protein-bound biotin in food is liberated first as biotinyl peptides and biocytin (biotinyl lysine) by the action of gastrointestinal proteases and then as free biotin by the action of intestinal biotinidase, which releases the vitamin from the epsilon-amide bond with lysine [1,2]. Animal studies indicate that biotin is absorbed in the jejunum and upper ileum by two mechanisms: one a saturable facilitated diffusion process that operates at low concentrations and the second a nonsaturable (linear) uptake operating at higher luminal concentrations of the vitamin [5,6]. Human studies indicate the presence of a high-affinity, low-capacity transport system that permits effective absorption of biotin at low concentrations in the succus entericus [7]. Preliminary evidence has been obtained that biotin is transported in the circulation by a biotin-binding glycoprotein [8]. Cellular uptake in rat hepatocytes did not appear to be energy-dependent [9]. Degradative turnover of biotin-dependent carboxylases in the tissues is thought to release biocytin and biotinyl peptides that can then be acted upon by biotinidase in plasma to release biotin for recycling back to the cell [10], accounting for the small daily requirement for the vitamin. Urinary excretion of biotin is generally less than the dietary intake, while the total of urinary and fecal excretion, because of bacterial synthesis in the gut, exceeds the amount ingested. Small amounts of biotin metabolites have been detected in urine that have not yet been identified [2].

DOSAGE FORMS

D-Biotin is available over the counter as tablets in 100-, 300-, and 600-mcg strengths. It is widely used in multivitamin formulations for both oral and parenteral administration.

CLINICAL STUDIES
Avidin-Induced Deficiency

A few cases have been reported in the literature of biotin deficiency induced by excessive consumption of raw eggs [2,11]. Erythematous exfoliative dermatitis and alopecia totalis occur, together with organic aciduria and a propensity toward ketosis. Withdrawal of raw eggs and administration of oral biotin, 1–10 mg/d, reverses the clinical and biochemical abnormalities. Doses of 5 or 10 mg/d may be used initially until skin lesions clear, after which dosage may be lowered.

Total Parenteral Nutrition

Until a few years ago, the infusates used for total parenteral nutrition (TPN) did not contain biotin. A number of reports appeared describing the onset of acute biotin deficiency in infants, children, and adults during long-term TPN [3,12–17]. Skin rash, alopecia, and neurologic symptoms were seen, as well as organic aciduria when measured. Modest biotin supplements, e.g., 60–100 mcg/d intravenously or 200 mcg/d orally, seem to reverse the biochemical abnormalities, but larger doses, for example 10 mg/d in one case [12], may be necessary for a time to obtain rapid clearance of the skin lesions. Commercial solutions for TPN now contain biotin.

Seborrheic Dermatitis and Leiner's Disease

If due to marginal biotin deficiency, seborrheic dermatitis in infants reponds readily to mg supplements of biotin orally [18–20]. One double-blind study gave negative results, perhaps because biotin deficiency may not have been the cause of the dermatitis [21].

Biotin-Responsive Multiple Carboxylase Deficiency

According to Thoene et al. [22], there appear to be two forms of this genetic disorder: the neonatal form, appearing in the first weeks of life with metabolic acidosis and ketosis and possible erythematous rash, and the juvenile form, appearing after 2–3 months with alopecia, erythematous rash, keratoconjunctivitis, ataxia or hypotonia or both, and metabolic acidosis with lactic acidosis. Genetic defects may occur in the carboxylases themselves or in the enzyme holocarboxylase synthetase that attaches biotin to the apocarboxylase to form the active holoenzymes.

The defect may also lie in the intestinal absorption mehanism, impairment leading to lack of sufficient biotin to activate the apocarboxylases adequately [22,23]. Response to 10 mg/d of biotin orally is good; the child must be maintained on high-dose biotin indefinitely. One case has been reported of biotin deficiency with neurologic and cutaneous manifestations but without organic aciduria; this child was maintained in good health with 10 mg/d of biotin [24].

Other cases of multiple carboxylase deficiencies have been described in which doses of 20 or 40 mg/d of oral biotin were necessary to obtain a good clinical result [25].

Prenatal diagnosis and treatment of multiple carboxylase deficiency has been described by Packman and his associates [26]. Amniocentesis was performed at 17 weeks in a woman whose first infant had been treated for the disorder. Elevated levels of methylcitrate and 3-hydroxyvalerate were found. At 23½ weeks gestation, well past the period of major organogenesis and teratogenicity risk, the mother was started on 10 mg/d oral biotin. A girl with normal growth parameters was born at term after an uncomplicated labor and delivery. The child was discharged on day 4 on breast milk and 40 mg/d oral biotin; growth and development were normal through a 3¼-year subsequent observation period. There was no apparent toxicity to mother or baby from biotin supplementation during the last 16 weeks of pregnancy despite 21- to 242-fold increases in serum biotin levels in the mother.

Methylcrotonyl-CoA Carboxylase Deficiency

Leonard et al.[27] reported four pediatric cases in which there was reduced activity of 3-methylcrotonyl-CoA carboxylase. Two children presented with metabolic acidosis, one in the neonatal period, the other with episodes of acidosis that started in the second year of life. In the other two children, neurologic symptoms were prominent, one having infantile spasms and the other showing developmental regression with skin rash and alopecia. Three of the four children showed good clinical response to high-dose biotin (5 or 10 mg/d) and restriction of protein intake, but the fourth, despite a good biochemical response, was left with a severe neurologic handicap.

Other cases of single carboxylase deficiencies have been reported for propionyl-CoA-carboxylase and for pyruvate carboxylase deficiency, leading to propionic acidemia and lactic acidosis, respectively, some of which have benefited from 10 mg/d oral biotin [1,2].

Biotinidase Deficiency

Another genetic disorder affecting biotin nutriture is biotinidase deficiency, in which there is inability to liberate biotin from biocytin and biotinylated peptides resulting from degradative turnover of carboxylases [1,2]. Since biotin cannot be recycled, the biotin requirement is increased. Ability to liberate biotin from biocytin and biotinyl peptides in the digestive tract during food digestion is also impaired. In addition, urinary excretion of biocytin and biotin appears to be increased in those with this genetic disorder [28].

The late-onset type of multiple carboxylase deficiency usually seems to be caused by biotinidase deficiency [29]. Signs and symptoms include seizures, skin rash, alopecia, ataxia, hearing loss, developmental delay, or metabolic decompensation that can terminate in coma and death [29,30]. A semi-quantitative colorimetric test is available for biotinidase activity employing dried-blood-impregnated filter papers that are used for neonatal metabolic screening [31]. Good clinical response is obtained by administration of 10 mg/d of oral biotin [29,30,32]. Hearing loss is frequent in these patients, so that patients should have their hearing evaluated early in life and monitored periodically throughout childhood [32]. Based on frequency of occurrence and difficulty of diagnosis in time to prevent irreversible sequelae, Wolf et al. [30] and Sweetman and Nyhan [1] recommend routine neonatal screening for biotinidase deficiency.

Hyperlipidemia

Biotin deficiency induced in humans by excessive consumption of uncooked egg white is accompanied by a rise in plasma cholesterol level [33]. Marshall and coworkers [34] carried out a 71-day double-blind placebo-controlled study of the effects of biotin supplementation on blood lipids of adults on ordinary diets, using a biotin dosage of 300 mcg t.i.d. Subjects were 40 healthy volunteers, 24 men and 16 women, aged 30–60 years, employed at the USDA Agricultural Research Center, Beltsville, Maryland. At the end of the study, plasma levels of total lipids, phospholipids, and triglycerides were at or below initial levels. Plasma biotin levels were elevated by supplementation, and there was a negative correlation between biotin levels and total plasma lipids. Russian workers have also reported some favorable effects of biotin (5 mg/d) on blood lipids in patients (aged 38–72 years) with arteriosclerosis and hyperlipidemia [35]. Although biotin effects on blood lipids appear modest, there seems sufficient evidence of possible efficacy to warrant further clinical studies.

Diabetes Mellitus

Marshall et al. [34] have suggested that, based on animal experiments indicating that biotin supplements elevated fasting insulin levels and improved the utilization of glucose, the value of biotin supplementation should be investigated in human diabetes mellitus. In a preliminary clinical study with a small number of insulin-dependent (type I) diabetics who were withdrawn from insulin therapy and then given either 16 mg/d of biotin orally or a placebo for 1 week, Coggeshall et al. [36] observed the expected rise in fasting blood glucose in the placebo subjects, but those on high-dose biotin experienced a significant decline in fasting blood glucose levels. During biotin therapy, plasma biotin concentrations were about 100 times pretreatment levels. Further studies are planned of this potentially important finding.

Sudden Infant Death Syndrome (SIDS)

Victims of SIDS are reported to have low hepatic concentrations of biotin [37,38]. Hood [39] has pointed out that infants below 12 months of age who receive a typical diet of 500 ml of milk supplemented with 200 g of a meat-based food and 200 g of a fruit-based food would receive only 7–14 mcg of biotin per day, as opposed to U.S. recommended intakes of 35 mcg/d for infants less than 6 months old and 50 mcg/d for infants 6–12 months old. Although the recommended intakes may have been set too high, further study of the relationship between marginal biotin status and SIDS has been recommended [1].

TOXICITY AND SIDE EFFECTS

No adverse effects have been reported from doses of biotin orally up to 40 mg/d in treatment of pediatric carboxylase deficiencies, and doses of 10 mg/d either orally or intramuscularly for periods exceeding 6 months. Also, no side effects were reported during intravenous administration of 5 mg biotin three times weekly for 6 weeks [2].

INTERACTIONS

Avidin of uncooked egg white forms a complex with biotin that renders the vitamin unavailable for intestinal absorption.

Long-term administration of sulfa drugs or antibiotics may decrease bacterial synthesis of biotin in the gut and increase the need for increased dietary or supplemental biotin [40].

Epileptics on long-term anticonvulsant therapy show reductions of mean blood biotin levels of about 50% [41], and regular biotin supplementation in these patients has been recommended [2].

REFERENCES

1. Sweetman L, Nyhan WL: Inheritable biotin-treatable disorders and associated phenomena. Ann Rev Nutr 6:317–343, 1986.
2. Bonjour J-P: Biotin. In: *Handbook of Vitamins*, LJ Machlin (ed). New York: Marcel Dekker, 1984, pp 403–435.
3. Mock DM, Baswell DL, Baker H, et al.: Biotin deficiency complicating parenteral alimentaton: Diagnosis, metabolic repercussions, and treatment. Ann NY Acad Sci 447:314–333, 1985.
4. Food and Nutrition Board: *Recommended Dietary Allowances*, 9th ed. Washington, DC: National Academy of Sciences, 1980.
5. Bowman BB, Selhub J, Rosenberg IH: Intestinal absorption of biotin in the rat. J Nutr 116:1266–1271, 1986.
6. Gore J, Hoinard C: Evidence for facilitated transport of biotin by hamster enterocytes. J Nutr 117:527–532, 1987.
7. Thoene JG, Lemons R, Baker H: Impaired intestinal absorption of biotin in juvenile multiple carboxylase deficiency. N Engl J Med 308:639–642, 1983.
8. Dakshinamurti K, Chalifour L, Bhullar RP: Requirement for biotin and the function of biotin in cells in culture. Ann NY Acad Sci 447:38–54, 1985.
9. Weiner DL, Wolf B: Biotin uptake and efflux in cultured rat hepatocytes: Implications for the treatment of biotinidase deficiency. Ann NY Acad Sci 447:435, 1985.
10. Heard GS, Grier RE, Weiner DL, et al.: Biotinidase—a possible mechanism for the recycling of biotin. Ann NY Acad Sci 447:400, 1985.
11. Sweetman L, Surh L, Baker H, et al.: Clinical and metabolic abnormalities in a boy with dietary deficiency of biotin. Pediatrics 68:553–558, 1981.
12. Kien CL, Kohler E, Goodman SI, et al.: Biotin-responsive in vivo carboxylase deficiency in two siblings with secretory diarrhea receiving total parenteral nutrition. J Pediatr 99:546–550, 1981.
13. Mock DM, DeLorimer AA, Liebman WM, et al.: Biotin deficiency: An unusual complication of parenteral alimentation. N Engl J Med 304:820–823, 1981.
14. Clinical nutrition case: Biotin deficiency as a complication of incomplete parenteral nutrition. Nutr Rev 39:274–277, 1981.
15. McClain CJ, Baker H, Onstad GR: Biotin deficiency in an adult during home parenteral nutrition. JAMA 247:3116–3117, 1982.
16. Innis SM, Allardyce DB: Possible biotin deficiency in adults receiving long-term total parenteral nutrition. Am J Clin Nutr 37:185–187, 1983.

17. Khalidi N, Wesley JR, Thoene JG, et al.: Biotin deficiency in a patient with short bowel syndrome during home parenteral nutrition. J Paren Ent Nutr 8:311–314, 1984.
18. Nisenson A: Seborrheic dermatitis of infants and Leiner's disease: A biotin deficiency. J Pediatr 51:537–548, 1957.
19. Messaritakis J, Kattamis C, Karabula C, Matsaniotis N: Generalized seborrhoeic dermatitis. Clinical and therapeutic data of 25 patients. Arch Dis Child 50:871–874, 1975.
20. Svejcar J, Homolka J: Experimental experiences with biotin in babies. Ann Paediatr 174:175–193, 1950.
21. Erlichman M, Goldstein R, Levi E, et al.: Infantile flexural seborrhoeic dermatitis. Neither biotin nor essential fatty acid deficiency. Arch Dis Child 56:560–562, 1981.
22. Thoene J, Baker H, Yoshino M, Sweetman L: Biotin-responsive carboxylase deficiency associated with subnormal plasma and urinary biotin. N Engl J Med 304:817–820, 1981.
23. Thoene J, Lemons R, Baker H: Impaired intestinal absorption of biotin in juvenile multiple carboxylase deficiency. N Engl J Med 308:639–642, 1983.
24. Swick HM, Kien CL: Biotin deficiency with neurologic and cutaneous manifestations but without organic aciduria. J Pediatr 103: 265–267, 1983.
25. Packman S, Sweetman L, Baker H, Wall S: The neonatal form of biotin-responsive multiple carboxylase deficiency. J Pediatr 99:418–420, 1981.
26. Packman S, Golbus M, Cowan MJ, et al.: Prenatal treatment of biotin-responsive multiple carboxylase deficiency. Ann NY Acad Sci 447:414–416, 1985.
27. Leonard JV, Seakins JWT, Bartlett K, et al.: Inherited disorders of 3-methylcrotonyl CoA carboxylation. Arch Dis Child 56:53–59, 1981.
28. Baumgartner ER, Suormala T, Wick H: Biotinidase deficiency associated with renal loss of biocytin and biotin. Ann NY Acad Sci 447:272–287, 1985.
29. Wolf B, Grier RE, Allen RJ, et al.: Phenotypic variation in biotinidase deficiency. J Pediatr 103:233–237, 1983.
30. Wolf B, Heard GS, Jefferson LG, et al.: Clinical findings in four children with biotinidase deficiency detected through a statewide neonatal screening program. N Engl J Med 313:16–19, 1985.
31. Heard GS, Secor McVoy JR, Wolf B: A screening method for biotinidase deficiency in newborns. Clin Chem 30:125–127, 1984.
32. Wolf B, Heard GS, Secor McVoy JR, et al.: Biotinidase deficiency. Ann NY Acad Sci 447:252–262, 1985.
33. Sydenstricker VP, Singal SH, Briggs AP, et al.: Observations on the "egg white injury" in man and its cure with a biotin concentrate. JAMA 118:1199–1200, 1942.
34. Marshall MW, Kliman PG, Washington VA, et al.: Effects of biotin on lipids and other constituents of plasma of healthy men and women. Artery 7:330–351, 1980.
35. Dokusova OD, Krivoruchenko IV: The effect of biotin on the blood cholesterol levels of atherosclerotic patients in idiopathic hyperlipidemia. Kardiologiia 12:113, 1972.
36. Coggeshall JC, Heggers JP, Robson MC, Baker H: Biotin status and plasma glucose in diabetics. Ann NY Acad Sci 447:389–392, 1985.
37. Johnson AR, Hood RL, Emery JL: Biotin and the sudden infant death syndrome. Nature 285:159–160, 1980.
38. Heard GS, Hood RL, Johnson AR: Hepatic biotin and the sudden infant death syndrome. Med J Aust 2:305–306, 1983.
39. Hood RL: The biotin content of infant foods. Nutr Rep Int 36:1039–1042, 1987.
40. Bonjour J-P: Biotin in man's nutrition and therapy—a review. Int J Vitam Nutr Res 47:107–118, 1977.
41. Krause KH, Bonjour J-P, Schmidt-Gayk, et al.: Vitamin status in patients on chronic anticonvulsant therapy. Int J Vitam Nutr Res 52:375–385, 1982.

13
Vitamin C

Introduction	202
Absorption, Metabolism, and Excretion	204
Dosage Forms	206
Clinical Studies	207
General Comment	207
The Common Cold	208
Herpes Labialis	209
Other Infections; Immune Function	210
Cardiovascular Disease	211
Ischemic heart disease	211
Fibrinolytic activity	211
Hypocholesterolemia	212
Hypertriglyceridemia	215
Other Clinical Studies	215
Hypertension	215
Diabetes mellitus	216
Pregnancy	217
The elderly	218
Cataract	221
Alcoholism	221
Periodontal disease	222
Bed sores	222
Surgery, burns, and trauma	223
Crohn's disease	223
Urine acidification	223
Allergy	224
Male infertility	224
Sickle cell anemia	224
Burns and frostbite	224
Cancer	224
Toxicity and Side Effects	227
Interactions	231
References	233

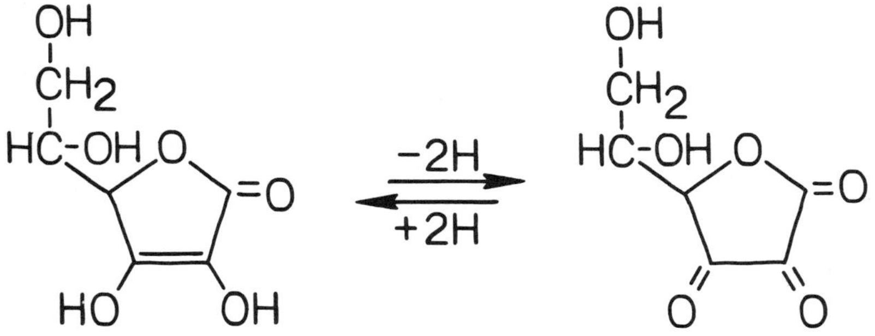

Fig. 1. Structural formulas of ascorbic acid (vitamin C) (left) and its primary oxidation product, dehydroascorbic acid (right).

INTRODUCTION

Vitamin C (L-ascorbic acid) in pure form is a white crystalline solid that is readily soluble in water. As an organic acid, its solutions have a tart acidic taste similar to that of citric acid. In the body it functions as an essential cofactor in several hydroxylation reactions. L-ascorbic acid and its primary oxidation product, dehydro-L-ascorbic acid, also form a reversible redox system (see Fig. 1) whereby the vitamin functions as a reducing agent in the organism, forming a part of the body's antioxidant defenses against such reactive oxygen species and free radicals as superoxide, hydroperoxy radical, singlet oxygen, hydroxyl radical, hydrogen peroxide, and peroxylipid radicals. These defenses include reducing agents and free radical quenchers from the diet, notably vitamins C and E and the carotenoids; intracellular enzymes such as catalase, glutathione peroxidase, and superoxide dismutase; soluble metabolites such as reduced glutathione, cysteine, bilirubin, and uric acid; and the circulating copper protein, ceruloplasmin. Ascorbic acid, which is insoluble in lipids but highly soluble in water, is of special importance in the circulation and in extracellular fluids, as well as in the aqueous phase within cells [1–3]. Excessive and chronic activation of phagocytes may lead to extracellular leakage of antimicrobial active oxygen species, which may cause various types of tissue damage and deterioration of immune response; ascorbic acid appears to deactivate these extracellular oxidants while protecting the phagocyte membrane and leaving intracellular generation of antimicrobial oxidants intact [4].

Oxidation of ascorbic acid proceeds stepwise by one-electron transfers, with ascorbate free radical (semidehydro-ascorbate) formed as an intermediate. Diliberto et al. [5] have presented evidence that dopamine-beta-hydroxylase, which catalyzes the final step in the biosynthesis of norepinephrine, uses ascorbic acid as the electron donor in reducing molecular oxygen to hydroxyl oxygen. An intracellular enzyme, semidehydroascorbate reductase, converts ascorbate radical back to ascorbate, utilizing NADH as the reducing agent [5,6]. Morré et al. [6] present evidence, from studies in Golgi apparatus isolates, that ascorbate free radical functions as an electron receptor in a redox system involving NADH-free ascorbate radical reductase, which is present in those cellular membranes most often involved in bulk translocations of cellular membranes and in membrane movements, e.g., plasma membrane, coated vesicles, secretory vesicles, and endocytic vacuoles. Their results, they conclude, are consistent with a role for this redox system in providing energy, via generation of a proton gradient or of a membrane potential, for membrane translocations and cell motility [6].

Rose [7] finds that ascorbate deactivates hydroxyl radical in aqueous medium, forming hydroxide ion and ascorbate free radical; the latter spontaneously disproportionates to form ascorbate and dehydroascorbate. Other free radicals may be quenched by ascorbate [3], and there is evidence that ascorbate, operating at the lipid-aqueous interface of cell membranes, can regenerate reduced tocopherol from vitamin E radical formed in the membrane when tocopherol reacts with peroxylipid radicals [8]. Further evidence that vitamin C protects cell membranes against lipid peroxidation is afforded by the fact that breath hydrocarbon levels of scorbutic guinea pigs, injected with carbon tetrachloride to stimulate peroxidation, are significantly lowered by pretreatment with ascorbic acid [9]. Thermal decomposition of peroxidized omega-3 and omega-6 unsaturated fatty acids yields ethane and pentane, respectively, which can be detected in expired breath by gas chromatography [10–12].

Other known actions of vitamin C include its role in maintenance of hepatic mixed-function oxidase activity; the 7-alpha-hydroxylation of cholesterol in the liver, which initiates the synthesis of bile acids; the hydroxylation of proline and lysine, required for production of stable cross-linked collagen; the biosynthesis of carnitine; the production of dopamine and norepinephrine from tyrosine; steroid metabolism in the adrenals; and regulation of histamine concentration in the blood [1]. Ascorbate serves as a cofactor for peptidyl-alpha-amidation enzyme, an enzyme present in secretory granules of nearly all neuroendocrine tissues that catalyzes the terminal COOH-amidation of neuroendocrine peptides by replacing terminal glycine residues with NH_2 [13]. Ascorbic acid markedly improves the absorption of nonheme iron ingested at the same meal [14].

Most mammals can biosynthesize their own ascorbic acid, but man and other primates cannot, so are dependent on dietary sources. The guinea pig is also unable to biosynthesize ascorbate and is commonly used as an animal model to study vitamin C nutrition. Dietary sources of the vitamin are primarily fruits and fresh vegetables, although in recent years various fruit-flavored beverages and even breakfast cereals have been fortified with synthetic vitamin C. The synthetic product is chemically identical to the naturally occurring vitamin.

Vitamin C deficiency disease is called scurvy. Only a small daily intake of the vitamin, of the order of 10 mg/day, is required to prevent clinical manifestations of scurvy. However, larger intakes are believed necessary to provide optimal ascorbate nurture and a reserve tissue storage. The current RDA is 60 mg/day, an intake easily met by usual American diets. A lower recommended intake of 40 mg/d for adult men and 30 mg/d for adult women has been proposed [15], but while young men appear to be able to maintain mean plasma ascorbate levels of 0.4 mg/dl (the dividing line between "adequate" and marginally deficient) on daily intakes of 40 mg [16], elderly men in good health appear to require 60 mg/d to maintain this plasma level [17]. Maximization of the ascorbate body pool (tissue saturation) requires an intake of 140–150 mg/d in both young [16] and elderly [17] men. Potential therapeutic uses of vitamin C have involved administration of daily doses many times larger.

ABSORPTION, METABOLISM, AND EXCRETION

Ascorbic acid is absorbed in the upper part of the small intestine by a saturable, Na^+-dependent transport process [18–20], and renal reabsorption is also by an active process [19]. Ascorbate exists in blood and body tissues mainly in reduced form; dehydroascorbate averages 5–7% of total vitamin C in the plasma of healthy individuals and 3–5% in their lymphocytes at vitamin C intakes ranging from usual dietary levels to 12 g/day [21]. Reduction of dehydroascorbic acid to ascorbic acid, both enzymatically and nonenzymatically, occurs in all tissues of experimental animals and is particularly strong in the adrenals and liver [22]. Urate in human plasma also appears to play a role in protecting the reduced form of vitamin C [23]. When a 1-g dose of vitamin C was administered to women in the form of dehydroascorbic acid, the vitamin was excreted in the urine almost entirely in the reduced state [24].

According to Hodges [25], a plasma ascorbate level above 0.6 mg/dl may be considered to reflect a well-nourished condition, while 0.40–0.59 mg/dl is in an adequate range, 0.10–0.39 mg/dl is marginal, and a value below 0.10 mg/dl reflects deficiency and will lead to scurvy. Most individuals on American diets, which tend to be rich in vitamin C-containing foods, have plasma levels in the range of 0.8–1.4 mg/dl [1]. Compared with plasma levels, tissue levels are generally far higher: several hundred times higher in pituitary, adrenal, thymus, corpus luteum, and retina; 10–100 times higher in brain, testicle, thyroid, small intestinal mucosa, lymph glands, lung, liver, spleen, white blood cells, pancreas, and salivary glands; and approximately ten times higher in kidney, muscle, and red cells [1,26]. Ascorbate enters the ventricular fluids via the choroid plexus and is actively transported against a concentration gradient into brain tissue, with especially high levels in the dentate gyrus of the hypothalamus [27]. Uptake of ascorbate into cells having a high ascorbate concentration is mediated by a stereospecific active transport mechanism [26]. The oxidized form, dehydroascorbate, is transported faster than the reduced form [26].

The cornea, lens, and aqueous humor of the eye are also strong concentrators of vitamin C [28,29]. Experiments by Varma et al. [30] indicate that vitamin C in the aqueous humor functions as a nonenzymatic scavenger of active oxygen species produced by light activation, thereby protecting cell membranes of the lens.

Catabolism of vitamin C in humans occurs by the irreversible hydrolysis of dehydroascorbic acid to 2,3-dioxo-L-gulonic acid, followed by oxidation to oxalic acid and 1-threonic acid, which, together with some ascorbate-2-sulfate, are excreted in the urine along with unmetabolized ascorbic acid [1].

The percentage of a single dose of ascorbic acid taken on an empty stomach that is absorbed decreases progressively as the dosage increases; according to one study, from 80–90% at 30 mg to about 50% at 1 g and 16% at 12 g [31]. Another study showed 75% absorption of a 1-g dose and 20% of a 5-g dose [32]. Of the amount absorbed from a 30-mg dose, about 7% is excreted in the urine as ascorbic acid and 93% as metabolites, while of that absorbed from total daily doses of 1–2 g, 87% is excreted as ascorbic acid and 13% as metabolites.

Reabsorption of ascorbate in the renal tubule is efficient at plasma levels below 0.8 mg/dl; as plasma ascorbate levels rise above 0.8 mg/dl, urinary ascorbate excretion rises at an accelerating rate, resulting in plateauing of plasma concentration usually somewhere in the range of 1–2 mg/dl, depending on the individual and the intake [15,16,33]. Melethil et al. [33] find the maximum amount absorbed from single doses as large as 20 g to be 500–600 mg.

When ascorbic acid was administered three times daily at meals, its urinary excretion was observed to increase with intake up to a total daily dose of 3 g, but was not further increased at 5 g/day [34]. Unabsorbed ascorbic acid appears to be metabolized by bacteria in the lower GI tract to carbon dioxide and unidentified products [35]. When administered even in gram quantities, no more than 5% of the dose could be recovered as unchanged ascorbic acid in the stool [35].

Kallner et al. investigated the pharmacokinetics of ascorbic acid in doses of 30–180 (4 x 45) mg/d under steady-state conditions in healthy nonsmoking men, measuring the time-course of radioactivity in plasma and urine after oral administration of a single dose of $(1-^{14}C)$ ascorbic acid [36]. Allowing for incomplete absorption and individual differences, they concluded that a daily intake of about 100 mg ascorbic acid would allow maintenance of a body pool of about 20 mg/kg bodyweight in 97.5% of the adult male population. In smokers, who have a higher metabolic turnover of vitamin C than nonsmokers, they calculate from similar studies that an intake of 140 mg/day would be required to maintain a comparable body pool [37]. Jacob et al. [16] calculate from studies in 11 nonsmoking young men that an average intake of 138 mg/d is required to maximize the body ascorbate pool. Smith and Hodges [38] estimated additional vitamin C needs for smokers from comparisons of dietary vitamin C intakes and serum ascorbate levels determined in NHANES II, which surveyed nutritional status of a nationwide population sample of 20,332 individuals. They found that, over the normal range of dietary intakes, smokers would need 60–65 mg/d more vitamin C than nonsmokers.

The efficiency of vitamin C absorption is increased by division of the daily dosage and by administration with meals. Absorption is particularly increased and prolonged by inclusion of the supplement in high-fat meals [39].

DOSAGE FORMS

Ascorbic acid is available over-the-counter (OTC) as tablets in 100-, 250-, 500-, and 1,000-mg strengths and in bulk as a powder. Orange-flavored chewable tablets are also marketed. Timed-release capsules and coated tablets are additional OTC forms.

Yung et al. [40] compared ascorbic acid absorption from four dosage forms (solution, tablet, chewable tablet, and timed-release capsule), each containing 1 g ascorbic acid, in four healthy young adults whose vitamin C stores had been saturated by oral administration of 1 g vitamin C daily for 2 weeks. With a separate trial of an intravenous dose of 1 g ascorbic acid, approximately 85% of the dose was recovered as ascorbic acid and its metabolites in a 24-hour urine collection. In contrast, when the oral forms were given as a single dose on an empty stomach, only about 30% of the dose was recovered in the urine from the solution and tablet forms and only 14% from the timed-release capsule. There was considerable intersubject variation in absorption, so that there appear to be good and poor absorbers of vitamin C supplements taken on an empty stomach. Whether the same is true when the vitamin is taken with meals is not known. Zetler et al. [41] obtained different results: 70% absorption of a 1-g dose in capsule form and essentially complete absorption of 1 g in a sustained-release capsule. Pharmaceutical formulation and timing of dosage are evidently important factors in extent of absorption, and much has still to be learned about the pharmacokinetics of the various vitamin C dosage forms.

Vitamin C is also available OTC as tablets containing rose hip solids or bioflavonoids. No advantage has been demonstrated for administration in such formulations. For individuals in whom the acidity of vitamin C may cause gastric discomfort, calcium ascorbate tablets may be employed. Ascorbyl palmitate, a fat-soluble form marketed as capsules, is promoted as a more slowly absorbed form, but again no advantage has been demonstrated. Pharmacokinetic data in human subjects are lacking for the various combinations, salts, and esters of ascorbic acid, as well as so-called "natural" vitamin C, for many of which special merit has been claimed over unmodified synthetic L-ascorbic acid.

Ascorbic Acid Injection, in concentrations of 100 or 250 mg/ml, is available in light-resistant, single-dose glass containers [42].

CLINICAL STUDIES
General Comment

Because of the limited intestinal absorption of vitamin C and its increasingly rapid renal excretion as plasma ascorbate levels rise above 0.8 mg/dl, it is difficult to elevate plasma levels above 1.3–1.4 mg/dl with high doses of vitamin C if they are administered only once daily. For example, in a large-scale, long-term study by Briggs [43], subjects receiving a 50-mg supplement once daily plus an average of 85 mg from diet had plasma ascorbate levels of 1.1 ± 0.3 mg/dl, while those ingesting a 1,000-mg supplement once daily plus 78 mg from diet had plasma levels elevated only to 1.3 ± 0.3 mg/dl. Tissue saturation, as reflected by platelet and leukocyte ascorbate levels, was apparently achieved at the 135 mg/day intake, as cellular levels were not further increased by the 1,078 mg/day intake.

It appears, therefore, that changes in cell biochemistry with vitamin C supplementation, e.g., in rates of enzymatic reactions, are most likely to be observed as ingestion of the vitamin progresses from marginal intakes of 20–30 mg/day to tissue saturation intakes of 100–150 mg/day, with little further change to be expected beyond this point. On the other hand, vitamin C actions of a pharmacologic nature that depend on an elevated concentration in such extracellular fluids as blood plasma, connective tissue fluid, and the humors of the eye may possibly be strengthened by high-dose administration, provided that proper attention is paid to maximizing absorption by giving multiple daily doses to compensate for accelerated urinary excretion of the vitamin.

It must be borne in mind that in patients whose tissue stores and plasma levels are depleted because of inadequate vitamin C intake over a long period or because of infection or surgical trauma, administration of 500 or 1,000 mg/day of ascorbic acid may be required for a week or two or until such time as normal plasma and leukocyte levels can be maintained with a smaller supplement or by the dietary intake.

The Common Cold

Enormous public interest developed in high-dose vitamin C for prophylaxis and therapy of the common cold following publication of Pauling's books on the subject [44,45]. Briggs [43] has critically reviewed the studies cited by Pauling as well as many controlled studies carried out since publication of his books and finds, in agreement with most other reviewers, that high-dose supplementation, i.e., 1–2 g/day, usually has only a small positive effect in lowering the incidence, shortening the duration, or ameliorating the symptoms of the common cold.

It is possible that favorable effects of supplementation occur primarily in those study subjects whose dietary intakes have been below tissue-saturating levels. For example, Baird et al. [46], studying incidence and symptoms due to colds over a 72-day period in 362 male and female students (17–25 years of age) randomly assigned to an 80-mg vitamin C supplement or placebo, observed a 14–21% reduction in cold symptoms in the supplemented group, together with an increase in the total number of symptom-free days. Since the average dietary vitamin C intake of the students was 50 mg/day, it appears that most of them were below tissue saturation initially and that the placebo group remained in that state, while in the supplemented group the total intake of 130 mg/day produced tissue saturation, actually documented by the investigators in terms of leukocyte ascorbate levels [46]. Briggs [43] carried out a multi-year study with 160 men and 368 women, half of whom were randomly assigned to receive a supplement of 50 mg ascorbic acid/day and half to one of 1,000 mg/day over observation periods of 3–6 months. She found no statistically significant differences between the two groups in incidence of colds or severity of symptoms. However, since dietary intakes averaged 85 mg/day in the 50-mg supplement group, their total intake of 135 mg/day was sufficient to provide tissue saturation comparable with that of the 1,000-mg group.

If the prophylactic value of vitamin C against the common cold is dependent on saturating the tissue ascorbate stores, it appears that a supplement of 100 mg/day, regularly taken, should be sufficient for the purpose even if the dietary intake is low. If dietary intake has been low for a protracted period, however, a 500-mg daily supplement might be advisable at onset of the common cold season, to be continued for several days to rapidly restore tissue levels, after which a smaller supplement should be satisfactory for maintenance.

There is evidence from in vitro, animal, and human studies that ascorbic acid is involved in regulation of histamine metabolism and in control of circulating levels of this vasoactive amine [47,48]. It is of interest that such cold symptoms as congestion and rhinorrhea, which are thought to be mediated at least in part by local histamine release, are the symptoms most often reported to be reduced in severity by vitamin C treatment [49–52].

Use of multigram quantities of ascorbic acid has been advocated for therapy of active colds, but evidence for efficacy obtained by controlled studies is lacking. Briggs [43] found supplements of 4 g/d to be no more effective than 200 mg/d, but it is not clear from her report whether these amounts were taken at one time or as divided doses. Controlled trials are needed of the possible therapeutic value during active colds of administering large oral doses, e.g., 500–1,000 mg, at hourly intervals through the day.

Herpes Labialis (Herpes Type I: Cold Sores)

Terezhalmy et al. [53] report successful use of a combination of ascorbic acid and water-soluble bioflavonoids in a group of patients suffering from recurrent attacks of herpetic lesions affecting the skin and mucous membranes of the mouth. Twenty episodes were treated with 200 mg ascorbic acid and 200 mg bioflavonoids administered three times daily for a total of 600 mg per day each. Another 20 episodes were treated with 200 mg ascorbic acid and 200 mg bioflavonoids administered five times daily. Ten episodes were treated with a lactose placebo. Treatment was continued for 3 days after recognition by the patient of the initial symptoms in the prodromal period. Patients receiving placebo had a mean duration of symptoms of 9.7 days, while those receiving ascorbic acid and bioflavonoids at 600 mg/day had a mean symptom duration of 4.2 days. The 1,000-mg/day regime was no better than the 600 mg/day dosage. The ascorbic acid-bioflavonoid treatment reduced vesiculation and prevented disruption of the vesicular membrane.

Therapy was most effective when initiated during the prodromal period, which lasted only about 12 hours from the first feelings of altered sensation to vesiculation. If started more than 12 hours after initial symptoms, therapy was much less effective.

The bioflavonoids are a large group of phenolic compounds widely distributed in the skin, peel, and outer layers of fruits and vegetables. Quercetin, rutin, and hesperidin are examples. They have been reported to decrease capillary fragility, but the evidence is controversial [54]. Studies are needed to determine whether vitamin C or the bioflavonoids cause the reported beneficial effects in recurrent herpes labialis, or whether the two agents act in concert. Since both are safe in chronic administration, the value of each, and of their combination, should be determined for long-term prophylaxis of recurrent herpetic lesions.

Other Infections; Immune Function

There is abundant evidence from animal and human studies that an adequate level of vitamin C nutriture is essential to normal immune function [43,47,55,56]. As opposed to a vitamin C deficiency state, adequacy of ascorbate nutrition improves a number of neutrophil functions, including chemotaxis, particulate ingestion, and lysozyme-mediated nonoxidative killing activity. Vitamin C also stimulates hexose monophosphate shunt activity in the neutrophil. In vitro and in some animal experiments, acorbic acid has displayed antiviral activity, but not against cold viruses [43,47]. High-concentration ascorbic acid stimulated interferon production in cultured mouse cells and in mice in vivo [57–59]. In elderly individuals, tissue-saturating doses (500 mg IM daily for 1 month) enhanced cell-mediated immune function as manifested by enhanced T-lymphocyte proliferative response in vitro and tuberculin skin hypersensitivity in vivo [60]. Ziemlanski et al. [61] found that healthy institutionalized Polish women aged 63–93 years, when given 200 mg vitamin C b.i.d. for 4 months, manifested significant increases in serum IgG, IgM, and complement component C3; these were similar to responses obtained earlier in middle-aged people by Vallance [62]. The initial mean plasma ascorbate level of the Polish subjects, 0.5 mg/dl, indicates low tissue saturation in many of them at the start of the experiment.

In the guinea pig, tissue-saturating intakes of vitamin C appear to produce higher plasma concentrations of the hydroxyproline-rich complement protein C1q than lower intakes [63]; C1q is essential for effective immune response, as it is the initiator of the complement cascade responsible for lysis of antibody-coated bacteria and viruses [64].

Plasma vitamin C levels are known to be lowered by infections and inflammatory disease such as rheumatic fever, rheumatoid arthritis, pneumonia, tuberculosis, and hepatitis [43,55], increasing the requirement to produce tissue saturation and normal plasma ascorbate levels. For example, tuberculosis patients may require as much as 250 mg vitamin C per day to achieve plasma levels similar to those of healthy normals [55].

Intravenous infusion of 5 or 10 g ascorbic acid per day has been reported to improve rate of recovery in acute hepatitis in both children and adults [65–68], but *oral* administration of 800 mg q.i.d. was ineffective in reducing the incidence of posttransfusion hepatitis in cardiac surgery patients [69]. In general, orally administered high-dose vitamin C seems to have had little effect on the course of infectious diseases [43]. However, precautionary administration orally of 100 mg t.i.d. or 500 mg once a day may be useful in preventing infection induced depletion of tissue and plasma ascorbate, which could be detrimental to immune function. Further studies are needed of the potential value, as well as possible adverse effects, of the high plasma ascorbate levels temporarily achievable by continuous intravenous infusion in therapy of viral and other infections.

Patrone and coworkers have demonstrated that ascorbic acid is important for normal human neutrophil function, specifically chemotaxis and bacterial killing [70,71]. In vitro, when neutrophils were treated with deactivating concentrations of a chemoattractant in the presence of ascorbic acid, they did not become deactivated, but retained normal chemotactic responsiveness; the ascorbate concentration in the medium was about that present in the neutrophil physiologically [70]. Clinically, this group has reported an association between recurrent infections, mainly of the skin, and impairment of neutrophil function [71]. In a small study with three patients, one with folliculitis, the other two with recurrent skin abscesses, they found that administration of 1–2 g/day of ascorbic acid for 3–6 weeks elevated chemotactic function from defective or low normal to fully normal and also restored normal bacterial killing (*Staphylococcus aureus*); simultaneously the skin infections cleared, and there was no recurrence in a subsequent year of vitamin C therapy. Plasma vitamin C levels before and after treatment were not reported; obviously confirmatory studies are needed.

Cardiovascular Disease

Ischemic heart disease (IHD). Gey [70] has reviewed the evidence from animal experiments that peroxidized polyunsaturated fatty acids can damage arterial and cardiac endothelium, provoke proliferation of smooth muscle cells, and inhibit arterial formation of prostacyclin; adequate plasma antioxidant activity provided by ascorbate and vitamin E can prevent or inhibit such damage. Gey and coworkers [71] now report results of surveys of mean plasma antioxidant levels (vitamins A, C, and E, beta-carotene, and selenium) in 703 healthy men aged 40–49 years in selected areas along a north-to-south gradient in Europe, corresponding to high, medium, and low incidence of IHD. They find that plasma ascorbate levels below 0.4 mg/dl and/or of cholesterol-standardized vitamin E levels below 0.9 mg/dl are associated with elevated risk of IHD mortality. The evidence, they conclude, justifies long-term intervention trials.

Fibrinolytic activity. Bordia and coworkers [74] noted that high single oral doses of ascorbic acid elevated fibrinolytic activity. Bordia [75] later reported results of administering 500-mg and 1-g oral doses b.i.d. for 6 months to patients with a past history of myocardial infarction. Of a total of 80 patients, about 41% received 2 g/day, 18% 1 g/day, and 41% no vitamin C supplement. All patients had normal or low-normal plasma ascorbate levels before supplementation. At the end of the study, in the group that received 1 g ascorbic acid b.i.d., mean plasma ascorbate had nearly doubled, fibrinolytic activity had increased by an average of 45%, and platelet adhesive index had dropped by 27%. In addition, mean total serum cholesterol had declined by 12%, with a significant decrease in LDL cholesterol and an increase in HDL cholesterol. The group receiving 500 mg ascorbic acid b.i.d. showed a moderate elevation of plasma ascorbate but no significant changes in fibrinolytic activity or plasma lipids. Elevated plasma ascorbate levels, not obtainable by dietary supply of the vitamin, appear essential for this physiological action. More recently, Bordia and Verma [76] showed that in coronary artery disease patients with hyperlipidemia, the platelet adhesiveness index and the platelet aggregation tendency were significantly decreased by 10 days of supplementation with 3 g/d of vitamin C, 1 g every 8 hours.

These results suggest the possible usefulness of oral high-dose vitamin C in the routine care of patients with cardiovascular disease, in prophylaxis of thrombophlebitis, in pre- and postoperative care of surgical patients, and in management of the chronically bedridden. Controlled studies with various dosage regimens are needed to evaluate these possibilities.

Hypercholesterolemia

It has been frequently demonstrated in guinea pigs that a marginal vitamin C intake, not low enough to cause scurvy or to retard normal growth and development, leads to an elevation of plasma cholesterol level simultaneously with a reduction of hepatic catabolism of cholesterol to bile acids [77–81]. The hepatic activity of cytochrome P-450-dependent cholesterol 7-alpha-hydroxylase, the enzyme that initiates biosynthesis of bile acids, appears to reach a maximum at tissue-saturating concentrations of ascorbic acid; at tissue ascorbate levels below this in the guinea pig, the rate of hepatic conversion of cholesterol to bile acids declines, and plasma cholesterol rises [77–81]. Jenkins [82,83] has repeatedly shown that cholesterol gallstones can readily be induced in guinea pigs fed cholesterol-containing diets at marginal intakes of vitamin C, but not at tissue-saturating intakes of the vitamin. In pregnant guinea pigs fed diets without added cholesterol and either 0.2 or 2.0 mg ascorbic acid per 100 g bodyweight per day, the gall bladder bile of the low-vitamin C group had more cholesterol and less bile acid than that of the higher-vitamin C group [84]. The lower dose of vitamin C was sufficient to ensure normal weight gain, appearance, mean fetal weight, and litter size, with no significant differences from pregnant animals receiving the higher dose of vitamin C.

Similar elevation of plasma cholesterol levels (17.5%) was observed in adult, initially vitamin C-adequate marmoset monkeys after being deprived of dietary vitamin C for 1 month, although at this time they were still excreting ascorbate (3–5 mg/100 ml urine), had no weight loss, and showed no clinical sign of deficiency [85].

A number of studies have indicated that tissue-saturating doses of vitamin C can cause significant lowerings of plasma cholesterol in hypercholesterolemic humans who are receiving less than optimal vitamin C in the diet. Much of this evidence was obtained in Czechoslovakia by Ginter and his associates, using daily dosages of 300–1,000 mg/day of vitamin C for periods up to 1 year. In one study with middle-aged and elderly men and women given 500 mg b.i.d. for 1 year, 60% of subjects showed substantial declines in plasma cholesterol, and the mean decline for the experimental group was 13%; 40% showed no significant change [86]. It should be noted that the study was initiated during a seasonal dietary deficit of vitamin C, the dietary intake averaging only 20 mg/day. Dobson et al. [87] have documented rises in serum cholesterol and falls in leukocyte ascorbate in elderly men and women in Scotland during low dietary vitamin C intakes in winter. Administration of 1 g/day of vitamin C for 12 months reduced mean plasma cholesterol levels by 14% and abolished seasonal changes in this parameter and in leukocyte ascorbate concentration. It is of interest that in contrast to young people, aged 29 ± 5 years, who showed a fall of 16% in mean serum cholesterol level within 2 months on the same dosage, the elderly subjects, aged 56 ± 3 years, required 6–12 months on this regimen before significant reductions in cholesterol levels occurred.

With a dosage of 500 mg ascorbic acid per day for 12 months, Ginter et al. [88] achieved a mean 20% reduction in serum cholesterol in 35 stabilized, permanently hypercholesterolemic maturity-onset diabetics, as well as a 10% decline in triglycerides. This was a double-blind, placebo-controlled study with 13 comparable diabetics serving as controls. Those on placebo showed no change in blood lipids. Vitamin C concentrations in the plasma and leukocytes of both experimental and control groups were significantly lower than in healthy normals at the start of the experiment, but were raised to high-normal levels by the vitamin C supplement.

There is no evidence that normolipemic or hyperlipemic subjects whose tissues are already saturated with vitamin C from their diet derive any further benefits in blood lipid levels from high-dose ascorbic acid supplementation. Ginter, in evaluating his group's experience, finds that little or no change in serum cholesterol levels was seen when initial levels of the sterol were low-to-normal, but if initial serum cholesterol concentration was in the range of 250–350 mg/dl, 10–20% reductions were regularly observed [89]; nonresponders could have been those whose bile acid synthetic capacity was greater and/or whose vitamin C requirement was lower than that of responders.

Young American adults, normolipemic and with tissue-saturating dietary intakes of vitamin C as evidenced by mean plasma ascorbate levels of 1.1–1.3 mg/dl at the start of the experiment, showed no response in blood lipids to administration of 1 g of ascorbic acid per day for several weeks [90,91]. In contrast, Kothari and Jain [92], in India, obtained substantial reductions in serum cholesterol in a 30-day study of the effects of supplementing healthy adult men with 1 g ascorbic acid per day. The subjects were 40 normolipemic physicians and residents, 20 aged 20–30 years and 20 aged 31–50 years. Each group was further subdivided randomly into ten subjects on vitamin C supplement and ten as controls. With vitamin C supplementation, mean serum ascorbate rose from 0.76 to 1.24 mg/dl in the younger subjects and from 0.74 to 1.22 mg/dl in the middle-aged group. Simultaneously, mean serum cholesterol levels declined from 204 to 177 mg/dl in the younger group and from 256 to 225 mg/dl in the older group, i.e., statistically significant declines of 10–15%. In this case, it appears that completion of tissue ascorbate saturation, evidenced by the rise of mean plasma ascorbate concentration from an initial 0.75 to about 1.2 mg/dl, had a favorable effect on serum cholesterol levels even in normolipemic subjects.

Ramirez and Flowers [93] compared leukocyte ascorbate levels in 101 patients with coronary artery disease with those of 49 patients without coronary artery disease, as demonstrated by coronary arteriography. The mean leukocyte ascorbate level was significantly and substantially lower in both male and female patients with coronary atherosclerosis than in those without arteriogram abnormalities. As reflected by leukocyte ascorbate concentrations, therefore, deficient tissue saturation was significantly more frequent in those with coronary artery disease than in those without.

The most direct evidence that intervention with high-dose vitamin C may lower risk of cardiovascular disease by an effect on blood lipids is afforded by the study of Jacques et al. [94] of the relation between plasma ascorbic acid levels and HDl- and total cholesterol levels in 680 noninstitutionalized elderly men and women in the Boston area. Age range was 60–99 years. Plasma ascorbate levels ranged from 0.18 to 2.63 mg/dl, with a mean of 1.3 mg/dl for women and 1.1 mg/dl for men, attesting to a generally good vitamin C status in this population. More than one-third of these individuals took a daily vitamin C supplement. Results established a positive correlation between plasma ascorbate and HDL cholesterol but not total cholesterol. The data suggest that each 0.1 mg/dl increase in plasma ascorbate level was associated with an increase of 1% in plaasma HDL cholesterol in those aged 60–69, of 0.5% in those aged 70–79, and of 0.01% in those aged 80 and over. Plasma ascorbate concentrations of those receiving 1,000 mg or more of vitamin C daily were about 1 mg/dl higher than those of individuals receiving less than 120 mg/d. The investigators point out that intervention trials with high-dose vitamin C would be justified by a potential 8–10% reduction in coronary heart disease mortality, which, from the Framingham studies, would be likely to follow an HDL-cholesterol elevation of 4–5 mg/dl [94].

Hypertriglyceridemia

Bobek et al. [95], treating three groups of male guinea pigs with different daily dosages of ascorbic acid, so that after 17 weeks their plasma ascorbate levels were 0.10, 0.40, and 1.17 mg/dl, respectively, found that the respective plasma triglyceride concentrations were 2.61, 1.92, and 1.13 μmole/liter. Vitamin C deficiency reduced the secretion rate of endogenous triglycerides by 25%, but the rate of removal of plasma triglycerides was much more strongly depressed, i.e., by 52%. The mechanism of ascorbic acid stimulation of triglyceride removal may involve in part an elevation of plasma lipoprotein lipase activity, demonstrated experimentally in guinea pigs, rabbits, rats, and baboons [95]. Physiological concentrations of sodium ascorbate have also been shown to increase the number of LDL receptors in cultured bovine arterial smooth muscle cells [96].

Vitamin C supplements have reduced plasma triglycerides in some studies with human subjects [88,97], but it is doubtful if any effect will be shown if the diet already provides tissue-saturating amounts of the vitamin. Wahlberg and Walldius [98] could find no effect on either triglycerides or cholesterol in nine stable type IV hypertriglyceridemia patients treated with 2 g ascorbic acid per day for a month. Data on dietary intakes and plasma levels of ascorbic acid were not obtained in this study, but the patients had received special instruction on lipid-lowering diets, which may have provided adequate vitamin C.

Geoly and Diamone [99] reported a striking 60–65% reduction in plasma triglycerides in two hemodialysis patients with severe hypertriglyceridemia when they were given 1,000 mg ascorbic acid t.i.d. for 2 months, *but plasma oxalate levels must be closely monitored in hemodialysis patients given even small supplements of vitamin C, as secondary hyperoxalemia may develop* [100].

If there is doubt as to the adequacy of the dietary vitamin C intake of the hypertriglyceridemic patient (with adequate renal function), the physician may wish to consider prescribing an ascorbic acid supplement of at least 100 mg daily. *High-dose vitamin C is contraindicated in patients with inadequate kidney function, as urinary excretion is the main route of disposal of high-dose ascorbic acid, as well as of its metabolite, oxalic acid.*

Other Clinical Studies

Hypertension. Yoshioka et al. [101] found an inverse correlation of serum ascorbic acid levels in 194 rural Japanese men, aged 30–39 years, with systolic and diastolic blood pressure. There were 20 individuals with readings of 140/90 mm Hg or higher in the 152 subjects with serum ascorbic acid concentrations of 0.8 mg/dl or less and only one with hypertension in the 42 subjects with serum ascorbic acid levels of 0.9 mg/dl or greater. It must be borne in mind, however, that dietary patterns resulting in a high vitamin C intake are also associated with relatively high intakes of other nutrients that may affect blood pressure, such as potassium and fiber.

Diabetes mellitus. Mann [102] hypothesized, based on a review of animal and human studies, that insulin is involved in the transport of vitamin C into insulin-sensitive tissues such as vascular endothelium, the retina, muscle, and adipose tissue, and further, that high blood glucose levels might exert a competitive inhibition on vitamin C transport. Damage to insulin-sensitive tissues could then result from local hypoascorbosis induced either by insulin lack or insensitivity or by high blood sugar levels or both. Subsequently Mann and Newton [103] reported that the uptake by erythrocytes of dehydroascorbic acid, the transportable form of vitamin C in the body, is strongly inhibited by glucose, mannose, and xylose; uptake of these nutrients in erythrocytes is by facilitated transport not dependent on insulin. Working with cultured cells from an insulin-sensitive tissue, fetal bovine heart endothelium, Verlangieri and Sestito [104] showed that uptake of ascorbic acid at physiologic concentrations was linearly related to log insulin concentration in the physiologic range of this hormone. In addition, they found that glucose at 180 mg/dl in the culture medium inhibited ascorbic acid uptake by about 80%. Chen et al. [105] demonstrated that hyperglycemia induced in normal human volunteers can cause depletion of ascorbate in mononuclear leukocytes (MNL). In an extension of these clinical studies [106], they found that hyperglycemia also depleted ascorbate in polymorphonuclear lymphocytes (PMN) and that prolonged hyperglycemia (maintained by glucose clamp technique) impaired chemotaxis in both MNL and PMN. The results may explain in part the increased susceptibility of diabetics to infections [106]. This evidence supports both aspects of Mann's hypothesis.

Verlangieri and coworkers [107,108] had previously shown that ascorbic acid supplementation reduces the extent of induced atherogenesis in the rabbit, even though it is an ascorbic acid-synthesizing species, and that the mechanism appears to be inhibition by ascorbate of degradation of glycosaminoglycans by the enzyme arysulfatase B in the extracellular matrix of the vascular endothelium [104,109]. Breakdown of intimal matrix due to localized ascorbate deficiency could contribute to the micro- and macroangiopathies associated with diabetes [104].

Chatterjee and coworkers found that diabetic patients tend to have plasma ascorbate levels significantly below those of normals and that vitamin C supplements raise these levels [110–112]. They also noted an increased ratio of dehydroascorbic acid to the reduced form in diabetics. The same disturbance of this ratio was reported to occur in experimental diabetes in laboratory rats and could be largely corrected by high-level vitamin C supplementation [113].

Subsequent clinical studies in the UK [114] and the US [115], however, failed to confirm any elevation of plasma dehydroascorbate levels in type I or type II diabetics.

Losert et al. [116] reported that high-dose vitamin C potentiated the glucose-lowering effect of exogenous insulin in both guinea pigs and healthy nondiabetic men without enhancing the antilipolytic action of insulin. In the animals, the rate of oxidation of administered ^{14}C-glucose was increased. In the human experiments, they evaluated the glucose-lowering effect of bovine insulin (0.05 U/kg) in subjects pretreated with 50 mg ascorbic acid per kg (about 3.5 g per 70-kg man) orally, obtaining a significant enhancement over the response without ascorbic acid treatment.

On the basis of this evidence, controlled clinical trials appear warranted in diabetics to assess the long-term benefits of maintaining high plasma ascorbate levels in order to overcome in some degree the competitive inhibition of transport of ascorbic acid into insulin-sensitive tissues by high blood glucose levels, and possibly to allow reduction of insulin dosage by potentiating the hormone's action. Dosages of 1,000 mg vitamin C b.i.d. or t.i.d. orally appear most likely to accomplish these objectives rather than ordinary dietary intakes; such doses could have, in addition, the potential advantages of increasing fibrinolytic activity and decreasing platelet adhesiveness [75,76]. Increased platelet aggregability has been reported in diabetes mellitus [117]. Sarji et al. [118] observed that ascorbate levels are reduced in insulin-dependent diabetics and that this may contribute to hyperaggregation.

Cox and Butterfield [119] reported that daily vitamin C supplements of 1,000 mg corrected the cutaneous capillary fragility of diabetic patients.

Pregnancy. Marginal states of vitamin C nutriture seem to be fairly common in pregnant women, often despite apparent adequacy of dietary intake. For example, in a group of 65 Irish women in the third trimester of pregnancy, whose vitamin C intakes from diet were calculated to be 90–150 mg/day, 6 had whole-blood ascorbate levels below 0.4 mg/dl, and 21 had levels below 0.8 mg/dl [120]; none were taking vitamin supplements. Clemetson [48] has reported wide variability in plasma ascorbate levels in a group of American women in the second and third trimester of pregnancy, even though all had been prescribed a daily multivitamin supplement containing 60 mg ascorbic acid. The cause of the interindividual variations is obscure, although Clemetson speculates that variable amounts of iron and copper in the food, drinking water, or prenatal vitamin-mineral supplements may catalyze varying extents of oxidation of vitamin C in the stomach.

The main concern in a low vitamin C status in late pregnancy relates to the possibility of premature separation of the placenta. In the Vanderbilt University cooperative study of maternal and infant nutrition [121], an association was observed between this complication and low serum ascorbate levels; the correlation was good with serum ascorbate concentration, but poor with dietary vitamin C intake. Clemetson noted a low plasma ascorbate level and elevated whole blood histamine in two patients after abruptio placentae and in one in whom partial separation of the placenta had occurred at 35 weeks. The latter patient was carried to term with cessation of bleeding by treatment with 333 mg sodium ascorbate and 20 mg rutin orally t.i.d. [48].

It would appear advisable for the physician to determine plasma ascorbate concentrations in pregnant patients and to recommend dietary measures and/or ascorbic acid supplements if the ascorbate concentration is much below 1.0 mg/dl, a level below which blood histamine levels may start to rise [48]. A supplement of 100 mg ascorbic acid b.i.d. or t.i.d. should ordinarily be sufficient when elevation of the plasma ascorbate reading cannot be accomplished by dietary means alone. If the patient is taking therapeutic iron, it would be best to take vitamin C at a different time of day [48].

Alcoholism can markedly impair the vitamin C status of mother and fetus during pregnancy. Norkus et al. [122] compared maternal plasma, cord blood plasma, and placental tissue ascorbate levels in alcoholic versus nonalcoholic mothers at delivery, finding marked decreases in all three measures in the alcoholics depsite comparable dietary vitamin C intakes and daily use of prenatal vitamin supplement.

The elderly. According to a large-scale longitudinal study carried out by Garry and his associates [123,124], elderly men and women in good health respond to vitamin C intake similarly to young adults. Dietary and supplemental intakes were related to plasma ascorbic acid levels in 270 free-living, healthy Americans of median age 72 years, all being older than 60. As in young people, plasma ascorbate levels increased rapidly with increasing dietary intake up to 100–150 mg/day after which only slight further increases occurred with the use of supplements; 57% of these individuals used supplements of 500–2,000 mg/day. For any given plasma ascorbate level, women required less vitamin C intake than men, probably because of a smaller bodily distribution volume [123,124]. At the RDA intake of 60 mg/d, the majority of elderly male subjects had plasma ascorbate levels below 0.4 mg/dl, suggesting marginal vitamin C status [124]. Garry et al. recommend as a prudent approach that the elderly receive a daily vitamin C intake that would allow plasma concentration to be maintained at or near 1.0 mg/dl. The daily intakes required to maintain this plasma level would be approximately 125 mg for healthy elderly men and 75 mg for healthy elderly women [124].

Although it is relatively easy to obtain tissue-saturating vitamin C intakes on the American diet, the physician needs to be aware that elderly individuals with very poor dietary habits may develop signs and symptoms of vitamin C deficiency, including follicular hyperkeratosis with coiled, fragmented corkscrewlike body hairs, perifollicular hermorrhages, petechiae, and ecchymoses [125]. Other signs may include swollen gums, malaise, weakness, lassitude, dyspnea, arthralgia, and bone pains. Sublingual petachiae and varicosities may also be a sign of prolonged marginal vitamin C nutriture in the elderly [126,127]. Price [125] recommends a dosage of 100 mg ascorbic acid t.i.d. to correct these conditions. The skin lesions usually clear in a matter of weeks, but severe arthralgias and peripheral neuropathies may require several months to resolve completely [128].

Reports of higher incidence of poor vitamin C status among the elderly have mostly originated from studies outside the United States. Schorah and his associates, for example, have reported the fairly common occurrence of very low plasma ascorbic acid levels among chronically sick elderly women in long-stay hospitals in England [129]; raising their intake to 90–120 mg/day resulted in plateauing of plasma levels at around 1.2–1.3 mg/dl. Schorah et al. [130] had previously reported encouraging clinical results from administration of 1 g ascorbic acid orally once a day in aged long-term institutionalized patients with initially low plasma and leukocyte ascorbate levels. A total of 94 patients were assigned randomly to vitamin C supplement or placebo in a double-blind protocol. After 2 months, the vitamin C group showed, relative to the placebo group, increased plasma and leukocyte ascorbate, and slightly but significantly increased body weight, plasma albumin, and plasma prealbumin. There was also a reduction in purpura and petechial hemorrhages in those receiving the vitamin C supplement. No changes in mood or mobility were noted.

Vir and Love [131] found that 40–50% of elderly individuals surveyed in Belfast, Northern Ireland, had plasma ascorbate levels below 0.3 mg/dl whether they lived at home, were hospitalized, or were in special residential accomodations; all subjects who were receiving multivitamin supplements had levels above 0.3 mg/dl. Bates et al. [132] observed low mean plasma ascorbate concentrations in elderly English men and women living at home, compared with the levels in healthy young adults, and found that plasma and leukocyte levels in the elderly could be elevated to those characteristic of young people by supplementation with 500 mg ascorbic acid per day.

Bates et al. [133] also noted in their studies a positive correlation between vitamin C status, as measured by plasma ascorbate, and plasma HDL-cholesterol level in elderly men but not in women. The same correlation was demonstrated in both men and women, aged 60–99 years, in the Boston area, with favorable effects on HDL-cholesterol levels resulting from self-prescribed use of high-dose vitamin C supplements [94].

In view of the free radical theory of aging [134], it is of interest that Sasaki et al. [135] determined, in studies of serum ascorbate radical by electron spin resonance (ESR) in 200 healthy Japanese male and female subjects, aged 12–90 years, whose only source of vitamin C was their diet, that the ESR intensity of ascorbate radical was positively correlated with the total ascorbic acid level and that both ascorbate radical and ascorbic acid concentration declined with age. Nishikimi [136] demonstrated that ascorbic acid is oxidized by superoxide radical, providing an extracellular defense against superoxide. Fessenden and Verma [137] showed that ascorbic acid quenches the hydroxyl radical, forming ascorbate free radical, while Bielski et al. [138] found that ascorbate radical is a relatively nonreactive species that decays mostly by disproportionation, thereby terminating propagation of free radical reactions. Since plasma ascorbic acid levels of older adults can be increased as readily as those of young people by vitamin C supplementation [123], there is no reason why the elderly cannot receive whatever protection may be afforded by this vitamin against free radical damage, the same as young people. Wartanowicz et al. [139] report that a daily vitamin C supplement of 400 mg given for 12 months to a group of elderly Polish subjects aged 60–100 years lowered their mean serum lipid peroxide levels by 13%; if 200 IU/d of vitamin E was also given, the peroxide level decreased 25%.

Improvement in immune capacity has been noted as a result of daily vitamin C supplementation in the elderly. Kennes et al. [58] found, in a placebo-controlled study with 20 healthy subjects of mean age 71 ± 7 years (8 males, 12 females), that supplementation with 500 mg/d of vitamin C for 1 month significantly enhanced the proliferative response of their T lymphocytes in vitro and their tuberculin skin hypersensitivity in vivo. Favorable trends in the concentrations of plasma proteins with immune functions (IgG, IgM, and C3) were noted in Polish women aged 63–93 years after 1 year of supplementation with 200 mg/d of vitamin C [59].

In sum, the evidence suggests that care of elderly patients should include determination of plasma ascorbic acid levels as well as evaluation of dietary intake. If plasma levels are not at or near 1.0 mg/dl, there may be risks of impaired immune response, inadequate defense against free radical damage, and increased liability to atherogenesis that can easily be avoided by recommendation of a 100-mg supplement once or twice daily. There are other possible benefits. Studies in young adult men [140] indicate that heat acclimation is enhanced by a 250-mg/d vitamin C supplement, while experiments in both laboratory animals [141] and adult men [142] suggest that a high vitamin C status improves acclimation to cold. The elderly adapt less well than the young to either high or low ambient temperatures, so that heat prostration during hot summer days and hypothermia during winter become real hazards to many of the aged. Pending future demonstration that the observations made in young people do or do not apply in the old, it would appear prudent to bring the elderly patient to a high vitamin C status.

Cataract. Jacques and coworkers [143], as part of a case-referral study of incidence of senile cataract and nutritional status in the Boston area, examined the eyes of 113 subjects aged 40–70 years and obtained dietary and supplement ingestion data. There was a significant inverse correlation between the prevalence of cataracts and the intakes of vitamin C and folic acid, measured between highest and lowest quintiles of intake.

Animal studies confirm the benefits of vitamin C in protection of the lens against damage that may lead to cataract formation. Varma [144] attributes genesis of senile cataract to damage by active oxidants and lipid peroxides generated largely by light activation, as well as by degradation products of lipid peroxides such as malondialdehyde and dienes. Hyperbaric oxygen therapy can induce cataracts in both animals [145] and humans [146]. Working with isolated rat lenses in culture media, Varma has demonstrated that ascorbic acid protects the lens membrane against oxygen-dependent photoinjury [144].

Rats provided with drinking water containing 1.2 g of ascorbic acid per liter were resistant to the cataract-inducing action of a 70% galactose diet, exhibiting only 6% of opacified lenses after 84 days, compared with 69% in controls [147]. Blondin et al. [148] compared the effect of UV light exposure in vivo in guinea pigs receiving either the minimum requirement of 2 mg/d or 50 mg/d of ascorbic acid, demonstrating that the higher intake markedly protected lens structural proteins and enzymes against photooxidation versus the lower intake.

Alcoholism. Chronic alcoholics frequently have a low vitamin C status, partly because of poor dietary habits and partly because of the adverse effects of alcohol consumption on vitamin C absorption [149]. Reduced plasma and leukocyte levels of the vitamin are especially common in those with alcoholic liver disease, including Laennec's cirrhosis [149,150]. Multivitamin supplementation, including high-dosage vitamin C, is recommended in the nutritional rehabilitation of alcoholics [1]. The initial deficiency may be difficult to correct by oral dosages of as much as 500 mg/day, requiring many days to several weeks in some patients [1]. Even intravenous therapy may require considerable time to achieve tissue and plasma saturation. Majumdar et al. [151] report results of treating 25 hospitalized English alcoholics intravenously with 500 mg ascorbic acid daily in an injectable multivitamin mixture providing at the same time 1,000 mg glucose, 250 mg thiamin hydrochloride, 50 mg pyridoxine hydrochloride, and 160 mg nicotinamide. Initial mean leukocyte ascorbate concentration was 12 ± 5 mcg/10^8 WBC (normal healthy controls, 20–50 mcg/10^8 WBC). After 5 days on IV therapy, mean leukocyte ascorbate had risen only to 19 ± 6 mcg/10^8 WBC, and only 9 of the 25 patients had leukocyte readings higher than 20 mcg/10^8 WBC. Majumdar et al. recommend prolonged replenishment with vitamin C in chronic alcoholics, with IV administration being the preferred route in resistant cases [151].

Vitamin C has been demonstrated in laboratory animals to protect against the anesthetic and lethal effects of acetaldehyde, the first intermediate formed in the metabolism of ethanol [1,152]. This suggests potential usefulness of IV ascorbic acid in emergency treatment of acute alcohol intoxication, but clinical reports on this possible application have not appeared as of this writing.

Periodontal disease. Using a double-blind experimental design, Aurer-Kozelj et al. [153] examined by electron microscopy the effect of ascorbic acid supplementation on periodontal tissue ultrastructure. The subjects were 11 men and 10 women (age range, 24–36 years) being treated in a Yugoslavian outpatient clinic for caries, but otherwise healthy. They were divided into two groups, 10 receiving placebo, the other 11 subjects 70 mg of ascorbic acid orally once a day. Initial plasma vitamin C levels of the 21 subjects were 0.18–0.30 mg/dl, reflecting their dietary intakes of 20–40 mg/day. Initial examination of interproximal gingival tissue showed that practically all subjects were affected by progressive periodontitis with marked changes in the connective tissue of the gingival lamina propria. After 6 weeks of supplementation, the mean plasma ascorbic acid concentration in the ascorbic acid group had risen from an initial 0.28 to 1.31 mg/dl, while the placebo group showed little change. Electron microscopy of gingival tissue showed in the experimental group, but not in the controls, connective tissue with signs of intensive cell activities characteristic of regeneration, including an increased number of collagen bundles in the periphery of fibroblasts, increased tonofibril content, and an increased number of desmosomes between adjacent cells. The authors considered these findings to represent an early phase in regeneration, implying that sustained ascorbic acid saturation or near-saturation is needed for optimal health of gingival tissue.

Bed sores (pressure sores). Taylor and associates [154] in England evaluated, in a double-blind, placebo-controlled trial, the effect of a twice-daily 500-mg dose of ascorbic acid on the healing of bed sores. Twenty surgical patients (8 males, 12 females) of ages 54–88 years (mean age 74.5 years) were assigned to the vitamin C or placebo group in order according to year of birth. Both groups received similar local therapy to the pressure areas. After 1 month, the leukocyte ascorbate concentration in the placebo group remained essentially unchanged, while that of the ascorbic acid group increased from a mean of 22 to 65.5 mcg/10^8 WBC (method of Denson and Bowers [155]). In the ascorbic acid group, there was a mean reduction of 84% in pressure-sore area after 1 month, compared with 42.7% in the placebo group. These patients had probably been on diets prior to surgery that failed to provide tissue ascorbate saturation, and the regular hospital diet consumed by both groups did not remedy this situation, although it probably met the British RDA for vitamin C and provided other essential nutrients in adequate amounts. Prescription of 500 mg ascorbic acid to be taken once or twice daily, both before and after surgery, should ensure against marginal vitamin C status as a contributing cause to bed sores. The vitamin C status of patients bedridden for other reasons should also be carefully monitored.

Surgery, burns, and trauma. There is a large literature on the use of supplemental ascorbic acid to aid in healing of wounds and burns; this has been reviewed by Irwin and Hutchins [53] and by Jaffe [1]. Oral dosages of 1,000–2,000 mg/day have been recommended pre- and postsurgery to assure optimal synthesis and crosslinking of collagen fibers in regenerating wound tissue. Abrahamian et al. [156] have described dosages and modes of administration of intravenous ascorbic acid in patients receiving long-term total parenteral nutrition.

Crohn's disease. Hughes and Williams [157] found significantly lower leukocyte ascorbate levels in Crohn's disease patients than in normal controls and concluded that this was due to the disease rather than to drug therapy or surgical treatment. They recommend vitamin C supplementation to correct marginal ascorbic acid nutriture in such patients, as persistence of a low ascorbate status could impair collagen formation, increasing the risk of spontaneous fistula formation. In addition to an otherwise adequate diet, a 100-mg supplement one or twice daily should ensure tissue saturation. If the patient is badly depleted, a daily 500-mg supplement may be required for 1 or 2 weeks to bring leukocyte and plasma levels up more rapidly, after which a lower dosage may be satisfactory.

Urine acidification. Oral vitamin C in dosages up to 6 or 8 g/d has often been used as a urine acidifier, although its effectiveness has been questioned. Naccarto et al. [158] found oral dosages of 1.0–6.0 g/d relatively ineffective in lowering urinary pH in 73 elderly patients receiving methenamine and with indwelling Foley catheters; effectiveness was increased by multiple daily dosage. In a short-term study (4 days), Nahata et al. [159] found doses up to 6 g/d to be ineffective in adults, and Barton et al. [160] reported that a single 2-g dose administered IV did not lower urinary pH in the following 3 hours.

On the other hand, Ekvall et al. [161] report the routine and successful use of daily dosages of 1,500–1,800 mg oral ascorbic acid, administered in four equal portions at about 4-hour intervals, to maintain urinary pH of 5.5–6 in myelomeningocele (spina bifida) patients aged 6–18 years. In healthy male volunteers, a dosage schedule of 1,000 mg five times daily for 5 days maintained urinary pH at 5.7–6.0; a dosage of 2,000 mg five times daily produced no further lowering of pH [162]. Others have reported regular and apparently successful use as a urine acidifier in paraplegic adults with a dosage of 1,000 mg ascorbic acid q.i.d. [163]. Multiple daily doses appear to be necessary, and efficacy as an acidifier apparently cannot be judged in short-term experiments.

Allergy. Zuskin and colleagues reported reduction by supplemental vitamin C of bronchoconstriction induced experimentally in humans by histamine [164], and occurring in industrial workers, by inhalation of hemp dust [165] and flax dust [166]. Vitamin C supplements did not appear of benefit in asthma patients [167]. High-dose vitamin C treatment significantly reduced anaphylactic deaths after IV injection of ovalbumin in mice previously sensitized to this protein [168,169].

Male infertility. Dawson and his associates [170] have found vitamin C supplements to be effective therapy in cases of male infertility characterized by sperm agglutination in excess of 25%. Their practice is to prescribe 1,000 mg/d for 10 days and then 500 mg/d through two menstrual cycles (60 days). Comparing 20 infertile men on this therapy with 10 who did not receive a vitamin C supplement, they found that at the end of 60 days all of the vitamin C group had impregnated their wives, while none of the unsupplemented group reported pregnancies. The vitamin C group showed increases in sperm count and motility and decreases in agglutination, abnormal sperm, and sperm precursors. Seminal plasma ascorbate levels were very low in these infertile men and could be elevated to normal by 1 week of 1,000 mg of vitamin C daily, but a supplement of 200 mg/d raised seminal plasma ascorbate level to only two-thirds of normal after 3 weeks.

Sickle cell anemia. Jain and Williams [171] find that sickle cell anemia patients tend to have below-normal plasma ascorbate levels, which may contribute to increased peroxidation of red cell membranes.

Burns and frostbite. Spillert and colleagues report that IV administration of ascorbic acid, followed by daily oral supplementation in drinking water, markedly reduced the severity of experimental burns [172] and tail frostbite [173] in mice. Comparable data for humans have not been reported.

Cancer. Cameron et al. [174] have reviewed the theoretical and hypothetical reasons why ascorbic acid might be expected to show anticancer properties and have summarized the available in vitro, animal, and human data on the subject as of 1979. Besides general elevation of immunocompetence, they hypothesize that high-dose vitamin C will increase the strength of the intracellular matrix and its resistance to penetration by invasive tumor cells. In addition, certain of the chemical characteristics of ascorbic acid, such as its antioxidant properties and its ability to inhibit synthesis of nitrosamines and detoxify certain other carcinogenic and mutagenic compounds, suggest the possibility that ascorbic acid supplements may have a useful place in prophylaxis and therapy of cancer.

In the prospective Basel population study relating plasma vitamin levels to subsequent cancer mortality, a statistically significant inverse association of plasma ascorbate levels was found with stomach cancer, but not cancer at other sites [175].

Studies of the effect of ascorbic acid on experimental cancer in vitro and in laboratory animals have given mixed results [174]. It is quite clear that ascorbic acid can reduce nitrite in a dose-related manner, being oxidized to dehydroascorbic acid in the process [176], thereby inhibiting or blocking the reaction of exogenous or endogenously produced nitrite with amines in ingested food to produce potentially carcinogenic nitrosamines [177–181]. High-dose vitamin C also partially inhibited liver damage produced by administration of preformed dimethylnitrosamine [179], an effect that may be the result of another chemical property of ascorbate, namely the ability to form an alternate anion form that reacts readily with alkylating agents such as nitrosamines [182].

Reddy et al. [183] found a lower incidence of colon and kidney tumors in rats given a single injection of the carcinogen 1,2-dimethylhydrazine (DMH) if their diets were supplemented with 0.25% or 1.0% sodium ascorbate, compared with controls on unsupplemented diet; however, if multiple weekly injections of DMH were given, incidence was the same in the two groups. Mirvish et al. [184] treated two groups of rats for 2 years with dietary morpholine and sodium nitrite in their drinking water with and without supplemental sodium ascorbate in their diets. Ascorbate delayed induction of liver tumors substantially and lowered incidence moderately, but did not lower the incidence of forestomach tumors [185]. High-dose ascorbic acid failed to postpone mammary tumor induction, reduce tumor incidence, or prolong survival time of rats treated with the carcinogen 7,12-dimethylbenzanthracene (DMBA), and also failed to inhibit the growth of transplantable mammary tumors [186]. Banic [187] actually noted a more rapid induction of tumors in adult guinea pigs receiving high-dose vitamin C by injection, compared with controls on regular diet, but after 584 days there was no difference between the two groups in tumor number. Kawasaki et al. [188] found that in rats on high dietary ascorbic acid, the incidence of gastric cancer induced by N-methyl-N'-nitro-N-nitroso-guanidine in their drinking water was not effectively inhibited; however, infiltrative growth was significantly repressed by heavy proliferation of connective tissue fibers around the tumors, as demonstrated by Asan-stain histologic examination. This evidence supports one aspect of the Cameron et al. hypothesis [174].

Results of human studies are similarly equivocal. Cameron and Pauling

[189–191] have reported significant prolongations of lifespan in Scottish patients with advanced cancer who were given 10 g ascorbic acid per day orally in divided doses, compared with similar patients who did not receive vitamin C supplements. Creagan and Moertel and their coworkers at the Mayo Clinic, on the other hand, observed no significant benefits in double-blind studies with randomized assignment of advanced cancer patients to placebo or 10 g ascorbic acid/d in five divided doses; neither patients with nor those without prior chemotherapy responded to vitamin C in terms of symptoms, performance status, appetite, weight, or survival time [192,193]. The studies of both groups may be criticized for failure to determine dietary vitamin C intake and plasma and leukocyte ascorbate levels initially and during therapy. Neither experimental protocol rules out the possibility that those cancer patients who are initially in low vitamin C status are the ones who may respond favorably to tissue-saturating or higher doses of ascorbic acid, while those who are already in high vitamin C status may not show any further response.

On the positive side, DeCosse and coworkers report that in trials running up to 2 years in patients with polyposis coli (familial polyposis), administration of 1 g ascorbic acid t.i.d. resulted in a significant reduction of polyp area as compared with placebo controls [194,195]; the dosage form was timed-release pellets in gelatin capsules (Ascorbicap®, ICN Pharmaceuticals). Bruce and Dion [196] report that fecal content of a partially characterized, isolatable mutagen often found in human stools was significantly reduced by inclusion of 400 mg ascorbic acid in the daily diet; the mutagen was not present in the food but was formed in the gut. Patients with recurrent polyposis given 400 mg each of vitamin C and alpha-tocopherol daily for 2 years appeared to have about a 20% reduction in polyp recurrence, compared with subjects on placebo, but more extensive trials are needed to confirm benefit [197].

In Czechoslovakia, Sram and coworkers determined that 1 g asorbic acid per day, 5 days per week for 3 months, significantly reduced chromosome aberrations in the peripheral blood lymphocytes of workers occupationally exposed to coal tar [198] and halogenated ethers [199], both of which are mutagenic and potentially carcinogenic. Ascorbic acid treatment significantly reduced chromatid and chromosome breaks and chromatid and chromosome exchanges. The subjects of both investigations were initially in low or relatively low vitamin C status, as evidenced by mean plasma ascorbate concentrations of about 0.2 mg/dl in one study [198] and 0.4–0.6 mg/dl in the other [199]. The vitamin C supplement raised mean plasma ascorbate levels to 1.1–1.6 mg/dl.

Wasserthal-Smoller et al. [200] carried out an epidemiologic study of the incidence of uterine cervical dysplasia in American women, noting that younger age, greater frequency of sexual intercourse, and younger age at first intercourse were positively associated with risk of cervical dysplasia. Among possible nutritional factors, low vitamin C intake was found to be an independent contributor to risk of severe cervical dysplasia when age and sexual activity variables were controlled. They point out that about 35% of U.S. women in their reproductive years have vitamin C intakes below 30 mg/day, and 68% have intakes below 88 mg/day (as opposed to tissue-saturating intakes of 100–150 mg/day). In this study, the mean daily vitamin C intake of 87 cervical dysplasia patients was 80 mg, while that of 82 dysplasia-free controls was 107 mg.

Schlegel [201] and his associates at Tulane University's Urology Clinic have explored the ability of excreted urinary ascorbic acid to prevent the oxidation and activation of carcinogenic aromatic amines. It was found that oxidation of an environmentally derived urinary contaminant, hydroxylated beta-naphthylamine, a known carcinogen, could be prevented by urinary ascorbic acid concentrations of 5 mg/dl or higher; a vitamin C intake of 300 mg/day or more produced such concentrations. Oxidation of the endogenously derived 3-hydroxyanthranilic acid, one of several potentially carcinogenic tryptophan metabolites normally found in urine, was also inhibited by ascorbic acid. Schlegel has therefore recommended that other urology centers join in the evaluation of the potential usefulness of ascorbic acid in preventing or inhibiting the development of bladder cancer. In their own clinic, since administration of ascorbic acid in moderate dosage is safe, they are routinely prescribing supplements of 500 mg t.i.d. for individuals at risk of recurrence of bladder cancer or those whose environmental conditions (including heavy smoking) might increase their risk of bladder carcinoma.

Overall, it appears likely that raising vitamin C status to tissue saturation may be advantageous in inhibiting cellular changes that lead to cancer in the first instance, as well as in increasing resistance to the growth or spread of established cancer. Whether or not elevation of ascorbate in plasma and other body fluids to unusually high levels by multiple daily dosages up to 10 g/day has any further benefits in prevention or treatment of cancer remains in question. A possible use for ascorbic acid as an adjunct to cancer therapy with Adriamycin® doxorubicin is suggested by the fact that intraperitoneal injection of high-dose ascorbic acid reduced lipid peroxide levels and cardiomyopathy in guinea pigs injected i.p. with this antitumor drug [202].

TOXICITY AND SIDE EFFECTS

Early hominids were confined to the tropics and lived mostly on vitamin C-rich fruits and vegetables, supplemented with animal protein from insects, frogs, and the like, and later by scavenging the kills of large predators [203]. With the advent of *Homo erectus*, a hunting-gathering way of life emerged that prevailed for most of humankind's existence until a few thousand years ago and still persists in a few areas remote from industrialized cultures.

Berlin and Markell [204] studied the nutritional and health status of the Aguaruna tribe of South American Indians, who live in a 22,000-km^2 tropical rain forest area in northcentral Peru, a habitat generally similar to that of early man. The Aguaruna are still a hunting culture (as of the time of the study, 1972–1975), but cultivate most of the rest of their food. Manioc, bananas, and plantains account for about 75% of calorie intake, but in all they cultivate 46 food plants, including roots, tubers, rhizomes, leaves, legumes, grains, and a large variety of fruits. Berlin and Markell measured dietary intakes of five closely studied families and calculated nutrient intakes from published food composition tables. Calculated within-family vitamin C intakes ranged, in the five families, from 350 to 870 mg/d per person, including both adults and children. Eaton and Konner [205], from available knowledge of the dietary habits of paleolithic *Homo sapiens*, estimate that the average vitamin C intake was close to 400 mg/d. Pauling [44] has calculated that 2,500 kcal/d derived solely from raw plant foods of the type available to early man would have provided 2.3 g of ascorbic acid (average of 110 foods). It could be argued, therefore, that vitamin C intakes of up to about 2 g/d can be considered within the physiologic range for the human species, even though only 100–150 mg/d is ordinarily needed to saturate body tissues, and the requirement to prevent scurvy is another order of magnitude lower still.

The most frequently expressed concern about high-dose ascorbic acid is that it might enhance urinary excretion of oxalic acid and increase the risk of calcium oxalate stone formation. Briggs [206] reported that in a group of 67 healthy adult volunteers treated with 4 g ascorbic acid daily for 7 days, three experienced 10- to 20-fold increases in oxalic acid excretion. However, other investigators found little or no increase in urinary oxalic acid of healthy subjects given ascorbic acid in dosages up to 4 g/d and increases of about twofold at dosages of 9 g/d [207–209]. One difficulty in carrying out such studies is the variation in urinary oxalic acid levels produced by the normally wide changes in dietary oxalate consumption, which may range from 70 to 980 mg/day, depending on the types of food consumed [210,211]; the experimental diet must therefore be closely controlled. In a carefully controlled study of oxalate excretion in healthy male volunteers given 2 g ascorbic acid five times daily, Schmidt et al. [162] found that mean urinary oxalate excretion was increased from about 50 mg to 87 mg/d (range 60–126 mg/d). The time-course of oxalate excretion revealed that after the third dose of 2 g ascorbic acid, the urinary oxalate excretion rate reached a plateau of 0.6 mcg/ml/min that was not exceeded with the additional 2-g doses of ascorbic acid. It appears, therefore, that in normal healthy adults, dosages of ascorbic acid as high as 10 g/d are not likely to increase significantly the risk of calcium oxalate stones.

Studies with radiolabeled ascorbic acid indicate that ascorbate contributes only a minor amount (about 5% in one patient studied in detail) to the several-fold enhancement above normal of oxalic acid excretion seen in primary hyperoxaluria [212,213]. One report, however, suggests that new stone formation may have been activated in five patients with a known history of calcium urolithiasis after they had taken 4 g or more of ascorbic acid daily for long periods [214]. Stone formation became active, following a long period of quiescence, 1–3 years after high-dose vitamin C supplementation had begun. *A history of calcium oxalate stone formation therefore dictates caution in the prescribing of vitamin C supplements.* Secondary hyperoxalemia has been reported in hemodialysis patients receiving supplemental vitamin C [100]; *ingestion of more than RDA amounts of ascorbic acid, including that provided by the diet, is contraindicated for dialysis patients.*

Stein et al. [215] noted an increase in urinary urate excretion with single doses of 4 g ascorbic acid in fasting subjects, but not with 2-g doses. Chronic administration of 2 g q.i.d. was also reported to increase clearance of uric acid [216]. Schmidt and coworkers, however, found no increase in uric acid excretion in healthy men receiving 2 g ascorbic acid five times daily for 5 days [162]. No cases of uric acid stones attributable to vitamin C ingestion have been reported.

Fears have been expressed that high vitamin C intakes might result in "rebound" scurvy after cessation of supplementation and return to usual dietary intakes. While there is some evidence of induction of an enhanced catabolism of ascorbate in maternal guinea pigs on high intakes and in their offspring [217,218], the effect is transient: upon return to a diet low or normal in vitamin C content, the enhanced ascorbate catabolism of the offspring returned to normal within 20 days [219]. Man is not known to have any catabolic system for ascorbate that is subject to comparable induction, the pathway to oxalate being saturable at very low output rates [162]. Controlled studies in men and women of various ages during prolonged intake of dosages as high as 4 g/d have failed to indicate any rebound scorbutic effect upon return to usual dietary intakes [86,220,221].

Effects on calcium metabolism have been investigated in view of the possible chelating action of ascorbic acid. Thornton [222] showed that injected ascorbic acid (10 mg/100 g/d) in the young chicken caused mobilization and excretion of calcium from bone during the initial 24 hours but had no further effect on continued treatment. Keith et al. [223] found elevated urinary calcium in female rats fed diets containing 2% ascorbic acid for 10 months. Weanling male guinea pigs fed diets with 8.7% ascorbic acid (equivalent to over 40 g/d for a human) showed reduced bone density after 6 weeks [224]. However, adult male chickens suffered no adverse effects in bodyweight, tibia ash, or tibia calcium content after 224 days on 1% dietary ascorbic acid (equivalent to about 5 g/d for a human [225]). In healthy men, Schmidt et al. [162] observed a slight reduction in urinary calcium excretion in subjects receiving 10 g/d (2 g ascorbic acid five times daily) for 5 days. No adverse effects on calcium or bone metabolism in humans have been reported at dosages of 10 g/d or less.

Erythrocytes of healthy volunteers receiving 5 g ascorbic acid per day showed in vitro an increased sensitivity to the lytic effects of hydrogen peroxide, but there was no evidence of hemolysis in vivo [226]. Excessive dosing with injected ascorbic acid may have contributed to the death of an elderly black man with glucose-6-phosphate dehydrogenase (G-6-PD) deficiency [227]. As part of treatment for second-degree burns of the hand, he was given 80 g ascorbic acid intravenously on each of 2 consecutive days, presumably with the thought that ascorbic acid would accelerate healing. The next day he became oliguric and showed extensive hemolysis. He was transferred in a comatose condition to a university medical center where he received hemodialysis, but his neurologic status steadily deteriorated, and he expired 3 weeks later. Why such enormous doses of ascorbic acid were used in the initial treatment is not disclosed, but they were large enough to subject the erythrocytes to considerable osmotic stress if not administered very slowly. The university investigators who supervised final treatment suggest that caution and appropriate testing be used before administration of large doses of ascorbic acid to individuals who might be susceptible to G-6-PD deficiency [227].

Patients with iron overload, such as those with beta-thalassemia and other hemolytic disorders, may experience scurvy because of iron-catalyzed destruction of ascorbate in the body, but vitamin C therapy should be undertaken with caution, as it may mobilize iron, cause rapid rises in transferrin saturation, serum ferritin, and the chelatable pool of iron; concomitant therapy with deferoxamine may be advisable to prevent iron toxicity [228]. Patients with hemochromatosis should avoid taking vitamin C supplements with meals, as vitamin C increases the absorption of iron from food [229].

Early studies with incompletely purified dehydroascorbic acid suggested that it might have diabetogenic properties when injected into laboratory animals, but Domke and Wais [230] could find no hyperglycemia or decreased glucose tolerance in rats injected IV with highly purified dehydroascorbic acid.

A female infant with congenital nephrotic syndrome experienced deterioration at 18–23 months of age to a point where long-term peritoneal dialysis was required [231]. During the period of deterioration, she had been receiving the very high dosage of 1.8–2.0 g ascorbic acid (200–250 mg/kg) daily. Although the decline in renal function may have been merely coincidental with ascorbic acid administration, nephrologists should be aware of the possibility of adverse effects of high-dose vitamin C in patients with compromised renal function.

Diarrhea and upper gastrointestinal-tract discomfort have been reported with high-dose vitamin C supplementation [1], but Creagan et al. [192] did not see any differences in incidence of such symptoms between placebo-treated and ascorbate-treated cancer patients receiving 2 g ascorbic acid five times daily.

Intravenous administration of ascorbic acid at slow rates in amounts of 5–10 g/d has been carried out without adverse incident [63–66]. Precautions to be used in intravenous use of ascorbic acid have been discussed by Abrahamian et al. [156].

INTERACTIONS

A number of drugs affect ascorbate metabolism, and conversely, high-dose vitamin C may affect the potency of certain drugs. Chronic therapy with aspirin and salicylates decreases the uptake of vitamin C in leukocytes [232], and studies in guinea pigs indicate reduced absorption of the vitamin when coadministered with aspirin [233]. It has been suggested that tissue ascorbate desaturation may contribute to aspirin-induced hematemesis [234]. Since ascorbic acid is metabolized in part to ascorbic acid sulfate, it may compete for sulfation with orally administered drugs that are conjugated (and inactivated) by sulfation in the first pass through the intestinal wall and liver [235].

Tests in female adult rhesus monkeys indicate a faster turnover of ascorbic acid when an oral contraceptive (mestramol-norethindrone) drug was taken, suggesting an increased requirement for vitamin C [236]. Conversely, high-dose vitamin C (1-g doses) increases the bioavailability of ethinylestradiol, probably by decreasing its sulfation in the intestinal wall [237], making supplementation in large doses questionable for oral contraceptive users.

In rats, high-dose ascorbic acid potentiated the antiamphetamine and cataleptogenic effects of haloperidol [238]. A young male manic-depressive patient being treated with fluphenazine was given 500 mg ascorbic acid b.i.d. to elevate his low plasma ascorbate level; this raised plasma ascorbate from 0.3 to 0.7 mg/dl within 6 days, but plasma fluphenazine level fell and symptoms worsened until vitamin C was discontinued; vitamin C may have induced higher levels of drug-metabolizing enzymes [239]. Guinea pigs treated with the appetite-suppressant drugs fenfluramine, mazindol, and diethylpropion showed significant reductions in brain ascorbate levels after 24 days; supplemental dietary vitamin C significantly inhibited the anorectic action of these drugs [240].

Simultaneous ingestion of alcohol (35 g) with 2 g ascorbic acid at a meal markedly depressed early absorption of the vitamin and kept plasma ascorbate levels at a reduced level for at least 24 hours, compared with the results of trials without ingestion of alcohol [241].

Therapeutic iron supplements should be taken at another time of day than oxidizable nutrients such as vitamins C and E to avoid iron-catalyzed oxidation in the GI tract. However, at the levels of iron occurring in foods, vitamin C of either food or supplemental origin is beneficial in increasing nonheme iron absorption, but Cook and Monsen [229] caution that supplemental vitamin C at mealtimes may cause excessive absorption of iron in patients who are subject to iron overload (idiopathic hemochromatosis, thalassemia major, sideroblastic anemia).

In animal studies, high-dose vitamin C has shown the ability to detoxify or reduce the adverse effects of a number of environmental toxins, including PCBs (polychlorinated biphenyls) [242,243], organophosphorus insecticides [244], and chlordane [225], suggesting that individuals occupationally or otherwise exposed to these pollutants should be kept in a high vitamin C status. In this connection, it is of interest that an epidemiologic survey by the Centers for Disease Control, Atlanta, Georgia, of 458 persons in Triana, Alabama, found positive correlations between serum PCB levels and serum cholesterol level and between serum PCB and blood pressure; these associations were independent of age, sex, body mass index, and social class [246]. High-dose oral ascorbic acid in laboratory animals also provides some degree of protection against sodium cyanide toxicity [241], against acetaldehyde [248], and against a combination of acetaldehyde, nicotine, and caffeine [249]. The protection of ascorbic acid against the carcinogenic action of nitrite and nitrosamines has been discussed in the Cancer section of this chapter [177–181].

Experimental animals were protected by high dietary vitamin C from much of the growth-depressing activity of toxic levels of salts of cobalt, selenium, vanadium, and cadmium [250]; mercury toxicity was increased by 0.2% dietary ascorbic acid but not by 1.0% ascorbic acid. Ascorbic acid in drinking water was as effective as EDTA in preventing lead toxicity in the rat, and the combination of vitamin C and EDTA was synergistically effective, especially in removing lead from the brain [251]. Ascorbic acid was also protective against nickel sulfate toxicity in rats [252].

In young adult men, use of vitamin C supplements (500 mg t.i.d.) for 64 days resulted in some reduction of serum copper levels and ceruloplasmin activity, with reversal on termination of supplementation [253]. The changes were within physiologic limits, but the data suggest that high-dose vitamin C might over the long term be adverse to copper nutritional status if the diet is marginal or low in copper (RDA 2–3 mg/d).

Experimental vanadium poisoning in animals can be ameliorated by ascorbic acid supplements, and Naylor and Smith, based on their findings of higher plasma vanadium levels in manic-depressives than in normal individuals, have evaluated ascorbic acid (1 g t.i.d.) alone and in combination with EDTA in treatment of manic-depressive patients, obtaining clinical improvement in some individuals, particularly with a combination of the two drugs [254].

An early report that vitamim B-12 might be unstable in the presence of high-dose vitamin C proved groundless, an error in analytical procedure having been responsible for this mistaken conclusion [255,256]. Individuals who had routinely taken 2–10 g/d of vitamin C for at least 2 years, along with a daily supplement of "therapeutic strength" multivitamins, were found to have elevated circulating levels of vitamin B-12 [257].

An in vitro study suggested that high levels of ascorbic acid in urine might interfere with the Cu-reduction glucose test, but subsequent trials in nondiabetics in vivo with 4–6 g/d of ascorbic acid [258], up to 27 g/d using the two-drop Clinitest® prodecure [259], and 4 g/d with methenamine in elderly patients with Foley catheters, again using two-drop Cu-reduction [260], failed to show false-positive tests for glucose.

Discontinuance of supplemental ascorbic acid for 48–72 hours is advisable before testing the feces for occult blood, as ascorbic acid in the stool may cause false-negative results in the Hemoccult® and Derman tests [261].

REFERENCES

1. Jaffe GM: Vitamin C. In: *Handbook of Vitamins: Nutritional, Biochemical, and Clinical Aspects*, L.J. Machlin (ed). New York: Marcel Dekker, 1984, pp 199–244.
2. Halliwell B: Oxidants and human disease: Some new concepts. FASEB J 1:358–364, 1987.
3. Machlin LJ, Bendich A: Free radical damage: Protective role of antioxidant nutrients. FASEB J 1:441–445, 1987.
4. Anderson R, Lukey PT: A biological role for ascorbate in the selective neutralization of extracellular phagocyte-derived oxidants. Ann NY Acad Sci 498:229–247, 1987.
5. Diliberto EJ Jr, Menniti FS, Knoth J, et al.: Ardrenomedullary chromaffin cells as a model to study the neurobiology of ascorbic acid: From monooxygenation to neuromodulation. Ann NY Acad Sci 498:28–53, 1987.
6. Morré DJ, Crane FL, Sun IL, Navas P: The role of ascorbate in biomembrane energetics. Ann NY Acad Sci 498:153–171, 1987.
7. Rose RR: Ascorbic acid protection against free radicals. Ann NY Acad Sci 498:506–508, 1987.
8. Niki E: Interaction of ascorbate and alpha-tocopherol. Ann NY Acad Sci 498:186–199, 1987.
9. Kunert K-J Tappel AL: The effect of vitamin C on in vivo lipid peroxidation in guinea pigs as measured by pentane and ethane production. Lipids 18:271–274, 1983.
10. Evans CD, List GR, Dolev A, et al.: Pentane from thermal decomposition of lipoxidase-derived products. Lipids 2:432–434, 1967.
11. Dillard DJ, Litov RE, Tappel AL: Effects of dietary vitamin E, selenium, and polyunsaturated fats on in vivo lipid peroxidation in the rat as measured by pentane production. Lipids 13:396–402, 1978.
12. Downey JE, Irving DH, Tappel AL: Effects of dietary antioxidants on in vivo lipid peroxidation in the rat as measured by pentane production. Lipids 13:403–407, 1978.

13. Glembotski CC: The role of ascorbic acid in the biosynthesis of the neuroendocrine peptides alpha-MSH and TRH. Ann NY Acad Sci 498:54–62, 1987.
14. Hallberg L, Brune M, Rossander-Hulthén L: Is there a physiological role of vitamin C in iron absorption? Ann NY Acad Sci 498:324–332, 1987.
15. Olson JA, Hodges RE: Recommended dietary intakes (RDI) of vitamin C in humans. Am J Clin Nutr 45:693–703, 1987.
16. Jacob RA, Skala JH, Omaye ST: Biochemical indices of human vitamin C status. Am J Clin Nutr 46:818–826, 1987.
17. VanderJagt DJ, Garry PJ, Bhagavan HN: Ascorbic acid intake and plasma levels in healthy elderly people. Am J Clin Nutr 46:290–294, 1987.
18. Stevenson NR: Active transport of L-ascorbic acid in the human ileum. Gastroenterology 67:952–956, 1969.
19. Mayersohn M: Ascorbic acid absorption in man— pharmacokinetic implications. Eur J Pharmacol 19:140–142, 1972.
20. Nelson EW, Lane H, Fabri PJ, Scott B: Demonstration of saturation kinetics in the intestinal absorption of vitamin C in man and the guinea pig. J Clin Pharmacol 18:325–335, 1978.
21. Yew M-L S: Megadose vitamin C supplementation and ascorbic acid and dehydroascorbic acid levels in plasma and lymphocytes. Nutr Rep Int 30:597–601, 1984.
22. Tsujimura M, Fukuda T, Kasai T: On the reduction of dehydroascorbic acid in guinea pig and rat. Nutr Rep Int 28:881–890, 1983.
23. Sevanian A, Davies KJA, Hochstein P: Conservation of vitamin C by uric acid in blood. J Free Rad Biol Med 1:117–124, 1985.
24. Tsujimura M, Fukuda T: Changes of vitamin C excreted into urine after oral administration of dehydroascorbic acid to human. Nutr Rep Int 31:1001–1008, 1985.
25. Hodges RE: Ascorbic acid. In: *Modern Nutrition in Health and Disease*, RS Goodhart, ME Shils (eds). Philadelphia: Lea & Febiger, 1980, pp 259–273.
26. Moser U: Uptake of ascorbic acid by leukocytes. Ann NY Acad Sci 498:200–215, 1987.
27. Oke AF, May L, Adams RN: Ascorbic acid distribution patterns in the human brain: A comparison with nonhuman mammalian species. Ann NY Acad Sci 498: 1–12, 1987.
28. Pirie A, Van Heyninger R: In: *Biochemistry of the Eye*, Springfield, IL: CC Thomas, 1956, pp 24–28.
29. Davson H In: *The Eye*. vol. 1, chapt. 2, The intra-ocular fluids. New York: Academic Press, 1962. p 90.
30. Varma SD, Kumar S, Richards RD: Light-induced damage to ocular cation pump: Prevention by vitamin C. Proc Natl Acad Sci USA 76:3504–3506, 1979.
31. Kuebler W, Gehler J: Zur Kinetik der enteralen Ascorbinsäure-Resorption zur Berechnung nicht dosisproportionaler Resorptionsvorgänge. Int J Vitam Nutr Res 40:442–453, 1970.
32. Hornig D, Vuilleumier J-P, Hartmann D: Absorption of large, single, oral intakes of ascorbic acid. Int J Vitam Nutr Res 50: 309–314, 1980.
33. Melethil S, Subrahmanyam MB, Chang CJ, Mason WD: Megadoses of vitamin C: A pharmacokinetic evaluation. Ann NY Acad Sci 498:491–493, 1987.
34. Angel J, Alfred B, Leichter J, et al.: Effect of oral administration of large quantities of ascorbic acid on blood levels and urinary excretion of ascorbic acid in healthy men. Int J Vitam Nutr Res 45:237–243, 1975.
35. Kellner A, Hornig D, Pelikka R: Formation of carbon dioxide from ascorbate in man. Am J Clin Nutr 41: 609–613, 1985.
36. Kallner A, Hartmann D, Hornig D: Steady-state turnover and body pool of ascorbic acid in man. Am J Clin Nutr 32:530–539, 1979.

37. Kallner AB, Hartmann D, Hornig DH: On the requirements of ascorbic acid in man: Steady-state turnover and body pool in smokers. Am J Clin Nutr 34:1347–1355, 1981.
38. Smith JL, Hodges RE: Serum levels of vitamin C in relation to dietary and supplemental intake of vitamin C in smokers and nonsmokers. Ann NY Acad Sci 498:144–152, 1987.
39. Yung S, Mayersohn M, Robinson JB: Ascorbic acid absorption in man: Influence of divided dose and food. Life Sci 28:2505–2511, 1981.
40. Yung S, Mayersohn M, Robinson JB: Ascorbic acid absorption in humans: A comparison among several dosage forms. J Pharm Sci 71: 282–285, 1982.
41. Zetler G, Seidel G, Siegers CP, Iven H: Pharmacokinetics of ascorbic acid in man. Eur J Clin Pharmacol 10:273–282, 1976.
42. *U.S. Pharmacopeia,* 21st revision. Rockville, MD: U.S. Pharmacoperial Convention, Inc., 1985.
43. Briggs M: Vitamin C and infectious disease: A review of the literature and the results of a randomized, double-blind, prospective study over 8 years. In: *Recent Vitamin Research,* MH Briggs (ed). Boca Raton, FL: CRC Press, 1984, pp 39–81.
44. Pauling L: *Vitamin C and the Common Cold.* San Francisco: W.H. Freeman & Co. 1970.
45. Pauling L: *Vitamin C, the Common Cold and the Flu,* San Francisco: W.H. Freeman & Co., 1976.
46. Baird IM, Hughes RE, Wilson HK, et al.: The effects of ascorbic acid and flavonoids on the occurrence of symptoms normally associated with the common cold. Am J Clin Nutr 32:1686–1690, 1979.
47. Stacpoole PW: Role of vitamin C in infectious disease and allergic reactions. Med Hypotheses 1:42–46, 1975.
48. Clemetson CAB: Histamine and ascorbic acid in human blood. J Nutr 110: 662–668, 1980.
49. Anderson TW, Reid DBW, Beaton GH: Vitamin C and the common cold: A double blind study. Can Med Assoc J 107: 503, 1972.
50. Coulehan JL, Reisinger KS, Rogers KD, Bradley DW: Vitamin C prophylaxis in a boarding school. N Engl J Med 290:6, 1974.
51. Wilson CWM, Loh HS: Ascorbic acid and upper respiratory inflammation. Acta Allerg 24:367, 1969.
52. Wilson, CWM, Loh HS: Common cold and vitamin C. Lancet 1:638, 1973.
53. Terezhalmy GT, Bottomley WK, Pelleu GB: The use of water-soluble bioflavonoid-ascorbic acid complex in the treatment of recurrent herpes labialis. Oral Surg 45:56–62, 1978.
54. Weininger J, Brigg GM: Bioflavoncids. In: *Modern Nutrition in Health and Disease,* 6th ed., RS Goodhart, ME Shils (eds). Philadelphia: Lea & Febiger, 1980, pp 279–281.
55. Irwin MI, Hutchins BK: A conspectus of research on vitamin C requirements of man. J Nutr 106:821–879, 1976.
56. Leibovitz B, Siegel BV: Ascorbic acid, neutrophil function and the immune response. Int J Vitam Nutr Res 48:159–164, 1978.
57. Siegel BV: Enhancement of interferon production by poly(rT).poly(rC) in mouse cell cultures by ascorbic acid. Nature 254:531–532, 1975.
58. Siegel, BV, Morton JI: Vitamin C and the immune response. Experientia 33:393–395, 1977.
59. Siegel BV, Morton JI: Vitamin C, interferon and the immune response. In: *Re-evaluation of Vitamin C,* A Hauck, G Ritzel (eds). Bern: Verlag Hans Huber, 1977, pp 245–265.
60. Kennes B, Dumont I, Brohee D, et al.: Effect of vitamin C supplements on cell-mediated immunity in old people. Gerontology 29:305–310, 1983.
61. Ziemlanski S, Wartanowicz M, Klos, A et al.: The effects of ascorbic acid and alpha-tocopherol supplementation on serum proteins and immunoglobulin concentrations in the elderly. Nutr Int 2:245–249, 1986.

62. Vallance S: Relationship between ascorbic acid and serum proteins of the immunological system. Br J Med 1:437, 1977.
63. Johnston CS, Cartee GD, Haskell BE: Effect of ascorbic acid nutriture on protein-bound hydroxyproline in guinea pig plasma. J Nutr 115:1089–1093, 1985.
64. Ziccardi RJ: The role of immune complexes in the activation of the first component of human complement. J Immunol 132:283–288, 1984.
65. Baur H, Staub H: Therapy of hepatitis with ascorbic acid infusions. Schweiz Med Wochenschr 84:595, 1954.
66. Kirchmair H: Epidemic hepatitis in children and its treatment with high doses of ascorbic acid. Dtsch Gesundheitwes 12:1525, 1957.
67. Kirchmair H: Treatment of epidemic hepatitis in children with high doses of ascorbic acid. Med Monatsschr 11:353, 1957.
68. Calleja HB, Brooks RH: Acute heptatitis treated with high doses of vitamin C. Ohio State Med J 56:821, 1960.
69. Knodell RG, Tate MA, Akl BR, Wilson JW: Vitamin C prophylaxis for posttransfusion hepatitis: Lack of effect in a controlled trial. Am J Clin Nutr 24:20, 1981.
70. Patrone F, Dallegri F, Lanzi G, Sacchetti C: Prevention of neutrophil chemotactic deactivation by ascorbic acid. Br J Exp Pathol 61:486–489, 1980.
71. Rebora A, Dallegri F, Patrone F: Neutrophil dysfunction and repeated infections: Influence of levamisole and ascorbic acid. Br J Dermatol 102:49–56, 1980.
72. Gey KF: On the antioxidant hypothesis with regard to arteriosclerosis. Bibl Nutr Dieta 37:53–91, 1986.
73. Gey KF, Stahelin HB, Puska P, Evans A: Relationship of plasma level of vitamin C to mortality from ischemic heart disease. Ann NY Acad Sci 498:110–123, 1987.
74. Bordia A, Paliwal K, Jain DKK, Kothari LK: Acute effect of ascorbic acid on fibrinolytic activity. Atherosclerosis 30:351, 1978.
75. Bordia AK: The effect of vitamin C on blood lipids, fibrinolytic activity and platelet adhesiveness in patients with coronary heart disease. Atherosclerosis 35:181–187, 1980.
76. Bordia A, Verma SK: Effect of vitamin C on platelet adhesiveness and platelet aggregation in coronary artery disease patients. Clin Cardiol 8:552–554, 1985.
77. Ginter E: Cholesterol: Vitamin C controls its transformation to bile acids. Science 179:702, 1973.
78. Ginter E, Nemec R, Cerven J, Mikus L: Quantification of lowered cholesterol oxidation in guinea pigs with latent vitamin C deficiency. Lipids 8:135–141, 1973.
79. Ginter E: Ascorbic acid in cholesterol and bile acid metabolism. Ann NY Acad Sci 258:410–421, 1975.
80. Bjorkhem I, Kallner A: Hepatic 7α-hydroxylation of cholesterol in ascorbate-deficient and ascorbate-supplemented guinea pigs. J Lipid Res 17:360–365, 1976.
81. Ginter E: Marginal vitamin C deficiency, lipid metabolism and atherogenesis. Adv Lipid Res 16:167–220, 1978.
82. Jenkins SA: Hypovitaminosis C and cholelithiasis in guinea pigs. Biochem Biophys Res Commun 77:1030–1035, 1977.
83. Jenkins SA: Biliary lipids, bile acids and gallstone formation in hypovitaminotic C guinea pigs. Br J Nutr 40:317–322, 1978.
84. Jenkins SA: Vitamin C status, serum cholesterol levels and bile composition in the pregnant guinea pig. Br J Nutr 43:95–100, 1980.
85. McIntosh GH, Richmond W, Himsworth RL: Vitamin C deficiency and hypercholesterolaemia in marmoset monkeys. Nutr Rep Int 23:237–244, 1981.
86. Ginter E, Cerna O, Budlovsky J, et al.: Effect of ascorbic acid on plasma cholesterol in humans in a long-term experiment. Int J Vitam Nutr Res 47:123–134, 1977.

87. Dobson HM, Muir MM, Hume R: The effect of ascorbic acid on the seasonal variations in serum cholesterol levels. Scott Med J 29:176–182, 1984.
88. Ginter E, Zdichynec B, Holzerova O, et al.: Hypocholesterolemic effect of ascorbic acid in maturity-onset diabetes mellitus. Int J Vitam Nutr Res 48:368–373, 1978.
89. Ginter E: Pretreatment serum-cholesterol and response to ascorbic acid. Lancet 2:958–959, 1979.
90. Khan AR, Seedarnee FA: Effect of ascorbic acid on plasma lipids and lipoproteins in healthy young women. Atherosclerosis 39:89–95, 1981.
91. Johnson GE, Obenshain SS: Nonresponsiveness of serum high-density lipoprotein-cholesterol to high-dose ascorbic acid administration in normal men. Am J Clin Nutr 34:2088–2091, 1981.
92. Kothari LK, Jain K: Effect of vitamin C administration on blood cholesterol level in man. Acta Biol Acad Sci Hung 28:111–114, 1977.
93. Ramirez J, Flowers NC: Leukocyte ascorbic acid and its relationship to coronary artery disease in man. Am J Clin Nutr 33:2079–2087, 1980.
94. Jacques PF, Hartz SC, McGandy RB, et al.: Vitamin C and blood lipoproteins in an elderly population. Ann NY Acad Sci 498:100–109, 1987.
95. Bobek P, Ginter E, Ozdin L, Mikus L: La cinetique des triglycerides chez le cobaye presentant une subcarence en vitamin C. Med Nutr 17:91–95, 1981.
96. Aulinskas TH, Van der Westhuyzen DR, Coetzee GA: Ascorbate increases the number of low density lipoprotein receptors in cultured arterial smooth muscle cells. Atherosclerosis 47:159–171, 1983.
97. Sokoloff B, Hori M, Saelhof C, et al : Effect of ascorbic acid on certain blood fat metabolism factors in animals and men. J Nutr 91:107–118, 1967.
98. Wahlberg G, Walldius G: Lack of effect of ascorbic acid on serum lipoprotein concentrations in patients with hypertriglyceridaemia. Atherosclerosis 43:283–288, 1982.
99. Geoly K, Diamond LH: Ascorbic acid and hypertriglyceridemia (Letter). Ann Intern Med 93:511, 1980.
100. Ono K: Secondary hyperoxalemia caused by vitamin C supplementation in regular hemodialysis patients. Clin Nephrol 26:239–243, 1986.
101. Yoshioka M, Matsushita T, Chuman Y: Inverse association of serum ascorbic acid level and blood pressure or rate of hypertension in male adults aged 30–39 years. Int J Vitam Nutr Res 54:343–347, 1984.
102. Mann GV: Hypothesis: The role of vitamin C in diabetic angiopathy. Perspec Biol Med 17:210–217, 1973.
103. Mann GV, Newton P: The membrane transport of ascorbic acid. Ann NY Acad Sci 258:243–252, 1975.
104. Verlangieri AJ, Sestito J: Effect of insulin on ascorbic acid uptake by heart endothelial cells: Possible relationship to retinal atherogenesis. Life Sci 29:5–9, 1981.
105. Chen MS, Hutchinson ML, Pecoraro RE, et al.: Hyperglycemia-induced intracellular depletion of ascorbic acid in human mononuclear leukocytes. Diabetes 32:1078–1081, 1983.
106. Pecoraro RE, Chen MS: Ascorbic acid metabolism in diabetes mellitus. Ann NY Acad Sci 498:248–258, 1987.
107. Verlangieri AJ, Bakos E: Influence of L-ascorbic acid on aortic mucopolysaccharides in cholesterol induced rabbit atherosclerosis. Fed Proc 35:661 (abstr. 2512), 1976.
108. Verlangieri AJ, Hollis T, Mumma R: Effects of ascorbic acid and its 2-sulfate on rabbit aortic intimal thickening. Blood Vessels 14:157–174, 1977.
109. Verlangieri AJ, Selvidge LA: Inhibition of arylsulfatase B with ascorbic acid: Possible influence on atherogenesis. Fed Proc 46:415A, 1987.

110. Chatterjee IB, Majumder AK, Nandi BK, Subramanian N: Synthesis and some major functions of vitamin C in animals. Ann NY Acad Sci 258:24–47, 1975.
111. Chatterjee IB, Banerjee A: Estimation of dehydroascorbic acid in blood of diabetic patients. Anal Biochem 98:368, 1979.
112. Som S, Basu S, Mukherjee D, et al.: Ascorbic acid metabolism in diabetes mellitus. Metabolism 30:572, 1981.
113. Yew M-LS: Ascorbic acid supplementation and induction of diabetes in rats. Nutr Rep Int 27:297–302, 1983.
114. Newill A, Habibzadeh N, Bishop N, Schorah CJ: Plasma levels of vitamin C components in normal and diabetic patients. Ann Clin Biochem 21:488–490, 1984.
115. Stankova L, Riddle M, Larned J, et al.: Plasma ascorbate concentrations and blood cell dehydroascorbate transport in patients with diabetes mellitus. Metabolism 33:347–353, 1984.
116. Losert VW, Vetter H, Wendt H: The influence of vitamin C (ascorbic acid) on glucose tolerance, insulin activity and insulin elimination. Arzneimittelforsch 30:21–22, 1980.
117. Kwaan HC, Colwell JA, Cruz S, et al.: Increased platelet aggregation in diabetes mellitus. J Lab Clin Med 80:236–246, 1972.
118. Sarji KE, Kleinfelder J, Brewington P, et al.: Decreased platelet vitamin C in diabetes mellitus: Possible role in hyperaggregation. Thromb Res 15:639–650, 1979.
119. Cox BD, Butterfield WJH: Vitamin C supplements and diabetic cutaneous capillary fragility. Br Med J 2:205, 1975.
120. Sharma SC, Molloy A, Walzman M, Bonner J: Levels of total ascorbic acid and histamine in the blood of women during the 3rd trimester of normal pregnancy. Int J Vitam Nutr Res 51:266–273, 1981.
121. Martin MP, Bridgforth E, McGanity WJ, Darby WJ: The Vanderbilt Cooperative Study of maternal and infant nutrition. J Nutr 62:201–224, 1957.
122. Norkus EP, Hsu H, Cehelsky MR: Effect of chronic alcohol intake during pregnancy on the vitamin C status of the mother and the newborn. Fed Proc 46:1192 A, 1987.
123. Garry PJ, Goodwin JS, Hunt WC, Gilbert BA: Nutritional status in a healthy elderly population: Vitamin C. Am J Clin Nutr 36:332–339, 1982.
124. Garry PJ, VanderJagt DJ, Hunt WC: Ascorbic acid intakes and plasma levels in healthy elderly. Ann NY Acad Sci 498:90–99, 1987.
125. Price NM: Vitamin C deficiency. Cutis 26:375–377, 1980.
126. Taylor GF, Eddy TP, Scott DL: A survey of 216 elderly men and women in general practice. J R Coll Gen Pract 21:267, 1971.
127. Eddy TP, Taylor GF: Sublingual varicosities and vitamin C in elderly vegetarians. Age Ageing 6:6–13, 1977.
128. Hodges RE: Scurvy. Nutr Prev Med (WHO Monogr Ser) 62:120, 1976.
129. Newton HMV, Schorah CJ, Habibzadeh N, et al.: The cause and correction of low blood vitamin C concentrations in the elderly. Am J Clin Nutr 42:656–659, 1985.
130. Schorah CJ, Tormey WP, Brooks GH, et al.: The effect of vitamin C supplements on body weight, serum proteins, and general health of an elderly population. Am J Clin Nutr 34:871–876, 1981.
131. Vir SC, Love AHG: Vitamin C status of institutionalized and non-institutionalized aged. Int J Vitam Nutr Res 48:274–280, 1978.
132. Bates CJ, Rutishauser IHE, Black AE, et al.: Long-term vitamin status and dietary intake of healthy elderly individuals. 2. Vitamin C. Br J Nutr 42:43–56, 1979.
133. Bates CJ, Mandal AR, Cole TJ: HDL cholesterol and vitamin C status. Lancet 2:611, 1977.

134. Harman D: Aging: A theory based on free radical and radiation chemistry. J Gerontol 11:298–300, 1956.
135. Sasaki R, Kurokawa T, Tero-Kubota S: Ascorbate radical and ascorbic acid level in human serum and age. J Gerontol 38:26–30, 1983.
136. Nishikimi M: Oxidation of ascorbic acid with superoxide anion generated by the xanthine-xanthine osidase system. Biochem Biophys Res Commun 63:463–468, 1975.
137. Fessenden RW, Verma NC: A time resolved electron spin resonance study of the oxidation of ascorbic acid by hydroxyl radical. Biophys J 24:93–100, 1978.
138. Bielski BHJ, Richter HW, Chan PC: Some properties of the ascorbate free radical. Ann NY Acad Sci 258:231–237, 1975.
139. Wartanowicz M, et al.: The effect of alpha-tocopherol and ascorbic acid on the serum peroxide level in elderly people. Ann Nutr Metab 28:186–191, 1984.
140. Strydom NB, Kotze HF, Van der Walt WH, Rogers GG: Effect of ascorbic acid on rate of heat acclimatization. J Appl Physiol 41:202–205, 1976.
141. Behrens WA, Madere R: Ascorbic acid in brown adipose tissue: Effect of cold acclimation and high intake of the vitamin. Experientia 37:63–64, 1981.
142. Livingstone SD: Effect of vitamin C on cold-induced vasodilatation. Lancet 2:319–320, 1976.
143. Jacques PF, Phillips J, Chylack LT, et al.: Vitamin intake and senile cataract. J Am Coll Nutr 6:435A, 1987.
144. Varma SD: Ascorbic acid and the eye with special reference to the lens. Ann NY Acad Sci 498:280–306, 1987.
145. Schocket SS, et al.: Induction of cataracts in mice by exposure to oxygen. Isr J Med Sci 8:1596–1601, 1972.
146. Palmquist B, Philipson B, Barr P: Nuclear cataract and myopia during hyperbaric oxygen therapy. Br J Ophthalmol 68:113–117, 1984.
147. Vinson JA, Possanza CJ, Drack AV: The effect of ascorbic acid on galactose-induced cataracts. Nutr Rep Int 33:665–558, 1986.
148. Blondin J, Baragi V, Schwartz ER, et al.: Dietary vitamin C delays UV-induced eye lens protein damage. Ann NY Acad Sci 498:460–463, 1987.
149. Bonjour JP: Vitamins and alcoholism. I. Ascorbic acid. Int J Vitam Nutr Res 49:434–441, 1979.
150. Brissot P, Deugneir Y, Le Treut, et al.: Ascorbic acid status in idiopathic hemochromatosis. Digestion 17:479–487, 1978.
151. Majumdar SD, Patel S, Shaw GK, et al.: Vitamin C utilization status in chronic alcoholic patients after short-term intravenous therapy. Int J Vitam Nutr Res 51:274–278, 1981.
152. Sprince H, Parker CM, Smith GG: Acetaldehyde, ascorbic acid, and catecholamine-regulating drugs: Data and hypothesis in relation to alcoholism and smoking. Nutr Rep Int 17:441–455, 1978.
153. Aurer-Kozelj J, Kralj-Klobucar N, Buzima R, Bacic M: The effect of ascorbic acid supplementation on periodontal tissue ultrastructure in subjects with progressive periodontitis. Int J Vitam Nutr Res 52:333–341, 1982.
154. Taylor TV, Rimmer S, Day B, et al.: Ascorbic acid supplementation in the treatment of pressure-sores. Lancet 2:544–546, 1974.
155. Denson KW, Bowers EF: The determination of ascorbic acid in white blood cells. A comparison of WBC ascorbic acid and phenolic acid excretion in elderly patients. Clin Sci 21:157–162, 1961.
156. Abrahamian V, Kaminski MV Jr, Santiago GC: Vitamin C supplementation of total parenteral nutrition formulas. J Paren Ent Nutr 7:465–469, 1983.

157. Hughes RG, Williams N: Leukocyte ascorbic acid in Crohn's disease. Digestion 17:272–274, 1978.
158. Naccarto DV, Bell CJ, Lamy PP: Appraisal of ascorbic acid for acidifying the urine of methenamine-treated geriatric patients. J Am Geriatr Soc 27:34–37, 1979.
159. Nahata MC, Shimp L, Lampman T, et al.: Effect of ascorbic acid on urine pH in man. Am J Hosp Pharm 34:1234–137, 1977.
160. Barton CH, Sterling, ML, Thomas R, et al.: Ineffectiveness of intravenous ascorbic acid as an acidifying agent in man. Arch Intern Med 141:211–212, 1981.
161. Ekvall S, Chen I-W, Bozian R: The effect of supplemental ascorbic acid on serum vitamin B-12 levels in myelomeningocele patients. Am J Clin Nutr 34:1356–1361, 1981.
162. Schmidt K-H, Hagmaier V, Hornig DH, et al.: Urinary oxalate excreation after large intakes of ascorbic acid in man. Am J Clin Nutr 34:305–311, 1981.
163. Afroz M. Bothinard B, Etzkorn J, et al.: Vitamins C and B-12. JAMA 232:246, 1975.
164. Zuskin E, Lewis A, Bouhuys A: Inhibition of histamine-induced airway constriction by ascorbic acid. J Allergy Clin Immunol 51:218–226, 1973.
165. Valic F, Zuskin E: Prevention of acute ventilatory capacity reduction in hemp workers by antihistamine, bronchodilator, and vitamin C. 17th International Congress on Occupational Health, Buenos Aires, 1972, abstracts, pp 108–109.
166. Valic F, Zuskin E: Pharmacological prevention of acute ventilatory capacity reduction in flax dust exposure. Br J Ind Med 30:381–384, 1973.
167. Kordansky D, Rosenthal R, Normal P: The effect of vitamin C on antigen-induced bronchospasm. J Allergy Clin Immunol 63:61–64, 1979.
168. Siegel B, Leibovitz B: The multifactorial role of vitamin C in health and disease. J Vitam Nutr Res, suppl. 23:9–22, 1982.
169. Leibovitz B, Siegel B: Ascorbic acid and the immune response. In: *Diet and Resistance to Disease*, M. Phillips, A. Baetz (eds). New York: Plenum Press, 1981, pp 1–25.
170. Dawson EB, Harris WA, Rankin WE, et al.: Effect of ascorbic acid on male fertility. Ann NY Acad Sci 498:312–323, 1987.
171. Jain SK, Williams DM: Vitamin C and sickle cell disease. Ann NY Acad Sci 498:484–486, 1987.
172. Hollinshead MB, Spillert CR, Lazaro EJ: The beneficial effects of ascorbic acid on murine burns. J Burn Care Res 6:50–54, 1985.
173. Spillert CR, Hollinshead MB, Lazaro EJ: Protective effects of ascorbic acid on murine frostbite. Ann NY Acad Sci 498:517–518, 1987.
174. Cameron E, Pauling L, Leibovitz B: Ascorbic acid and cancer: A review. Cancer Res 39:663–681, 1979.
175. Stähelin HB, Gey KF, Brubacher G: Plasma vitamin C and cancer deaths: The prospective Basel study. Ann NY Acad Sci 498:124–131, 1987.
176. Basu TK, Weiser T, Dempster JF: An in vitro effect of ascorbate on the spontaneous reduction of sodium nitrite concentration in a reaction mixture. Int J Vitam Nutr Res 54:233–236, 1984.
177. Mergens WJ, Vane FM, Tannenbaum SR: In vitro nitrosation of methapyrilene. J Pharm Sci 68:827–832, 1979.
178. Newmark HL, Osadca M, Araujo M, et al.: Stability of ascorbate in bacon. Food Technol 28:28–31, 1974.
179. Kamm JJ, Dashman T, Conney AH, Burns JJ: Protective effect of ascorbic acid on hepatotoxicity caused by sodium nitrite plus aminopyrine. Proc Natl Acad Sci USA 70:747–749, 1973.

180. Kamm JJ, et al.: The effect of ascorbate on amine-nitrite hepatotoxicity. In: *N-Nitroso Compounds in the Environment*. Lyon: International Agency for Research on Cancer, 1974, pp 200–204.
181. Tannenbaum SR, Wishnok JS: Inhibition of nitrosamine formation by ascorbic acid. Ann NY Acad Sci 498:354–363, 1987.
182. Edgar JA: Ascorbic acid and biological alkylating agents. Nature 248:136–137, 1974.
183. Reddy BS, Hirota N, Katayama S: Effect of dietary sodium ascorbate on 1,2-dimethylhydrazine- or methylnitrosourea-induced colon carcinogenesis in rats. Carcinogenesis 3:1097–1099, 1982.
184. Mirvish SS, Pelfrene AF, Garcia H, Shubik P: Effect of sodium ascorbate on tumor induction in rats treated with morpholine and sodium nitrite, and with nitrosomorpholine. Cancer Lett 2:101–108, 1976.
185. Mirvish SS, Salmasi S, Cohen SM, et al.: Liver and forestomach tumors and other forestomach lesions in rats treated with morpholine and sodium nitrite, with and without sodium ascorbate. J Natl Cancer Inst 71:81–85, 1983.
186. Abul-Hajj YJ, Kelliher M: Failure of ascorbic acid to inhibit growth of transplantable and dimethylbenzanthracene induced rat mammary tumors. Cancer Lett 17:67–73, 1982.
187. Banic S: Vitamin C acts as a cocarcinogen to methylcholanthrene in guinea pigs. Cancer Lett 11:239–242, 1981.
188. Kawasaki H, Morishige F, Tanaka H, Kimoto E: Influence of oral supplementation of ascorbate upon the induction of N-methyl-N'-nitro-N-nitrosoguanidine. Cancer Lett 16:57–63, 1982.
189. Cameron E, Campbell A: The orthomolecular treatment of cancer. II. Clinical trial of high-dose ascorbic acid supplements in advanced human cancer. Chem Biol Interact 9:285–315, 1974.
190. Cameron E, Pauling L: Supplemental ascorbate in the supportive treatment of cancer: Prolongation of survival times in terminal human cancer. Proc Natl Acad Sci USA 73:3685–3689, 1976.
191. Cameron E, Pauling L: Supplemental ascorbate in the supportive treatment of cancer: Reevaluation of prolongation of survival times in terminal human cancer. Proc Natl Acad Sci USA 75:4538–4542, 1978.
192. Creagan ET, Moertel CG, O'Fallon JR, et al.: Failure of high-dose vitamin C (ascorbic acid) therapy to benefit patients with advanced cancer. N Engl J Med 301:687–690, 1979.
193. Moertel CG, Fleming TR, Creagan ET, et al.: High-dose vitamin C versus placebo in the treatment of patients with advanced cancer who have had no prior chemotherapy. N Engl J Med 312:137–141, 1985.
194. DeCosse JJ, Adams MB, Kuzma JF, et al.: Effect of ascorbic acid on rectal polyps of patients with familial polyposis. Surgery 78:608–612, 1975.
195. Bussey HJR, DeCosse JJ, Deschner EE, et al.: A randomized trial of ascorbic acid in polyposis coli. Cancer 50:1434–1439, 1982.
196. Bruce WR, Dion PW: Studies relating to a fecal mutagen. Am J Clin Nutr 33:2511–2512, 1980.
197. McKeown-Eyssen GE, Holloway V, Jazmaji V, et al.: A randomized trial of vitamin C and E supplementation on recurrence of colorectal polyps. Fed Proc 46:883A, 1987.
198. Sram RJ, Dobias L, Pastorkova A, et al.: Effect of ascorbic acid prophylaxis on the frequency of chromosome aberrations in the peripheral lymphocytes of coal-tar workers. Mutat Res 120:181–186, 1983.

199. Sram RJ, Samkova I, Hola N: High-dose ascorbic acid prophylaxis in workers occupationally exposed to halogenated ethers. J Hyg Epidemiol Microbiol Immunol 27:305–318, 1983.
200. Wassertheil-Smoller S, Romney SL, Wylie-Rosett J, et al.: Dietary vitamin C and uterine cervical dysplasia. Am J Epidemiol 114:714–724, 1981.
201. Schlegel JU: Proposed uses of ascorbic acid in prevention of bladder carcinoma. NY Acad Sci 258:432–437, 1975.
202. Fujita K, Shinpo K, Yamada K, et al.: Reduction of Adriamycin toxicity by ascorbate in mice and guinea pigs. Cancer Res 42:309–316, 1982.
203. Shipman P: The ancestor that wasn't. The first hominids were timid scavengers, not fearless hunters. Sciences 25:43–48, 1985.
204. Berlin EA, Markell EK: An assessment of the nutritional and health status of an Aguaruna Jivaro community, Amazonas, Peru. Ecol Food Nutr 6:69–81, 1977.
205. Eaton SB, Konner M: Paleolithic nutrition. N Engl J Med 312:283–289, 1985.
206. Briggs M: Vitamin C-induced hyperoxaluria. Lancet 1:154, 1976.
207. Lamden MP, Chrystowski GA: Urinary oxalate excretion by man following ascorbic acid ingestion. Proc Soc Exp Biol Med 85:190–192, 1954.
208. Takenouchi K, Aso L, Kawase K, et al.: On the metabolites of ascorbic acid, especially oxalic acids eliminated in urine, following the administration of large amounts of ascorbic acid. J Vitaminol 12:49–58, 1966.
209. Takiguchi H, Furuyama S, Shimazono N: Urinary oxalic excretion by man following ingestion of large amounts of ascorbic acid. J Vitaminol 12:307–312, 1966.
210. Hagler L, Herman RH: Oxalate metabolism. II. Urinary oxalate and the diet. Am J Clin Nutr 26:882–889, 1973.
211. Watts RWE: Studies on the urinary excretion of oxalate by normal subjects. Clin Sci 16:405, 1957.
212. Atkins GL, Dean BM, Griffin WJ, et al.: Primary hyperoxaluria. The relation between ascorbic acid and the increased urinary excretion of oxalate. Lancet 2:1096–1097, 1963.
213. Atkins GL, Dean BM, Griffin WJ, et al.: Quantitative aspects of ascorbic acid metabolism in patients with primary hypoeroxaluria. Clin Sci 29:305–314, 1965.
214. Smith LH: Risk of oxalate stones from large doses of vitamin C. N Engl J Med 298:856, 1978.
215. Stein HB, Hasan A, Fox IH: Ascorbic acid-induced uricosuria. A consequence of megavitamin therapy. Ann Intern Med 84:385, 1976.
216. Berger L, Gerson TD, Yu T: The effect of ascorbic acid on uric acid excretion with a commentary on the renal handling of ascorbic acid. Am J Med 62:71–76, 1977.
217. Norkus EP, Rosso P: Changes in ascorbic acid metabolism of the offspring following high intake of this vitamin in the pregnant guinea pig. Ann NY Acad Sci 258:401–409, 1975.
218. Nandi BK, Majumder AK, Halder K: Effects of high intake of vitamin C by guinea pigs in pregnancy and lactation on the tissue levels of the vitamin in their offspring. Int J Vitam Nutr Res 47:200–205, 1977.
219. Norkus EP, Rosso P: Effects of maternal intake of ascorbic acid on the postnatal metabolism of this vitamin in the guinea pig. J Nutr 111:624–630, 1981.
220. Ludvigsson J, Hansson L-O, Stendahl O: The effect of large doses of vitamin C on leukocyte function and some laboratory parameters. Int J Vitam Nutr Res 49:160–165, 1979.
221. Omaye ST, Skala JH, Jacob RA: Plasma ascorbic acid in adult males: Effects of depletion and supplementation. Am J Clin Nutr 44:257–264, 1986.
222. Thornton PA: Influence of exogenous ascorbic acid on calcium and phosphorus metabolism in the chick. J Nutr 100:1479–1485, 1970.

223. Keith MO, Shad BG, Nera EA, Pelletier O: The effect of high ascorbic acid and iron intake on the renal excretion of oxalate, calcium, and iron and on the kidneys of rats. Nutr Rep Int 10:357–369, 1974.
224. Bray DL, Briggs GM: Decrease in bone density in young male guinea pigs fed high levels of ascorbic acid. J Nutr 114:920–928, 1984.
225. Sifri M, Kratzer FH, Norris LC: Lack of effect of ascorbic and citric acids on calcium metabolism of chickens. J Nutr 107:1484–1492, 1977.
226. Mengel CE, Greene HL Jr: Ascorbic acid effects on erythrocytes. Ann Intern Med 84:490, 1976.
227. Campbell GD, Steinberg MH, Bower JD: Ascorbic acid-induced hemolysis in G-6-PD deficiency. Ann Intern Med 82:810, 1975.
228. Cohen A, Cohen IJ, Schwartz E: Scurvy and altered iron stores in thalassemia major. N Engl J Med 304:158–160, 1981.
229. Cook JD, Monsen ER: Vitamin C, the common cold, and iron absorption. Am J Clin Nutr 30:235–241, 1977.
230. Domke I, Weis W: Reinvestigation of the diabetogenic effect of dehydroascorbic acid. J Vitam Nutr Res 53:51–60, 1983.
231. Reznik VM, Griswold WR, Brams MR, Mendoza SA: Does high-dose ascorbic acid accelerate renal failure? N Engl J Med 302 1418–1419, 1980.
232. Ovesen L: Drugs and vitamin deficiency. Drugs 18:278, 1979.
233. Ioannides C, Stone AN, Breacker PJ, Basu TK: Impairment of absorption of ascorbic acid following ingestion of aspirin in guinea pigs. Biochem Pharmacol 31:4035–4038, 1982.
234. Coffey G, Wilson CWM: Ascorbic acid deficiency and aspirin-induced haematemesis. Br Med J Jan. 25: 208, 1975.
235. Houston JB, Levy G: Modification of drug biotransformation by vitamin C in man. Nature 255:78–79, 1975.
236. Weininger J, King JC: Effect of oral contraceptive agents on ascorbic acid metabolism in the rhesus monkey. Am J Clin Nutr 35:1408–1412, 1982.
237. Back DJ, Orme MLE: Oral contraceptives and ascorbic acid metabolism. Am J Clin Nutr 37:330, 1983.
238. Rebec GV, Centore JM, White LK, Alloway KD: Ascorbic acid and the behavioral response to haloperidol: Implications for the action of antipsychotic drugs. Science 227:438–440, 1985.
239. Dysken MW, Cumming RJ, Channon RA, Davis JM: Drug interaction between ascorbic acid and fluphenazine. JAMA 241:2008, 1979.
240. Odumosu A: Anoretic drugs and vitamin C: Role in appetite and brain ascorbic acid in guinea pigs. Int J Vitam Nutr Res 51:247–253, 1981.
241. Fazio V, Flint DM, Wahlqvist ML: Acute effects of alcohol on plasma ascorbic acid in healthy subjects. Am J Clin Nutr 34:2394–2396, 1981.
242. Kato N, Okada T, Takenaka Y, Yoshida A: Ameliorative effect of dietary ascorbic acid on PCB toxicity in guinea pigs. Nutr Rep Int 15:125–130, 1977.
243. Kato N, Kawai K, Yoshida A: Effect of dietary level of ascorbic acid on the growth, hepatic lipid peroxidation, and serum lipids in guinea pigs fed polychlorinated biphenyls. J Nutr 111:1727–1733, 1981.
244. Chakraborty D, Bhattacharyya A, Mazumdar K, et al.: Studies on L-ascorbic acid metabolism in rats under chronic toxicity due to organophosphorus insecticides: Effects of supplementation of L-ascorbic acid in high doses. J Nutr 108:973–980, 1978.
245. Chatterjee K, Banerjee SK, Tiwari R, et al.: Studies on the protective effects of L-ascorbic acid in chronic chlordane toxicity. Int J Vitam Nutr Res 51:254–265, 1981.

246. Kreiss K, Zack MM, Kimbrough RD, et al.: Association of blood pressure and polychlorinated biphenyl levels. JAMA 245:2505–2509, 1981.
247. Sprince H, Smith GC, Parker CM, Rinehimer DA: Protection against cyanide lethality in rats by L-ascorbic acid and dehydroascorbic acid. Nutr Rep Int 25:463–470, 1982.
248. Sprince H, Parker CM, Smith CG: Acetaldehyde, ascorbic acid, and catecholamine-regulating drugs: Data and hypothesis in relation to alcoholism and smoking. Nutr Rep Int 17:441–455, 1978.
249. Sprince H, Parker CM, Smith GC: Lethal synergy of acetaldehyde with nicotine, caffeine, or dopamine in rats: Protection by ascorbic acid, cysteine, and anti-adrenergic agents. Nutr Rep Int 23:43–54, 1981.
250. Hill CH: Studies on the ameliorating effect of ascorbic acid on mineral toxicities in the chick. J Nutr 109:84–90, 1979.
251. Goyer RA, Cherian MG: Ascorbic acid and EDTA treatment of lead toxicity in rats. Life Sci 24:433–438, 1979.
252. Chatterjee K, Chakraborty D, Majumdar K, et al.: Biochemical studies on nickel toxicity in weanling rats—influence of vitamin C supplementation. Int J Vitam Nutr Res 49:264–275, 1979.
253. Finley EB, Cerklewski FL: Influence of ascorbic acid supplementation on copper status in young adult men. Am J Clin Nutr 37:553–556, 1983.
254. Naylor GJ, Smith AHW: Vanadium: A possible aetiological factor in manic-depressive illness. Psychol Med 11:249–256, 1981.
255. Hogenkamp HPC: The interaction between vitamin B-12 and vitamin C. Am J Clin Nutr 33:1–3, 1980.
256. Marcus M, Prabhudesai M, Wassef S: Stability of vitamin B-12 in the presence of ascorbic acid in food and serum: Restoration by cyanide of apparent loss. Am J Clin Nutr 33:137–143, 1980.
257. Baker H, Pauling L, Frank O: Mega-ascorbate taken with other vitamins permits elevation of circulating vitamins including B-12 in humans. Nutr Rep Int 23:669–677, 1981.
258. Nahata MC, McLeod DC: Noneffect of oral ascorbic acid on urinary copper reduction glucose test. Diabetes Care 1:34, 1978.
259. Smith D, Young WW: Effect of large dose ascorbic acid on the two-drop clinitest determinations. Am J Hosp Pharm 34:1347, 1977.
260. Nahata MC, McLeod DC: Lack of effect of ascorbic acid, hippuric acid and methenamine (urinary formaldehyde) on the copper-reduction glucose test in geriatric patients. J Am Geriatr Soc 28:230–233, 1980.
261. Jaffe RM, Kasten B, Young DS: MacLowry JD: False-negative stool occult blood tests caused by ingestion of ascorbic acid (vitamin C). Ann Intern Med 83:824–826, 1975.

SECTION II: PHARMACOLOGY OF THE TRACE ELEMENTS

14
Chromium

Introduction	247
Absorption, Metabolism, and Excretion	248
Dosage Forms	250
Clinical Studies	250
Hyperglycemia, Glucose Intolerance, and Insulin Resistance	250
Blood Lipids	251
The Elderly	252
Pregnancy	252
Athletics and Exercise	252
Toxicity and Side Reactions	253
Interactions	253
References	253

INTRODUCTION

Chromium, initially demonstrated to be an essential micronutrient for laboratory animals, was not definitely established as essential for humans until acute signs and symptoms of abnormal glucose metabolism were observed to develop in two patients who had received long-term TPN (total parenteral nutrition) without chromium supplementation. In both animals and man, acute deficiency of this element causes insulin resistance and failure to utilize glucose [1–3].

Chromium potentiates the action of insulin when it is present in an organically complexed form known as "glucose tolerance factor" or GTF. GTF is present in the diet and can also be synthesized in the body from inorganic Cr^{3+} compounds supplied in the diet or as a supplement [1–3]. Although the structure of GTF is not fully known, one biologically active form, isolated from brewer's yeast, was found to contain nicotinic acid and amino acids complexed with chromium [4]. GTF has been demonstrated in animal studies to potentiate the action of both exogenous and endogenous insulin in terms of glucose tolerance, insulin-induced hypoglycemia, utilization of glucose for glycogen formation, cell transport of amino acid analogs, and utilization of amino acids for protein synthesis [3].

In human deficiency resulting after long-term TPN, the first reported case developed failure to utilize glucose as manifested by hyperglycemia, depressed respiratory quotient, and weight loss; there was relative insensitivity to exogenous insulin. These adverse findings, as well as peripheral neuropathy, were completely reversed by an intravenous supplement of 250 mcg/d of Cr^{3+} for 2 weeks, followed by a maintenance supplement of 20 mcg/d in the TPN infusate [5]. A similar case was later reported by Freund et al. [6], with the added symptom of a metabolic encephalopathy-like confusional state; again, complete reversal of adverse signs and symptoms was accomplished by chromium supplementation.

Occurrence of a marginal chromium deficiency state appears to have been established in children suffering from protein calorie malnutrition in Jordan, Nigeria, and Turkey [2], but otherwise there is no clear-cut evidence of deficient dietary intakes of chromium by humans. However, marginal dietary deficiencies or defects in utilization of dietary chromium have been suspected among the elderly, in diabetics, and in individuals with coronary artery disease [1–3].

Good food sources of chromium include brewer's yeast, spices, vegetable oils, unrefined sugar, liver, and kidney [2]. No RDA has been established, but the National Research Council has promulgated the range 50–200 mcg/d as a safe and adequate intake [7]. Many healthy adults are consuming less than this amount without apparent ill effects [8,9], and Offenbacher et al. [10] found that two elderly men, ages 62 and 66 years, were in apparent chromium balance on dietary intakes of 38 mcg/d. Wallach [1] has summarized the many difficulties hampering a fuller understanding of the metabolism and clinical importance of chromium, namely lack of reliable means of assessing chromium nutriture, difficulties in analysis of chromium in biological fluids and tissues, and different forms of chromium in food including that arising from cooking and food-processing equipment and eating utensils, with inadequate knowledge of the composition and bioavailability of these various forms.

ABSORPTION, METABOLISM, AND EXCRETION

Anderson and Kozlovsky [9] examined dietary chromium absorption in 10 men and 22 women (age range 22–65 years) and found the percent absorption to be inversely related to the dietary intake, varying from about 2% at an intake of 10 mcg/d to 0.5% at an intake of 40 mcg/d. Offenbacher et al. [10] observed an absorption of about 1.8% in two men ingesting 38 mcg/d of dietary chromium. Both groups found a relatively constant urinary excretion of chromium if intake was constant. At an intake of 38 mcg/d, urinary output was 0.3 mcg/d [10].

Absorbed chromium circulates as unbound Cr^{3+} and as Cr^{3+} bound to transferrin and other plasma proteins and also as GTF-chromium. Trivalent Cr is accumulated by kidney, skeleton, liver, spleen, lung, and large intestine, with other tissues being less active in accumulation [1]. It is thought that GTF probably concentrates in tissues that are insulin-responsive [1]; in laboratory animals, administered GTF is also accumulated by the liver [2]. Laboratory mice accumulated inorganic chromium in endocrine and exocrine pancreas [11].

Turnover of tissue chromium and probably release of Cr^{3+} by degradation of GTF occurs, which, together with absorption from the diet, maintains homeostasis of the body Cr^{3+} pool, from which a relatively constant fraction is excreted daily by the kidneys. It has not been conclusively established whether there is any reabsorption in the kidney, but recent work of Wallach and Verch [12] in laboratory rats suggests that chromium conservation is compatible with proximal reabsorption, but may be explainable equally well by lack of glomerular filtration or by binding to a specific Cr-binding substance.

Since 98% or more of dietary chromium is not absorbed, most ingested chromium passes through the gut and is excreted in the feces [10].

Studies by Anderson and coworkers [13–15] in 76 healthy free-living adults, aged 21–69 years, who were not taking supplements or medications, found that they had a mean serum Cr of 0.13 ng/ml (range 0.05–0.56) and a mean urinary excretion of 0.20 ng/ml daily (range 0.05–0.58). These values did not differ by sex. Application of a glucose load (1 g/kg) after overnight fasting increased mean urinary Cr excretion significantly, but in an unpredictable fashion that could not be correlated with Cr status. After 3 months on a chronic chloride supplement providing 200 mcg Cr per day, mean serum Cr increased to 0.38 ng/ml in both men and women, while urinary Cr increased about fivefold to a mean of 1.13 ng/ml. After supplementation, there was no increase in urinary Cr excretion in response to a glucose load. In another study by this group in which 37 healthy male and female volunteers took part, Cr excretion was compared when the subjects consumed diets for 12 weeks containing either 35% of calories as complex carbohydrates and 15% as simple sugars, or 15% as complex carbohydrates and 35% as simple sugars [16]. All diets contained about 16 mcg Cr per 1,000 kcal. Compared with the 15% sugar diet, the diet containing 35% simple sugars increased urinary Cr losses 10–300% for 27 of the 37 subjects. The authors conclude that Cr losses stimulated by high-sugar diets, coupled with marginal Cr intakes, could lead to a degree of Cr deficiency great enough to impair glucose and lipid metabolism [16].

DOSAGE FORMS

GTF-chromium is available over the counter as a high-Cr brewer's yeast prepared by culturing yeast in a high-chromium growth medium. This product is marketed as tablets containing 100 or 200 mcg of elemental Cr each. Chromium is also marketed as Cr^{3+} chelated with amino acids in the form of tablets, furnishing 100 mcg elemental Cr each. Chromic chloride ($CrCl_3$) is available through pharmacies.

Using as their criterion the maximum decrease in fasting blood glucose level produced in six men and one woman, aged 22–42 years, by administration of a single dose of 100 mcg Cr in various forms, Vinson and Hsiao [17] determined that high-Cr yeast and a Cr^{3+}-EDTA complex were more active than either inorganic Cr^{3+} (chromic chloride) or conventional brewer's yeast. Whether such differences are of significance in chronic daily administration, where serum Cr builds up to a rather constant plateau [13–15], appears doubtful.

CLINICAL STUDIES
Hyperglycemia, Glucose Intolerance, and Insulin Resistance

Mertz [3] has summarized the progress of developing chromium deficiency as follows: a first phase involving mild insulin resistance, which is compensated by elevation of circulating insulin levels and an increased insulin secretion in response to glucose loading; a second phase with abnormal glucose tolerance and disturbances of lipid metabolism; and a final phase of marked insulin resistance producing a diabetes-like syndrome. Insulin resistance and impaired glucose tolerance may have a number of causes, but Mertz recommends that chromium deficiency be considered as a possible contributory factor in all clinical situations in which an unexplained insulin resistance develops in patients [3].

Wallach [1] has summarized and evaluated clinical studies carried out over the period 1978–1982 on the effects of chromium supplements, usually 150–250 mcg/d as $CrCl_3$ or brewer's yeast, on blood glucose and insulin levels prevailing during oral glucose tolerance testing (OGTT). In general, healthy adults with a normal OGTT had no change in glucose tolerance, whil diabetics and individuals with abnormal OGTTs responded variably but with average decreases of 13–17% in blood glucose levels, and one study on long-term Cr supplementation noted a 17% decrease in the circulating level of glycosylated hemoglobin. For most subjects, plasma insulin levels were not significantly affected, but Riales and Albrink [18] noted small declines in insulin levels during OGTT after 12 weeks of Cr supplementation in adults who had had normal glucose levels together with elevated insulin levels prior to supplementation, suggesting that insulin sensitivity had been improved by Cr.

Clinical trials since 1982 have shown the same general trends, again with wide individual variations. Anderson et al. [19] reported that in 20 of the 76 healthy adults they studied who had serum glucose concentrations ≥ 100 mg/dl at 90 minutes after glucose loading (1g/kg), a 3-month Cr (200 mcg/d) supplementation period produced a significant mean 15% decline of serum glucose levels at 90 min of the OGTT; fasting glucose levels also decreased significantly in this group. Those whose 90-min glucose levels were above the fasting concentration but below 100 mg/dl were not affected by Cr supplementation. Interestingly, those whose 90-min glucose concentrations were below their fasting levels showed a significant increase (from a mean of 71 to 81 mg/dl) in the 90-min level after Cr supplementation, suggesting the possibility that Cr may be beneficial in reactive hypoglycemia.

Saner et al. [20] reported an improvement in glucose tolerance with Cr supplementation in patients with Turner's syndrome, a genetic disorder in which there is a high incidence of diabetes. Uusitupa et al. [21] evaluated the effect of 200 mcg/d of Cr^{3+} ($CrCl_3$) given for 6 weeks to 10 noninsulin-dependent (type II) diabetics aged 37–68 years; OGTT glucose values were slightly but significantly depressed, indicating improved insulin sensitivity. Vinson and Bose [22] treated 23 adults, comprising 6 normals, 5 hyperglycemics, 7 type I diabetics receiving insulin, and 5 type II diabetics taking oral antidiabetic drugs, with 218 mcg/d of Cr in high-Cr brewer's yeast for 6 months; only the hyperglycemic group showed a statistically significant improvement in OGTT. In five healthy elderly but glucose-intolerant subjects, given 200 mcg/d of Cr^{3+} ($CrCl_3$) for 12 weeks, Potter et al. [23] found a small improvement in glucose tolerance, significant only at the 60-min point of the OGTT, and a small but significant increase in glucose utilization during hyperglycemic clamp studies.

Blood Lipids

In the majority of cases in which Cr supplementation has improved glucose tolerance, there has also been an improvement in blood lipid profile, with increases in HDL-cholesterol level and/or in the ratio of HDL cholesterol to total cholesterol [1]. In addition, both normal healthy adults and type II diabetics have shown this favorable blood lipid response to Cr supplementation even when there has been no change in their glucose tolerance [1]. Elwood et al. [23] reported that 8 weeks of supplementation with 48 mcg/d of Cr as high-chromium brewer's yeast caused a significant decrease (24–26 mg/dl) in mean total cholesterol levels in both normolipidemic and hyperlipidemic adults, as well as rises in HDL-cholesterol levels and HDL- to total cholesterol ratios. Vinson and Bose [22] observed a significant decrease in LDL cholesterol and an increase in HDL cholesterol, as well as improved glucose tolerance, in hyperglycemic but nondiabetic adults after 6 months of supplementation with 218 mcg/d of Cr as high-chromium brewer's yeast. However, a few studies have noted little or no change in total blood lipid profile in normal adults [19], type II diabetics [21], or elderly glucose-intolerant subjects [23] after 1.5–3 month periods of Cr supplementation. The reasons for these discrepancies in the results of different trials are obscure.

The Elderly

Tissue chromium concentration appears to decline with age [1–3], but most of the decline occurs in the first decade of life, with fluctuations thereafter [1]. Further studies by modern analytical methods are needed to clarify the picture. Studies of radio-Cr uptake in young and old rats by Wallach and Verch [22] showed a marked difference in tissue distribution of an injected dose of ^{51}Cr in old animals compared with that in young rats. Plasma ^{51}Cr levels were generally higher and tissue ^{51}Cr levels generally lower in the old rats, except for spleen ^{51}Cr, which increased. Skeletal ^{51}Cr was markedly decreased in the old rats.

Some studies of Cr supplementation in institutionalized elderly individuals found that oral glucose tolerance was improved in 40–50% of those with abnormal glucose tolerance [25,26], but Potter et al. [23] noted only minimal improvements in five glucose-intolerant subjects aged 54–79 years (mean 66 years).

Pregnancy

Saner [27], investigating hair Cr concentration in pregnant Turkish women by serial sampling of newgrown hair, found that if dietary chromium was low, the hair Cr content declined with advancing pregnancy; hair Cr was lower in multiparous than in nulliparous women. In pregnant Canadian women, Shapcott et al. [28] found much lower Cr levels in pubic hair at delivery than in nonpregnant women, but 1 year later the median Cr level had doubled, indicating that the low values seen at delivery were only a temporary phenomenon in this group. It is inferred from such evidence and from the relatively high Cr tissue levels of infants that a considerable transfer of Cr occurs from the mother to the fetus, and this is confirmed in laboratory studies of the placental transport of ^{51}Cr in pregnant rats [29]. If there is doubt about the adequacy of dietary chromium intake of the pregnant patient, Cr supplements have been recommended [2].

Athletics and Exercise

Anderson et al. [30] observed an almost fivefold increase in urinary Cr excretion in long-distance runners and attributed this to increased mobilization of Cr incident to the increased glucose uptake of exercising muscle. This suggests that individuals engaging in regular strenuous exercise may be at risk of depleting body Cr stores and should ensure an increased intake either by dietary measures or by use of a Cr supplement.

TOXICITY AND SIDE REACTIONS

Chromium salts appear to have relatively low toxicity in animals and man [30–32]. No toxicity has been reported from the use of high-Cr brewer's yeast as a supplement.

INTERACTIONS

Stoecker and Oladut [33] obtained evidence that in guinea pigs dietary chromium deficiency potentiated the hyperglycemic and hypercholesterolemic effects of vitamin C deficiency, suggesting that Cr deficiency may impair tissue uptake of ascorbic acid as it does that of glucose.

REFERENCES

1. Wallach S: Clinical and biochemical aspects of chromium deficiency. J Am Coll Nutr 4:107–120, 1985.
2. Saner G: The metabolic significance of dietary chromium. Nutr Int 2:213–220, 1986.
3. Mertz W: Clinical and public health significance of chromium. In: *Clinical, Biochemical, and Nutritinal Aspects of Trace Elements,* AS Prasad (ed). New York: Alan R. Liss, 1982, pp 315–323.
4. Toepfer EW, Mertz W, Polansky MM. et al.: Preparation of chromium-containing material of glucose tolerance factor activity from brewer's yeast extracts and by synthesis. J Agric Food Chem 25:162–166, 1977.
5. Jeejeebhoy KN, Cher RG, Marliss EB, et al.: Chromium deficiency, glucose intolerance, and neuropathy reversed by chromium supplementation in a patient receiving long-term total parenteral nutrition. Am J Clin Nutr 30:531–538, 1977.
6. Freund H, Atamian S, Fischer JE: Chromium deficiency during total parenteral nutrition. JAMA 241:496–498, 1979.
7. Food and Nutrition Board: *Recommended Dietary Allowances,* 9th ed. Washington, DC: National Academy of Sciences, 1980.
8. Bunker VW, Lawson MS, Delves HR, Clayton BE: The uptake and excretion of chromium by the elderly. Am J Clin Nutr 39:792–802, 1984.
9. Anderson RA, Kozlovsky AS: Chromium intake, absorption, and excretion of subjects consuming self-selected diets. Am J Clin Nutr 41:1171–1183, 1985.
10. Offenbacher EG, Spencer H, Dowling HJ, Pi-Sunyer FX: Metabolic chromium balances in men. Am J Clin Nutr 44:77–82, 1986.
11. Barggren PO, Flatt PR: Effects of trivalent chromium administration on endogenous chromium stores in lean and obese hyperglycaemic (ob/ob) mice. Nutr Rep Int 31:213, 1985.
12. Wallach S, Verch RL: Radiochromium distribution during saline diuresis. J Am Coll Nutr 5:299–304, 1986.
13. Anderson, RA, Polansky MM, Bryden NA, et al.: Urinary chromium excretion of human

subjects: Effects of chromium supplementation and glucose loading. Am J Clin Nutr 36:1184–1193, 1982.
14. Anderson RA, Polansky MM, Bryden NA, et al.: Effects of chromium supplementation on urinary Cr excretion of human subjects and correlation of Cr excretion with selected clinical parameters. J Nutr 113:276–281, 1983.
15. Anderson RA, Bryden NA, Polansky MM: Serum chromium of human subjects: Effects of chromium supplementation and glucose. Am J Clin Nutr 41:571–577, 1985.
16. Kozlovsky AS, Moser PB, Reiser S, Anderson RA: Effects of diets high in simple sugars on urinary chromium losses. Metabolism 35:515–518, 1986.
17. Vinson JA, Hsiao KH: Comparative effect of various forms of chromium on serum glucose: An assay for biologically active chromium. Nutr Rep Int 32:1–7, 1985.
18. Riales R, Albrink MJ: Effect of chromium chloride supplementation on glucose tolerance and serum lipids including high-density lipoprotein of adult men. Am J Clin Nutr 34:2670–2678, 1981.
19. Anderson RA, Polansky MM, Bryden NA, et al.: Chromium supplementation of human subjects: Effects on glucose, insulin, and lipid variables. Metabolism 32:894–899, 1983.
20. Saner G, Yuzbasiyan V, Neyzi O, et al.: Alterations of chromium metabolism and effect of chromium supplementation in Turner's syndrome patients. Am J Clin Nutr 38:574–578, 1983.
21. Uusitupa MIJ, Kumpulainen JT, Voutilainen E, et al.: Effect of inorganic chromium supplementation on glucose tolerance, insulin response, and serum lipids in noninsulin-dependent diabetics. Am J Clin Nutr 38:404–410, 1983.
22. Vinson JA, Bose P: The effect of a high chromium yeast on the blood glucose control and blood lipids of normal and diabetic human subjects. Nutr Rep Int 30:911–918, 1984.
23. Potter JF, Levin P, Anderson RA, et al.: Glucose metabolism in glucose-intolerant older people during chromium supplementation. Metabolism 34:199–204, 1985.
24. Elwood JC, Nash DT, Streeten DHP: Effect of high-chromium brewer's yeast on human serum lipids. J Am Coll Nutr 1:263–274, 1982.
25. Streeten DHP, Gerstein MM, Marmor BM, et al.: Reduced glucose tolerance in elderly human subjects. Diabetes 14:579, 1965.
26. Doisy RJ, Streeten DHP, Freiberg JM, et al.: Chromium metabolism in men and biochemical effects. In: *Trace Elements in Human Health and Disease,* AS Prasad (ed). New York: Academic Press, 1976, 2:79.
27. Saner G: The effect of parity on maternal hair chromium concentration and the changes during pregnancy. Am J Clin Nutr 34:853–855, 1981.
28. Shapcott D, Cloutier D, Demers PP, et al.: Hair chromium at delivery in relation to age and number of pregnancies. Clin Biochem 13:129–131, 1980.
29. Wallach S, Verch RL: Placental transport of chromium. J Am Coll Nutr 3:69–74, 1984.
30. Levy LS: Carcinogenic and mutagenic activity of chromium containing materials. Br J Cancer 32:254, 1975.
31. Krisha G, Mathur JS, Mehrotra SK, et al.: Blood and urine concentrations of chrome in chrome industry workers. Indian J Med Res 63:1357, 1975.
32. Mertz W, Roginski EE, Reba RC: Biological activity and fate of trace quantities of intravenous chromium in the rat. Am J Physiol 209:489, 1965.
33. Stoecker BJ, Oladut WK: Effects of chromium and ascorbate deficiencies on glucose tolerance and serum cholesterol of guinea pigs. Nutr Rep Int 32:399–405, 1985.

15
Copper

Introduction	255
Absorption, Metabolism, and Excretion	257
Dosage Forms	258
Clinical Uses	259
The Infant	259
Pregnancy and Lactation	259
The Elderly	260
Hypercholesterolemia	260
Abnormal Glucose Tolerance	260
Heart Beat Irregularities	261
Familial Benign Copper Deficiency	262
Other Possible Uses	262
Toxicity and Side Effects	263
Interactions	263
References	264

INTRODUCTION

Copper is an essential nutrient because it is required as a constituent or cofactor in many enzymes [1–4], including:

1. Ceruloplasmin (ferroxidase I), capable of oxidizing ferrous ion mobilized from iron stores to ferric ion for binding to plasma transferrin.

2. Ferroxidase II, a lipoprotein with ferrous ion-oxidizing properties similar to but more potent than those of ceruloplasmin.

3. Tyrosinase, involved in conversion of tyrosine to dopamine and required for synthesis of melanin pigment.

4. Dopamine-beta-hydroxylase, required for synthesis of norepinephrine.

5. Lysyl oxidase, essential for the formation of mature cross-linked collagen and elastin.

6. Amine oxidases, involved in the catabolism of norepinephrine, epinephrine, serotonin, histamine, and various diamines.

7. Cytochrome oxidase, essential for energy generation.

8. Cytosolic superoxide dismutase (SOD), a copper- and zinc-dependent enzyme that catalyzes dismutation of the superoxide anion and protects against peroxidative damage.

The activity of copper in these enzymes depends on its redox functions as it shifts back and forth between two valence states, Cu II and Cu I [1].

There is suggestive evidence from animal studies that copper may be required for the activation of two enzymes essential for the regulation of plasma lipid levels, namely plasma lecithin:cholesterol acyltransferase [5,6] and endothelial lipoprotein lipase (activity measured after release into plasma by heparin injection) [7].

The blue copper protein, ceruloplasmin, accounts for the majority of circulating copper. Frieden [8] has reviewed the many functions of this plasma protein. Besides its copper transport functions, ceruloplasmin directly mobilizes iron into the plasma for binding to transferrin and provides the major molecular link between copper and iron metabolism [8,9]. In addition, it is, because of its redox properties, the most prominent plasma antioxidant, inhibiting deleterious oxidation of polyunsaturated fatty acids and other oxidizable substrates. It scavenges superoxide radicals, serves as an acute-phase reactant (an endogenous modulator) of the inflammatory response, and may be involved in regulating plasma levels of epinephrine and sertotonin [8]. Vitamin A, specifically retinoic acid, appears to have a regulatory role in the synthesis of ceruloplasmin [10].

The total amount of copper in the adult human body is 80–150 mg, of which about one-third is in liver and brain at high tissue concentrations and one-third in muscle (at low concentrations offset by large mass) [1–3]. Studies of serum copper levels in men and women in The Netherlands showed the following: men 18–60 years old, mean 109, range 75–150 mcg/dl; men, >60 years, mean 124, range 100–155 mcg/dl; women, 18 years and older, mean 137, range 105–185 mcg/dl [9]. These values are probably typical for adults consuming Western-type diets and suggest considerable individual variability in the status of copper nutriture. Similar mean values and ranges for plasma copper were found in healthy elderly men and women living in England [11].

Severe copper deficiency in animals results in hypocupremia, microcytic anemia, neutropenia, reduced immune status, depressed levels of neurotransmitters, failure of myelination, reduced pigment in hair and skin, connective tissue and cardiac defects, bone defects, and hypercholesterolemia [1–4]. Thymus weights decrease, and antibody titers and natural killer-cell cytotoxicity are markedly suppressed by acute copper deficiency in young rats [12]. Acute copper deficiency in humans is rare and has been reported only in premature infants, manifesting at 3 months of age or older, with findings of normocytic (rather than microcytic) hypochromic anemia, poor iron absorption, osteopenia with scurvy-like cupping of the metaphyses, and pathologic fractures [13,14]; inadequate copper in formula or in TPN solutions, together with low copper stores at birth due to prematurity, were precipitating causes. Mild or subclinical copper deficiency in humans appears to be a definite possibility, however, as copper is low or absent in staple foods constituting a large part of Western-type diets, e.g., sugar and other simple carbohydrates, refined flour and cereal products, and fats and oils. Marginal copper deficiency in adults has been suggested to be a contributory factor in hypercholesterolemia, abnormal glucose tolerance, cardiovascular disease, and cardiac arrythmias, based on evidence obtained in both animal and human studies [15].

The recommended safe and adequate intake of copper for adults is 2–3 mg/d [16]. Diets that include liver and legumes easily provide this intake, but lack of these in the diet usually causes the daily intake to drop below 2 mg [3]. Oysters are also an excellent source of copper. About 75% of the diets of military personnel were found to provide less than 2 mg/d [17]. The minimum daily requirement of men consuming typical American foods appears to be about 1.5 mg/d [18,19]. Many American diets provide less than this amount [19], but it must be recognized that since the liver can store copper, occasional higher-than-habitual daily intakes to replenish liver stores may enable an individual to maintain an adequate copper status. Healthy male subjects maintained for 3–6 months on controlled diets providing 0.8–1.0 mg/d of copper developed biochemical signs of deficiency, including decline of total serum copper and ceruloplasmin concentrations, reduced erythrocyte SOD activity, rising serum cholesterol level, abnormal glucose tolerance, and decreased plasma enkephalins [19–23]. In the largest of these studies [22], 4 of the 24 subjects developed heart beat irregularities and had to be removed from study for repletion with supplemental copper. A 90-day study by Turnlund et al. [24], comparing groups of young men on intakes of 2, 0.8, and 8.0 mg/d of copper, indicated no statistically significant differences between groups with respect to plasma copper, ceruloplasmin, or superoxide dismutase activity, but trends toward slightly lower plasma copper and ceruloplasmin after 42 days on 0.8 mg/d of dietary copper suggest that this intake is marginal for healthy young men.

ABSORPTION, METABOLISM, AND EXCRETION

Dietary copper of plants is believed to be largely of inorganic form, while in animal protein foods it is usually present as cuproproteins. Inorganic (free ionic) copper is absorbed in the stomach and the proximal duodenum under acidic conditions, while cuproprotein copper, after digestion, is absorbed mostly below the pancreatic duct, probably as chelates with amino acids [25]. Metallothionein, an inducible protein rich in sulfhydryl groups, binds the copper in the mucosal absorbing cells and releases it gradually to the circulation, where it binds to albumin and amino acids [1,2]. Metallothionein is also present in the liver, and its complex with copper is believed to serve as a copper store for synthesis of ceruloplasmin and cuproenzymes [2]. Bremner [26] has recently reviewed the role of metallothionein in the hepatic metabolism of copper, including its presumed involvement in the uptake, storage, and transfer of copper. Measurements of copper absorption from food vary, most being in the range of 40–60% [2].

Transport of copper in the circulation is largely in the form of ceruloplasmin, but substantial amounts are associated with another plasma protein still undergoing identification and with albumin and amino acids [27]. Small amounts are present as cuproenzymes in red and white cells [2].

The major excretory route for copper is via the bile, with only 0.5–3% of daily intake being excreted in the urine. Surface losses (sweat, shed cells, hair growth) are estimated at a mean of 0.34 mg/d in adult men [28].

Derangements of copper metabolism occur in two genetic disorders: Menkes' syndrome and Wilson's disease. In Menkes' syndrome, there is defective absorption of copper, accumulation of the metal in the intestinal mucosa, and greatly reduced levels of copper in the blood, brain, and liver. The specific metabolic defects are poorly understood; injection of copper salts or chelates does not improve the copper status or prevent the invariably fatal outcome [1–4]. Wilson's disease is characterized by excessive accumulation of copper in liver, brain, and kidney, leading ultimately to cirrhosis, neurologic disorders, and kidney failure; the cause is believed to be an abnormal metallothionein with a very high binding affinity for copper [1–4]. The disease has been successfully controlled by administration of copper-chelating agents, notably D-penicillamine, to increase urinary excretion of the metal [2], and more recently by administration of pharmacologic doses of zinc, which depresses absorption of ingested copper (see chapter on Zinc, pp. 285–319).

There is also a copper-accumulation disorder reported from India called Indian childhood cirrhosis, in which hepatic copper is elevated; a genetic origin is only one of several speculative causes suggested [1–3]. Chronic cholestasis, as in primary biliary cirrhosis, may also result in hepatic accumulation of copper [1–3].

DOSAGE FORMS

Copper in amino acid-chelated form is available over the counter as tablets containing 2.5 mg elemental copper. Copper sulfate is available in pharmacies.

CLINICAL USES
The Infant

About 80% of fetal copper stores, much of which is in the liver, is accumulated during the last trimester [29]. Whether term or premature, however, the newborn infant has low total copper and ceruloplasmin levels relative to adult values, probably due to incomplete maturation of hepatic production of ceruloplasmin; immaturity of copper absorption is another possibility [30]. On formula providing 0.4 mg of copper per liter, which supplied a copper intake of 70–90 mcg/kg bodyweight at 120–150 kcal/kg, term infants reached adult serum levels of copper and ceruloplasmin within 1 month, while in preterm infants these increases were delayed—up to 3 or 4 months in very premature infants [30]. If these increases are further delayed, consideration should be given to a somewhat higher level of copper supplementation [30].

Studies of the copper status of exclusively breast-fed Finnish infants showed that adult levels of serum copper and ceruloplasmin were not attained until 6 months of age, but no signs of copper deficiency were observed [31]. Maternal copper supplementation (2 or 4 mg/d) did not change the results.

Chilean clinicians noted a prevalence of 30% of hypocupremia in marasmic infants being repleted on a cow's milk-cornstarch-sugar-vegetable oil formula; supplementation with 80 mcg/kg/d of copper as copper sulfate was effective in preventing hypocupremia [32,33].

Urinary copper losses are greatly increased in infants receiving total parenteral nutrition and could contribute to development of copper deficiency [34]. Wilmore et al. [35] have recommended supplementation of the TPN formula for infants with sufficient copper to provide 0.22 mg/kg/d.

Pregnancy and Lactation

Pregnancy is associated with a marked increase, almost a doubling, of serum copper and ceruloplasmin levels, and this persists during the early postpartum period, returning to normal nonpregnant levels by 6 weeks postpartum [30,36]. At delivery, there is a strong correlation between maternal and cord blood copper concentration, but the maternal level is about 5 times that of cord blood [36].

Results differ between different investigators as to whether ordinary dietary levels of copper are sufficient for the pregnant woman. Turnlund et al. [37], studying copper retention in women during the second half of their pregnancy, noted apparent positive balances at intakes of 1.44 and 2.53 mg copper per day from animal and plant protein diets, respectively, with calculated total retentions adequate to provide for the needs of the fetus. On the other hand, from their studies Taper et al. [38] concluded that dietary copper intakes were usually not adequate and that retention sufficient to provide for fetal needs without some depletion of maternal copper stores was probably not possible without a supplement. A supplement of 2 mg/d, particularly during the latter part of pregnancy when the major part of fetal copper is being transferred [29], should help to protect the mother's copper status. Diet histories should be obtained by a dietitian and evaluated for copper intake.

During lactation, milk copper concentration was observed to decline from a mean level of 0.34 mg/d in colostrum to 0.12 mg/d at 9 months, with wide interindividual variations [31]. If evaluation of the maternal diet indicates a copper intake below the recommended 2–3 mg/d, a 2-mg/d supplement may be prescribed to protect maternal copper status during lactation. Maternal supplementation at 2 or 4 mg/d does not appear to affect copper concentration in breast milk [31].

The Elderly

Although dietary copper intakes below the existing U.S. recommendations [16] appear to be adequate for healthy active elderly people [39], Bunker et al. [40] found that housebound elderly patients (70–85 years of age) with stable chronic diseases, who ate self-selected diets, were in significant negative copper balance.

Hypercholesterolemia

In experimental animals, copper deficiency is known to produce hypercholesterolemia [1–4,41], as well as depressed plasma lecithin:cholesterol acyltransferase [5,6] and lipoprotein lipase [7] activities. Mild copper deficiency induced in a male human subject by 15 weeks of feeding a diet supplying 0.83 mg copper per day caused a linear rise in plasma cholesterol from 204 to 232 mg/dl; repletion with a copper supplement during the following 39 days reduced plasma cholesterol to 198 mg/dl [20]. A recent large study involving 24 male volunteers, aged 21–57 years, showed that when maintained for 11 weeks on diets containing 0.36 mg Cu/1,000 kcal, they experienced significant increases in LDL cholesterol and significant decreases in HDL cholesterol, as compared with readings obtained on their pretest self-selected diets providing a mean 0.57 mg Cu/1,000 kcal or on a repletion diet providing 1.41 mg Cu/1,000 kcal [42]. Klevay et al. [43] report significant declines in plasma HDL cholesterol in 8 women, aged 18–36 years, maintained for 135 days on a diet made up of conventional foods supplying 0.6 mg Cu/d.

Abnormal Glucose Tolerance

Laboratory rats fed copper-deficient diets develop impaired glucose tolerance [41,44], with decreased insulin binding [45]. The effects are magnified if the major dietary carbohydrate is sucrose, glucose, or fructose and reduced if the carbohydrate is supplied by starch [45,46]. The percentage of glycosylated hemoglobin is markedly increased in copper-deficient rats [47]. In rats rendered diabetic by streptozotocin treatment and kept on a copper-deficient diet, the metabolism of injected glucose was improved more by injecting a combination of insulin and copper than by injecting either insulin or copper alone [48].

Klevay and coworkers [21] studied the effect of mild copper deficiency, induced by feeding a diet supplying 0.78 mg copper daily for 150 or 175 days, on plasma glucose clearance in two healthy young men. Intravenous glucose tolerance was significantly decreased by copper deficiency, even though the deficiency produced was not great enough to cause significant hematological changes. After repletion with 6 mg copper per day for 30 days, glucose clearance was restored to initial levels or better.

Whether mild copper deficiency is a significant contributor to impaired glucose tolerance in individuals consuming Western-type diets is not known, but such diets often have copper contents as low or almost as low as in the Klevay experiment and are often high in simple sugars, which may aggravate copper deficiency. It therefore appears advisable to assess the dietary copper intake of patients with glucose intolerance or type II diabetes and to prescribe a copper supplement if their habitual intakes are below the recommended range of 2–3 mg/d.

Heart Beat Irregularities

During the latter part of a 15-week depletion period with a single adult male subject, using a diet providing 0.83 mg copper daily. Klevay et al. [20] observed an abnormal EKG pattern that disappeared on repletion with a 2.5-mg/d copper supplement. Reiser et al. [22], studying 24 male subjects receiving 1.03 mg dietary copper per day over an 11-week period, observed development of acute tachycardia in one, aged 26 years, after 7 weeks, and in another, aged 33 years, at 10 weeks. A third subject, aged 57 years, showed an intermittent second-degree heart block at 11 weeks. These cardiac irregularities disappeared on repletion with 3 mg/d of supplemental copper.

Disappearance of chronic ventricular premature beats with copper gluconate supplementation (4 mg elemental copper per day) was reported by Spencer [49] in three instances: a man, aged 57 years, with a lifelong history of premature beats; a young woman, aged 19 years, who had experienced several premature beats per minute for several years; and a man, aged 48 years, with duration of the condition not given. Resolution of the abnormality was rapid (2 weeks in the case of the young woman), and the premature beats did not recur as long as the supplement was continued. Copper status and dietary copper intake were not assessed.

Since the supplement of 4 mg/d of copper as the gluconate is innocuous and since its effect appears to be rapid, the physician may wish to try such copper supplementation in patients with tachycardia or arrythmias even if dietary copper intake appears adequate. Individual copper requirements are known to vary, and these patients may possibly require more than the generally recommended intake of 2–3 mg/d. Confirmatory clinical studies are needed.

Familial Benign Copper Deficiency

Hungarian clinicians [50] describe a congenital copper deficiency syndrome in a 21-month-old- boy hospitalized for repeated seizures and failure to thrive. He had blond curly hair of normal microscopic appearance, mild hypochromic anemia, low serum iron and copper with normal ceruloplasmin levels, and spurring of the femora and tibiae. Muscle tone, EEG, and mental development were normal. The abnormalities were corrected with daily supplements of 7.5 mg of oral copper as copper sulfate syrup. No iron was given. Over the ensuing 20 months, it was determined that if this copper dosage was reduced or discontinued, serum copper level dropped and seizures again occurred. This child, his mother, and his maternal uncle had subnormal serum copper levels, curly hair subject to thinning in adulthood, and facial seborrhea. The normal mental development, normal ceruloplasmin levels, and response to copper supplementation distinguish the condition from Menkes' syndrome.

Other Possible Uses

The acute-phase response in infections, rheumatoid arthritis, cancer, and possibly other stresses causes release of ceruloplasmin and copper complexes from the liver, and this may result in a two- or threefold increase in blood copper concentrations. Sorenson [51] postulates that this increase fulfills a physiologic need. He and his coworkers [52,53] have investigated the pharmacologic properties of copper complexed with antiarthritic drugs and with various amino acids, finding many of them to be effective antiinflammatory, antiulcer, anticonvulsant, anticancer, and antidiabetic agents in laboratory animals. Products of this kind have not yet been marketed. There have been no reports of efficacy of ordinary forms of supplemental copper for these applications.

Although serum ceruloplasmin is a positive acute-phase reactant after stress, Boosalis et al. [54] found depressed ceruloplasmin and total serum copper levels in 23 adult patients hospitalized with second- and third-degree burns. Urinary excretion of copper was also elevated. Loss of ceruloplasmin from thermal-wound sites may be an important factor in severe burns of large area. Whether repletion by copper supplementation may prove useful in treatment of burn patients is a subject for future investigation. Burn patients are highly susceptible to infection, and copper deficiency is known to impair immune function [12,55,56].

Copper deficiency in young adult rats has been observed to lead to hypertension, and Klevay has suggested that this aspect of marginal copper-deficiency requires further investigation in human studies [57].

Dreith et al. [58] propose that an adequate copper status, because of the antioxidant properties of copper-containing enzymes, may be important in cancer patients undergoing treatment with Adriamycin®. They find that in mice of good copper status, Adriamycin® did not produce a decrease in cardiac reduced-glutathione level, but rather a 20% increase, whereas in mice of low copper status, cardiac content of reduced-glutathione decreased 20%, thereby permitting an increase in peroxidation of cardiac lipids.

TOXICITY AND SIDE EFFECTS

The initial toxic level of inorganic copper in adult men when ingested long-term has been estimated at 10–15 mg/d [2]. However, Pratt et al. [59] administered 10 mg/d of elemental copper as copper gluconate for 12 weeks to three men and four women patients with back pain; seven comparable patients received a placebo in a double-blind experimental design. The seven subjects receiving 10 mg/d of copper (in addition to their normal dietary intake) experienced no change in the level of copper in serum, urine, or hair. There was also no change in serum zinc or magnesium, nor in levels of hematocrit, triglycerides, SGOT, GPT, LDH, total cholesterol, or alkaline phosphatase. The incidence of heartburn, nausea, and diarrhea was the same in the copper-supplement groups as in the placebo group. There was no effect on back pain, spine motion, or trunk muscle strength.

The signs and symptoms of acute copper poisoning, as by ingestion of an excessive amount of copper sulfate, are similar to those observed in other metal poisonings and are probably mostly due to a direct irritant effect on the gastric mucosa: metallic taste, ptyalism, epigastric pain, nausea, vomiting, and diarrhea [2,3]. There may also be intravascular hemolysis, jaundice, hemoglobinuria, and renal failure. Copper toxicity and its treatment are discussed by Bremner [60]. Industrial exposure to copper, copper dust, copper oxide fumes, and other copper products may result in respiratory symptoms and contact dermatitis [3].

INTERACTIONS

Copper absorption is depressed by high intakes of zinc, apparently because excessive zinc stimulates increased production of metallothionein in the intestines, which has a greater binding affinity for copper than for zinc; the bound copper is returned to the intestinal lumen with mucosal cell turnover [18]. It is uncertain whether zinc intakes at the high end of the physiologic range (around 20 mg/d) have any effect when copper intake is adequate, but there could be adverse effects if copper intake is low [18]. Prasad et al. [61] found that hypocupremia developed in sickle cell anemia patients being treated with 150 mg/d of zinc as zinc acetate, but this was readily corrected by a daily supplement of 1 or 2 mg of copper per day as copper sulfate [61].

Very high levels of vitamin C in the diets of monkeys and other animal species tend to depress serum copper and ceruloplasmin levels, especially if the copper intake is marginal [62]. Milne et al. [63] report that in eight women, aged 18–36 years, a low Cu intake (0.6 mg/d), combined with a high vitamin C intake (1,500 mg/d), for 42 days did not markedly affect commonly measured indices of copper and iron metabolism, except for a 20% reduction in ceruloplasmin diamine oxidase activity. Subjects were repleted with 2 mg Cu/d for 37 days. In laboratory rats, high iron content in the diet worsened the degree of copper deficiency developed on a low-copper diet [64].

Patients who are taking supplemental zinc, high-dose vitamin C, or therapeutic iron should probably also be taking 2–3 mg of copper daily, particularly if their dietary copper intake is low and/or their intake of sucrose and other simple sugars, which can depress copper status [22,45,46,64], is high.

REFERENCES

1. Solomons NW: Biochemical, metabolic, and clinical role of copper in human nutrition. J Am Coll Nutr 4:83–105, 1985.
2. Mason KE: A conspectus of research on copper metabolism and requirements of man. J Nutr 109:1979–2066, 1979.
3. Williams DM: Clinical significance of copper deficiency and toxicity in the world population. In: *Clinical, Biochemical, and Nutritional Aspects of Trace Elements,* AS Prasad (ed). New York: Alan R. Liss, 1982, pp 277–299.
4. O'Dell BL: Biochemical basis of the clinical effects of copper deficiency. In: *Clinical, Biochemical, and Nutritional Aspects of Trace Elements,* AS Prasad (ed). New York: Alan R. Liss, 1982, pp301–313.
5. Lau BWC, Klevay LM: Plasma lecithin: cholesterol acyltransferase in copper-deficient rats. J Nutr 111:1698–1703, 1981.
6. Harvey PW, Allen KGD: Decreased plasma lecithin: cholesterol acyltransferase activity in copper-deficient rats. J Nutr 111:1855–1858, 1981.
7. Lau BWC, Klevay LM: Postheparin plasma lipoprotein lipase in copper-deficient rats. J Nutr 112:928–933, 1982.
8. Frieden E: Caeruloplasmin: A multifunctional metalloprotein of vertebrate plasma. In: *Biological Roles of Copper,* Ciba Foundation Symposium 79. Amsterdam: Elsevier/North-Holland, Excerpta Medica, 1980, pp 93–124.

9. Schreurs WHP, Klosse JA, Muys T, Haesen JP: Serum copper levels in relation to sex and age. Int J Vitam Nutr Res 42:68–74, 1982.
10. Barber EF, Cousins RJ: Induction of ceruloplasmin synthesis by retinoic acid in rats: Influence of dietary copper and vitamin A status. J Nutr 117:1615–1622, 1987.
11. Bunker VW, Hinks LJ, Lawson MS, Clayton BE: Assessment of zinc and copper status of healthy elderly people using metabolic balance studies and measurement of leucocyte concentrations. Am J Clin Nutr 40:1096–1102, 1984.
12. Koller LD, Mulhern SA, Frankel NC, et al.: Immune dysfunction in rats fed a diet deficient in copper. Am J Clin Nutr 45:997–1006, 1987.
13. Ashkenazi A, Levin S, Djaldetti M, et al: The syndrome of neonatal copper deficiency. Pediatrics 52:525, 1973.
14. Heller RM, Kirchner SG, O'Neill JA, et al.: Skeletal changes of copper deficiency in infants receiving prolonged total parenteral nutrition. J Pediatr 92:947, 1978.
15. Klevay LM: Changing patterns of disease: Some nutritional remarks. J Am Coll Nutr 3:149–158, 1984.
16. National Research Council: *Recommended Dietary Allowances,* 9th ed. Washington, DC: National Academy of Sciences, 1980.
17. Milne DB, Schnakenberg DD, Johnson HL, Kuhl GL: Trace mineral intake of enlisted military personnel. J Am Diet Assoc 76:41, 1980.
18. Klevay LM, Reck SJ, Jacob RA, et al.: The human requirement for copper. I. Healthy men fed conventional American diets. Am J Clin Nutr 33:45–50, 1980.
19. Sandstead HH: Copper bioavailability and requirements. Am J Clin Nutr 35:809–814, 1982.
20. Klevay LM, Inman L, Johnson LK, et al.: Increased cholesterol in plasma in a young man during experimental copper depletion. Metabolism 33:1112–1118, 1984.
21. Klevay LM, Canfield WK, Gallagher SK, et al.: Decreased glucose tolerance in two men during experimental copper depletion. Nutr Rep Int 33:371–382, 1986.
22. Reiser S, Smith JC Jr, Mertz W, et al.: Indices of copper status in humans consuming a typical American diet containing either fructose or starch. Am J Clin Nutr 42:242–251, 1985.
23. Bhathena SJ, Recant L, Voyles NR, et al.: Decreased plasma enkephalins in copper deficiency in man. Am J Clin Nutr 43:42–46, 1986.
24. Turnlund JR, Keen CA, Acord LL: Indices of copper status at three levels of dietary copper. Am J Clin Nutr 45:854A, 1987.
25. Ashmead HD, Graff DJ, Ashmead HH. In: *Intestinal Absorption of Metal Ions and Chelates.* Springfield, IL: CC Thomas, 1985, pp 74, 215–218.
26. Bremner I: Involvement of metallothionein in the hepatic metabolism of copper. J Nutr 117:19–29, 1987.
27. Weiss KC, Linder MC: Copper transport in rats involving a new plasma protein. Am J Physiol 249:E77–E88, 1985.
28. Jacob RA, Sandstead HH, Munoz JM, et al.: Whole body surface loss of trace metals in normal males. Am J Clin Nutr 34:1379–1383, 1981.
29. Shaw JCL: Trace elements in the fetus and young infant. II. Copper, manganese, selenium, and chromium. Am J Dis Child 134:74–81, 1981.
30. Hillman LS: Serial serum copper concentrations in premature and SGA infants during the first 3 months of life. J Pediatr 98:305–308, 1981.
31. Salmenperä L, Perheentupa J, Pakarinen P, Siimes MA: Cu nutrition in infants during prolonged exclusive breast feeding: Low intake but rising serum concentrations of Cu and ceruloplasmin. Am J Clin Nutr 43:251–257, 1986.

32. Castillo-Duran C, Fisberg M, Valenzuela A, et al.: Controlled trial of copper supplementation during the recovery from marasmus. Am J Clin Nutr 37:898–903, 1983.
33. Uauy R, Castillo-Duran C, Fisberg M, et al.: Red cell superoxide dismutase activity as an index of human copper nutrition. J Nutr 115:1650–1655, 1985.
34. Tyrala EE, Brodsky NL, Auerbach VH: Urinary-copper losses in infants receiving free amino acids. Am J Clin Nutr 35:542–545, 1982.
35. Wilmore DW, Groff DBR, Bishop HC, Dudrick SJ: Total parenteral nutrition in infants with catastrophic intestinal anomalies. J Pediatr Surg 4:181, 1969.
36. Vir SC, Love AHG, Thompson W: Serum and hair concentrations of copper during pregnancy. Am J Clin Nutr 34:2382–2388, 1981.
37. Turnlund JR, Swanson CA, King JC: Copper absorption and retention in pregnant women fed diets based on animal and plant proteins. J Nutr 113:2346–2352, 1983.
38. Taper LJ, Oliva JT, Ritchey SJ: Zinc and copper retention during pregnancy: The adequacy of prenatal diets with and without dietary supplementation. Am J Clin Nutr 41:1184–1192, 1985.
39. Bunker VW, Hinks LJ, Lawson MS, Clayton BE: Assessment of zinc and copper status of healthy elderly people using metabolic balance studies and measurement of leucocyte concentrations. Am J Clin Nutr 40:1096–1102, 1984.
40. Bunker VW, Hinks LJ, Stansfield MF, et al.: Metabolic balance studies for zinc and copper in housebound elderly people and the relationship between zinc balance and leukocyte zinc concentrations. Am J Clin Nutr 46:353–359, 1987.
41. Klevay LM: Copper and ischemic heart disease. Biol Trace Elem Res 5:245–255, 1983.
42. Reiser S, Powell A, Yang C-Y, Canary JJ: Effect of copper intake on blood cholesterol and its lipoprotein distribution in men. Nutr Rep Int 36:641–649, 1987.
43. Klevay LM, Milne DB, Mullen LM, Lukaski HC: Decreased HDL cholesterol associated with negative copper balance. Fed Proc 46:909A, 1987.
44. Hassel CA, Marchello JA, Lei KY: Impaired glucose tolerance in copper-deficient rats. J Nutr 113:1081–1083, 1983.
45. Fields M, Ferretti RJ, Smith JC Jr, Reiser S: Effect of copper deficiency on metabolism and mortality in rats fed sucrose or starch diets. J Nutr 113:1335–1345, 1983.
46. Fields M, Ferretti RJ, Smith JC Jr, Reiser S: Impairment of glucose tolerance in copper-deficient rats: Dependency on the type of dietary carbohydrate. J Nutr 114:393–397, 1984.
47. Klevay LM: An increase in glycosylated hemoglobin in rats deficient in copper. Nutr Rep Int 26:329, 1982.
48. Fields M, Reiser S, Smith JC Jr: Effect of copper or insulin in diabetic copper-deficient rats. Proc Soc Exp Biol Med 173:137–139, 1983.
49. Spencer JC: Direct relationship between the body's copper/zinc ratio, ventricular premature beats, and sudden coronary death. Am J Clin Nutr 32:1184–1185, 1979.
50. Mehes K, Petrovicz E: Familial benign copper deficiency. Arch Dis Child 57:716–718, 1982.
51. Sorenson JRJ: Copper complexes: A physiologic approach to treatment of chronic diseases. Compr Ther 11:49–64, 1985.
52. Sorenson, JRJ, Oberley LW, Crouch RK, et al.: Pharmacologic activities of copper compounds in chronic diseases. Biol Trace Elem Res 5:257–273, 1983.
53. Sorenson JRJ, Kishore V, Pezeshk A: Newer pharmacologic activities of copper complexes. Nutr Res Suppl 1:S-457–S-463, 1985.
54. Boosalis MG, McCall JT, Solem LD, et al.: Serum copper and ceruloplasmin levels and urinary copper excretion in thermal injury. Am J Clin Nutr 44:899–906, 1986.
55. Prohaska JR, Lukasewycz OA: Copper deficiency suppresses the immune response of mice. Science 213:559–561, 1981.

56. Prohaska JR, Downing SW, Lukasewycz OA: Chronic dietary copper deficiency alters biochemical and morphological properties of mouse lymphoid tissues. J Nutr 113:1583–1590, 1983.
57. Klevay LM: Hypertension in rats due to copper deficiency. Nutr Rep Int 35:999–1005, 1987.
58. Dreith D, Zidenberg-Cherr S, Keen CL: The influence of Cu status of Adriamycin (ADR) induced lesions in mice. Fed Proc 46:950A, 1987.
59. Pratt WB, Omdahl JL, Sorenson JRJ: Lack of effects of copper gluconate supplementation. Am J Clin Nutr 42:681–682, 1985.
60. Bremner I: Heavy metal toxicities. Q Rev Biophys 7:75–124, 1974.
61. Prasad AS, Brewer GJ, Schoomaker EB, Rabbani P: Hypocupremia induced by zinc therapy in adults. JAMA 240:2166–2168, 1978.
62. Milne DB, Omaye ST, Amos WH Jr: Effect of ascorbic acid on copper and cholesterol in adult cynomolgus monkeys fed a diet marginal in copper. Am J Clin Nutr 34:2389–2393, 1981.
63. Milne DR, Klevay LM, Hunt JR: Effects of copper (Cu) intake and vitamin C supplements on copper and iron nutriture in women Fed Proc 46:908A, 1987.
64. Johnson MA, Hove SS: Development of anemia in copper-deficient rats fed high levels of dietary iron and sucrose. J Nutr 116:1225–1238, 1986.

16
Selenium

Introduction	269
Absorption, Metabolism, and Excretion	271
Dosage Forms	272
Clinical Studies	273
Cancer	273
Cardiovascular Disease	274
The Elderly	275
Chronic Disease	276
Alcoholism	276
Parenteral Nutrition	276
Diabetic Retinopathy	277
Acne Vulgaris	277
Toxicity and Side Effects	278
Interactions	279
References	279

INTRODUCTION

Selenium (Se) is required for synthesis of Se-dependent glutathione peroxidase (GSHPx), in which it is a covalently bound constituent. The Se is introduced into the enzyme, as it is being biosynthesized, in the form of selenocysteine, an analog of cysteine in which Se takes the place of sulfur. GSHPx is widely distributed in the body, being particularly abundant in the kidney, liver, pancreas, and erythrocyte, but in all cells its function, as an essential part of the antioxidant defenses of the body, is to inactivate hydrogen peroxide and peroxidized polyunsaturated fatty acids by converting them to water and hydroxyl derivatives. The enzyme employs reduced glutathione as a reducing agent to accomplish these reactions; the resultant oxidized glutathione is recycled back to the reduced form by the action of another enzyme, glutathione reductase. The primary function of Se-dependent GSHPx is believed to be the protection of cellular membranes, including those within the cell, from oxidative damage; in this respect it complements the action of vitamin E [1–4].

Selenium deficiency due to inadequate Se in the diet has been reported only from China and is known there as Keshan disease [5,6]. The disorder, manifesting in childhood as a selenium-responsive cardiomyopathy, occurs in a region where the soil is extremely low in Se. In this region, the intake from locally grown produce may be as low as 10 mcg/d. In similar areas, a disorder termed Kaschin-Beck disease, characterized by necrosis of cartilage and dystrophy of skeletal muscle, has been suggested as another manifestation of severe Se deficiency [6]. On the other hand, there are other areas in China where the soil is so high in Se that intakes from food may reach 3,200–6,690 mcg/d, resulting in distinct toxic symptoms [6,7]. There are other parts of the world where soils are low in Se, but not low enough to produce any clinical signs of deficiency; these include New Zealand and parts of Finland [8,9]. In the United States and Canada, there are areas of low-Se soils in the Pacific northwest, the southeast Atlantic coastal region, and around the Great Lakes, but interstate and import shipments of food tend to level out dietary Se intakes, which usually fall in the range of 62–216 mcg/d and are considered adequate [3].

Symptoms of Se deficiency have also been observed in patients undergoing long-term total parenteral nutrition without Se supplementation; muscle weakness, pain, and tenderness were noted that responded quickly to Se supplementation [8]. Feldman's group in Augusta, Georgia, a low-Se region, observed that a low Se status was common in hospitalized middle-aged patients with protein-calorie malnutrition, particularly among black males [10]. Decreased Se concentrations in red cells and serum occur rather commonly in patients with alcoholic cirrhosis or other chronic liver disease, gastrointestinal disease, cancer, and chronic renal failure [10,11]. Boyne and Arthur [12] report that marginal Se status in mice markedly increased the rapidity of initiation and the severity of infections induced by IV injection of the yeast organism, *Candida albicans,* a yeast known to cause serious and often fatal infections in immunologically suppressed humans.

Levander and coworkers [13] determined that adult North American men required about 54 mcg/d of Se to replace fecal and urinary losses and maintain their usual plasma levels and body stores; assuming 80% absorption, the dietary intake should be about 70 mcg/d. In New Zealand, Stewart et al. [14] found that young women could remain in Se balance on intakes of about 24 mcg/d, although with mean plasma Se levels less than one-half those of North American men. New Zealanders typically have lower blood levels of Se and GSHPx than North Americans or Australians, owing to the low Se content of indigenous foods [8]. The U.S. National Research Council has established the range of 50–200 mcg/d as a safe and adequate Se intake for adults, 10–40 mcg/d for infants, and 20–120 mcg/d for children [15]. These intakes are readily met by most North American diets, with animal protein foods (meat, poultry, fish) and cereal grain foods being the principal Se sources [16].

ABSORPTION, METABOLISM, AND EXCRETION

The predominant form of Se in wheat, soybeans, and probably many other plant foods appears to be selenomethionine (Se-Met), although some vegetables may contain large amounts of selenate. Selenium-enriched yeast also contains Se-Met. In meat and other animal proteins, Se is present mostly as selenocysteine (Se-Cys), together with small amounts of Se-Met, selenotrisulfide, selenopersulfides, and metal selenides [16,17].

Laboratory studies show that Se-Met is absorbed in the duodenum by active transport, while selenate is absorbed actively by the ileal mucosa, utilizing the same transport system as sulfate. Selenite appears to be absorbed by passive diffusion throughout the tract, and probably some selenate is absorbed passively in the duodenum [18–20]. Se-Cys does not appear to be absorbed by active transport [21].

Selenate, selenite, and Se-Met can be converted in animal tissues to Se-Cys, which is used in the biosynthesis of glutathione peroxidase. Some Se-Cys or Se-Met appears to be used for synthesis of non-GSHPx proteins, e.g., plasma protein P (found in the rat and rhesus monkey) and Se-proteins in the tissues (found in rat testes, kidney, liver, and sperm). The functions of these Se-proteins are not known, although plasma protein P may be involved in Se storage and transport [22].

It is definitely known that Se-Met can be incorporated directly into hemoglobin and non-GSHPx proteins in plasma and probably into newly synthesized proteins in the liver and other tissues [23–26]. When human subjects of initially low Se status are given 100–200 mcg/d Se largely in the form of Se-Met, as found in wheat or Se-enriched yeast, higher ciculating levels of Se are reached than when they receive Se as selenite or selenate, even though platelet and red cell GSHPx activities reach the same plateau with each type of Se supplement [27–30]. In one study, plasma Se levels plateaued within 4 weeks when sodium selenate (200 mcg Se/d) was the supplement, while in subjects receiving the same amount of Se from wheat or Se-rich yeast, plasma Se was higher and still increasing when the experiment ended at 11 weeks [29]. Since many populations normally have Se-Met intakes this large or larger while maintaining steady plasma and total circulating Se concentrations, an equilibrium is evidently reached at each intake level, over the long term, between synthesis and degradation of proteins containing Se-Met.

Plateauing of red cell GSHPx levels has variously been reported to occur when plasma Se concentration reaches 70 ng/ml [10], 110 ng/ml [29], and 140 ng/ml [31].

Thomson and Robinson [32] have recently reported, based on studies of sodium selenite and sodium selenate absorption and urinary excretion in young New Zealand women, that selenate is much more rapidly absorbed than selenite, confirming animal studies [18–20]. There is also much greater urinary excretion of selenate than selenite (dosage, 1 mg Se in water) during the first 12 hours after administration, but the 5-day total recovery from urine and feces was similar for selenite-Se (60%) and selenate-Se (68%). This is in contrast to results with a 1-mg dose of Se as Se-Met, which was mainly retained, with only 26% of the dose recovered in urine and feces [33].

The total body content of Se varies with the habitual intake and has been estimated at 6 mg for New Zealand women [34] and about 15 mg for North American adults [35].

With high-dose (1 mg Se) administration, selenate is absorbed better than selenite (about 90% versus 60%) [32]. A number of human studies indicate that Se-Met and other organically bound forms of Se found in foods are also better absorbed than selenite and produce earlier rises in blood Se and GSHPx levels [16,30,33,36,37], but in long-term administration, selenite, selenate, and Se-Met are equivalent with regard to efficacy in raising GSHPx activity, although Se-Met produces higher total blood Se values.

Selenium is excreted in the urine and feces. In studies of Se utilization in pregnant and nonpregnant women ingesting about 150 mcg/d of food Se, of which about 80% was absorbed, Swanson et al. [38] observed urinary excretions of 83–124 mcg/d, accounting for 75–80% of total Se output; urinary Se excretion was slightly but not significantly less in pregnant than in nonpregnant women. Fecal Se ranged from 24 to 48 mcg/d, with no significant difference between groups. Selenium compounds present in urine include trimethylselenonium ion, an unidentified organoselenium metabolite, and selenite and selenate (if these inorganic forms are ingested as supplements or are present in foods) [32].

DOSAGE FORMS

Selenium supplements are available over the counter as tablets of Se-enriched yeast providing 50, 100, or 200 mcg Se. Sodium selenate is available as tablets providing 100 mcg of Se.

Selenious Acid Injection (Selenitrace®, Armour) is supplied in 10-ml vials containing 40 mcg Se per ml.

CLINICAL STUDIES
Cancer

By far the most active research interest in the potential of selenium as a clinically useful supplement is in connection with its ability experimentally to inhibit both the initiation and promotion phases of chemically induced carcinogenesis [39]. Selenium is also effective in inhibiting virally induced and transplantable tumors of mice [40]. Interest in laboratory investigations of the relation of Se nutriture to cancer was originally aroused by Shamberger and Frost's 1969 publication [41] calling attention to an inverse epidemiologic correlation between blood Se levels and cancer death rates in certain regions of the United States. Subsequent epidemiologic observations have verified the inverse correlation of Se blood levels or dietary Se intake with age-adjusted cancer mortality rates on a worldwide basis [42], in the United States by region [43] and by county [44], and in China by region [45]. The literature has become very extensive on experimental, clinical, and epidemiologic evidence that Se intake and incidence of most types of cancer are inversely related; pre-1985 studies have been reviewed by Combs and Clark [46]. Some more recent reports include that by Salonen et al. [47] associating general cancer death risk with low Se status in an epidemiological study in Finland; one by Borek et al. [48] showing that Se and vitamin E separately inhibited both radiogenic and chemically induced cell transformation in vitro; and a large population study by Fex et al. [49] in Sweden, involving analysis of plasma samples obtained several years prior to death by cancer, which showed low plasma Se to be a risk factor for cancer death in middle-aged men who lived in the same area. One epidemiologic study, however, noted an apparent positive risk association of plasma Se with squamous-cell lung cancer risk [50].

The mechanism(s) by which Se might inhibit development and progress of cancer remain unclear, although many possible explanations, in addition to the antioxidant activity of GSHPx, have been suggested, all of which have some experimental justification: enhancement of immune response by high Se intakes [51], facilitation of DNA repair [52,53], antimutagenic activity [54], blocking of DNA synthesis and proliferation of rapidly dividing cells [55–57], and depression of reverse transcriptase activity [58].

Several caveats have been expressed regarding the practicality of selenium supplementation as a means of reducing the incidence of human cancer. Nearly all of the laboratory animal studies showing efficacy have been done with near-toxic dosages of Se. Furthermore, almost all investigators have used as their Se source sodium selenite or selenious acid, forms not commonly found in food. There are a few exceptions, for example the study by Schrauzer et al. [59] showing that dietary supplementation with 1 ppm Se as Se-enriched yeast (equal to about 500 mcg Se in a day's human consumption of 500 g of food) reduced the incidence of mammary tumors in a mouse strain that is susceptible to genesis of spontaneous mammary tumors.

Burk [60] has pointed out that Se-Met is probably the only ingested form of Se that can markedly elevate blood Se levels over and above that due to GSHPx content, and furthermore that plant foods are the major dietary source of Se-Met. High blood levels of Se may therefore be largely an indicator of high intake of Se-rich plant foods. Since plant foods may contribute to reduced cancer incidence in a number of ways, e.g., as sources of vitamin C, folic acid, carotenes, fiber, etc., the epidemiologic data relating cancer incidence inversely with Se intake and blood Se levels may not be as valid as generally supposed.

Nevertheless, the total weight of evidence has inspired institution of several double-blind clinical trials to evaluate sodium selenite, sodium selenate, and Se-enriched brewer's yeast as potential agents for reducing the risk of cancer. Daily supplemental dosages are being varied by individual in the range of 50–200 mcg Se. The studies are being conducted in populations at extremely high risk of cancer by: Cornell University and the Institute of Cancer, Chinese (People's Republic of China) Academy of Medical Science (IC-CAMS) for primary liver cancer; the U.S. National Cancer Institute and IC-CAMS for cancer of the espohagus; and Cornell University for nonmelanoma skin cancer [61]. These investigations should do much to settle the question of the value of selenium supplementation as a means of reducing the incidence of human cancer.

Cardiovascular Disease

Shamberger et al. [62] reported an inverse correlation between blood Se levels or Se intakes and heart mortality in 19 U.S. states. Some epidemiologic studies have failed to confirm this correlation [3], but the large prospective case-control study of Salonen et al. [63] in a low-Se area of Finland showed a two- to threefold greater risk of fatal or nonfatal myocardial infarction during a 7-year follow-up period in subjects whose serum Se concentration was below 45 ng/ml than in those whose serum Se was above that level. Concentrations below 45 ng/ml are in a range in which blood and tissue levels of GHSPx are subnormal [10], which might predispose the heart to injury from oxygen stress [64]. In 106 hospitalized patients undergoing coronary arteriography, highest mean plasma Se values (mean 136 ng/ml) were seen in patients without coronary blockage and lowest (mean 105 ng/ml) in those with atheromatous blockage in all three coronary arteries [65]. Platelet GSHPx activity was found to be significantly lower in recent heart attack patients than in controls [66]. On the negative side, a recently published Netherlands report on a 9-year follow-up study of cardiovascular mortality in 10,532 individuals concluded that there were no significant associations between serum Se, vitamin A and vitamin E levels, and CVD mortality [67].

Arachidonate-induced respiratory distress and platelet aggregation in mice were aggravated in a selenium-deficient group, as compared with a selenium-supplemented group; aortic biosynthesis of the vasodilative hormone prostacyclin (or prostacyclin-like substances) was inhibited by deficiency, probably because of an increased amount of hydroperoxide derivatives of arachidonate [68]. Schiavon et al. [69] found that Se added to the culture medium of pig aortic endothelial cells enhanced GSHPx activity and concurrently increased prostacyclin production, probably by preventing inhibition of prostacyclin synthetase by lipid hydroperoxides. When six healthy human volunteers (four males, two females, 26–55 years) were given supplemental Se for 6 weeks (10 mcg/kg/d as sodium selenite), their platelet GSHPx activities approximately doubled, and their bleeding time almost doubled. The investigators suggest that supplemental Se may exert an antithrombotic effect by promoting vascular prostacyclin synthesis [69].

The evidence developed so far provides ample justification for further clinical investigation of supplemental Se in prevention of atherosclerotic heart disease.

The Elderly

Tolonen et al. [70] in Finland carried out a double-blind trial of high-level Se and vitamin E supplementation for 1 year in 30 elderly individuals in an old people's home, one-half receiving the two supplements, and one-half placebo. Mean age of both groups was 76 years. Based on encouraging results previously obtained with high-dose Se and vitamin E therapy of the rare genetic disease, juvenile neuronal ceroid lipofuscinosis [71], the supplemented group was given 1,720 mcg/d of Se as sodium selenate in two doses, 400 mg/d (600 IU) of d-alpha-tocopherol in two doses, and 45 mcg of Se as Se-enriched yeast. The patients were assessed every 2 months independently by two nurses using the Sandoz Clinical Assessment Geriatric Scale [72]. The supplemented group showed statistically significant improvements in measures of depression, anxiety, self-care, mental alertness, emotional lability, motivation and initiative, hostility, interest in the environment, fatigue, anorexia, and general impression. A distinct improvement of the general condition was noticed only after 2 months and continued up to the end of the 1-year study.

Mean whole blood Se and plasma alpha-tocopherol concentrations were the same in both groups initially and were not at deficient levels (blood Se 124 ng/ml, plasma alpha-tocopherol 0.87 mg/dl). In the supplemented group, mean plasma tocopherol increased almost threefold, blood Se about 50%, and blood GSHPx activity 40%. Urinary Se increased from a mean of 24.5 mcg/l to

1,060 mcg/l, as expected when Se is administered as selenate [32]. No side effects were noted despite the high dosages employed.

The results are intriguing and deserve further investigation to see if the favorable results can be confirmed in other aged populations and to evaluate other dosages and forms of Se, as well as to separate the effects of vitamin E and selenium.

Chronic Disease

Low Se status has been observed in many chronic disease states, particularly when there is weight loss or protein-calorie malnutrition reflective of low food intake, poor dietary patterns, or malabsorption. Depressed plasma Se and red cell GSHPx levels have been observed in a sizable proportion of hospitalized malnourished patients with liver disease, gastrointestinal disorders (such as inflammatory bowel disease, celiac disease, and Crohn's disease), pancreatitis, congestive heart failure, renal failure, cancer, and neuropsychiatric disorders [10,73–76]. Dworkin et al. [77] compared Se status of 15 stable cystic fibrosis outpatients with that of 13 age-matched controls and noted, besides a general undernourished state (low serum albumin and decreased weight-for-height and weight-for-age), that the CF patients had reductions of approximately 30% in whole blood, plasma, and RBC Se levels versus controls. The Se status of such patients should be assessed and supplementation begun as part of a total nutritional rehabilitation regimen if their status is low or marginal.

Alcoholism

Low plasma Se levels are commonly observed in alcoholics even when liver structure and function appear normal, probably because of low dietary intake and possibly because of malabsorption induced by alcohol [10,78]. Chronic alcohol intake is associated with increased free radical production, which may contribute to progression of liver disease, and Tanner et al. [79] report finding elevations of free radical activity in the blood of alcoholics, associated with significant reductions of Se and vitamin E levels. Plasma Se is depressed by liver disease and is most reduced in decompensated alcoholic cirrhosis [80]. The withdrawal of alcohol and resumption of good dietary habits will restore the Se status if the liver is normal [78], but the alcoholic with liver dysfunction may require a selenium supplement.

Parenteral Nutrition

Depressed blood Se and red cell GSHPx levels have frequently occurred in both adults and children maintained for long periods on total parenteral nutrition [81–85]. Long-term home parenteral nutrition (HPN) is also associated with declining Se status, and many of these patients have low serum Se levels prior to starting HPN [86,87]. Defective granulocyte function [83] and proximal muscle weakness [87] were noted. Diverted gut secretions in HPN add to loss of Se [86]. Brown et al. [87] treated an adult woman patient on HPN with 400 mcg/d of selenious acid intravenously; muscle strength was regained in 6 weeks, and red cell GSHPx returned to normal in 3 months. Cohen et al. [85] supplemented the TPN infusate of three children with gastrointestinal disease with 240 mcg/d of Se as sodium selenite; plasma GSHPx activity returned to normal in 4–5 weeks, while red cell GSHPx took 3–4 months to recover.

Premature infants have a lower Se concentration in serum than full-term infants and children. Amin et al. [88] describe decline of serum Se in preterm infants, especially in one infant with respiratory distress syndrome requiring TPN during the first few weeks of life. Supplementation of the parenteral fluids with 3 mcg Se/kg/d as sodium selenite prevented the decline in serum Se level.

Diabetic Retinopathy

Long-term consumption of high-sucrose diets by rats produces vascular damage in the retina similar to that seen in human diabetic retinopathy [89,90]. Thornber and Eckhert [91] have demonstrated that this occurs with the semipurified diets used because of their low Se levels. Supplementation with 0.1 mg Se/kg of diet significantly reduced the retinal capillary damage observed after 10 months as compared with that on unsupplemented diet. The number of acellular collapsed capillary strands within arterial-venous capillary networks was significantly reduced. The unsupplemented diet contained approximately 0.13 mg of Se per kg (equivalent to 65 mcg Se in a day's human consumption of 500 g of food, dry weight) and the supplemented diet 0.23 mg/kg (equivalent to 115 mcg in 500 g food). The change in Se status therefore involved a range of dietary Se concentrations very similar to that occurring in the United States and other western countries. Since diabetic retinopathy is a major cause of blindness, this laboratory lead should be vigorously followed up in human studies.

It should be noted that best results were obtained with a supplement combination of Se with chromium, which improved glucose tolerance in these nondiabetic rats, and a modest supplement (5% of diet) of corn oil, which supplies polyunsaturated fatty acids. However, the added Se was essential for protection of the retinal capillaries from damage [91].

Acne Vulgaris

Based on analytical data showing that blood levels of GSHPx were low in male patients suffering from acne vulgaris, Michaelsson and Edqvist [92] carried out a preliminary open study in 29 acne patients. They were given 200 mcg Se and 10 mg vitamin E succinate (a small nutritional amount of this vitamin) twice daily for 6–12 weeks. Response to this treatment is described as good, especially in patients with pustular acne and initial low GSHPx levels. Beneficial effects were usually accompanied by a gradual increase in erythrocyte GSHPx activity. Five of the 29 patients who did not improve with Se therapy showed either unchanged or decreased GSHPx levels during therapy. The authors conclude from this preliminary study that Se deficiency may be one of several contributing factors in inflammatory acne, but further clinical data are needed [92].

TOXICITY AND SIDE EFFECTS

Selenium is not as toxic as was formerly supposed, but distinctions must be drawn between the forms in which it is supplied. Sodium selenite was administered to an adult woman in her TPN solution at 400 mcg Se per day for 3 months [86], and to three small children (8–23 months of age) at 240 mcg Se per day for 4 months [84], without any indication of toxicity. Thomson et al. [93] reported laboratory tests on 23 New Zealanders, most over 60 years of age, who had been self-administering 500–3,000 mcg of Se per day, mainly as selenite, for 6 months to as long as 15 years "for relief of muscular pain," apparently without ill effects; tissue Se levels were high, but blood GSHPx levels were no higher than in individuals taking 100 mcg Se per day. However, Yang et al. [7] noted incipient toxicity, manifested as thickened, fragile fingernails and a garlic odor in dermal secretions, in an employee who had been taking 1,000 mcg of Se per day as sodium selenite for 2 years; upon cessation of the supplement, nail growth became smooth.

Sodium selenate in high dosage appears less likely to lead to toxic symptoms because it is more rapidly excreted in the urine than selenite [32]. Tolonen et al. [70] observed no side effects in elderly patients given 1,720 mcg of Se per day as sodium selenate (8 mg/d of $Na_2SeO_4 \cdot H_2O$) for 1 year.

In China, no Se toxicity appears to occur with food Se, which is largely Se-Met, at daily intakes as great as 750 mcg Se [7]. In one area studied, the usual daily Se intake averaged 4,990 mcg (range 3,200–6,690 mcg). Selenosis became manifest, however, only after failure of the rice crop due to drought, causing greater consumption of high-Se vegetables and maize and fewer protein foods. Signs of poisoning included loss of hair and nails, lesions of the skin (redness, swelling, blisters), and peripheral neuritis. There was rapid recovery upon change of diet to low-Se foods, except for the neuritis, which took longer to resolve [7].

According to Donaldson, as quoted by Clark and Combs [61], some U.S. cancer patients have been given doses as high as 2,000 mcg Se/d as Se-enriched brewer's yeast, and in spite of attaining blood Se levels close to 1 ppm, they showed no clinical toxicity or biochemical abnormalities.

Dosages of Se in the range of 50–200 mcg/d orally appear safe for chronic use regardless of whether they are administered as sodium selenite, sodium selenate, Se-Met, or Se-enriched yeast.

INTERACTIONS

Vitamin E and Se have been found to spare one another in animal experiments designed to explore minimum requirements. Excessively high Se intakes are less toxic in vitamin E-adequate than in vitamin E-deficient animals [3].

In the young chick, supplemental sodium selenite in the diet decreased the toxicity of the urinary antimicrobial drug, nitrofurantoin [94]. Adequate Se status also appears to reduce the toxicity of the herbicide, paraquat [3,95].

Selenium decreases the toxicity of certain heavy metals, such as mercury and cadmium [3]. On the other hand, arsenic, zinc, and lead have been found to reduce the anticancer efficacy of Se in animal cancer models [96–98]. Clinically, this could be a problem in cancer patients taking supplemental zinc.

An apparent antagonistic action of vitamin C on the anticancer activity of Se was reported by Jacobs and Griffin [99]. Dissolved in the drinking water of rats treated with chemical carcinogens, vitamin C (1.2%) and sodium selenite (4 ppm Se) each inhibited colon and liver tumor incidence when given separately, but not when present in the drinking water simultaneously. No such antagonism has been reported between vitamin C and food Se or Se-Met. It is known that ascorbic acid can reduce selenite to elemental Se [100] and that selenite is incompatible with parenteral solutions that contain both copper and ascorbic acid [101].

REFERENCES

1. Sunde RA, Hoekstra WG: Structure, synthesis and function of glutathione peroxidase. Nutr Rev 38:265–273, 1980.
2. Chow CK: Nutritional influence on cellular antioxidant defense systems. Am J Clin Nutr 32:1066–1081, 1979.
3. Levander OA: Selenium: biochemical actions, interactions, and some human health implications. In: *Clinical, Biochemical and Nutritional Aspects of Trace Elements,* AS Prasad (ed). New York: Alan R. Liss, 1982, pp 345–368.
4. Stadtman TC: Specific occurrence of selenium in enzymes and amino acid tRNAs. FASEB J 1:375–379, 1987.
5. Chen X, Yang G, Chen J, et al.: Studies on the relations of selenium and Keshan disease. Biol Trace Elem Res 2:91–107, 1980.
6. Yang, Guang-Qi: Research on selenium-related problems in human health in China. In:

Selenium in Biology and Medicine, GF Combs Jr., et al. (eds). New York: Van Nostrand Reinhold, 1987, Vol. A, pp 9–32.
7. Yang G, Wang S, Zhou R, Sun S: Endemic selenium intoxication of humans in China. Am J Clin Nutr 37:872–881, 1983.
8. Robinson MF: Clinical effects of selenium deficiency and excess. In: *Clinical, Biochemical, and Nutritional Aspects of Trace Elements,* AS Prasad (ed). New York: Alan R. Liss, 1982, pp 325–343.
9. Westermarck T, Raunu P. Kirjarinto M, Lappalainen L: Selenium content of whole blood and serum in adults and children of different ages from different parts of Finland. Acta Pharmacol Toxicol 40:465–475, 1977.
10. Smith DK, Teague J, McAdam PA, et al.: Selenium status of malnourished hospitalized patients. J Am Coll Nutr 5:243–252, 1986.
11. Miller L, Mills BJ, Blotcky AJ, Lindeman RD: Red blood cell and serum selenium concentrations as influenced by age and selected diseases. J Am Coll Nutr 4:331–341, 1983.
12. Boyne R, Arthur J: The response of selenium-deficient mice to *Candida albicans* infection. J Nutr 116:816–822, 1986.
13. Levander OA, Sutherland B, Morris VC, King JC: Selenium balance in young men during selenium depletion and repletion. Am J Clin Nutr 34:2662–2669, 1981.
14. Stewart RDH, Griffiths NM, Thomson CD, et al.: Quantitative selenium metabolism in normal New Zealand women. Am J Clin Nutr 33:303–323, 1980.
15. National Research Council: *Recommended Dietary Allowances,* 9th ed. Washington, DC: National Academy of Sciences, 1980.
16. Young VR, Janghorbani M: Selenium bioavailability with reference to human nutrition. Am J Clin Nutr 35:1076–1088, 1982.
17. Beilstein MA, Whanger PD: Deposition of dietary organic and inorganic selenium in rat erythrocyte proteins. J Nutr 116:1701–1710, 1986.
18. Ardüser F, Wolffram S, Scharrer E: Active absorption of selenate by rat ileum. J Nutr 115:1203–1208, 1985.
19. Wolffram S, Ardüser, Scharrer E: In vivo absorption of selenate and selenite by rats. J Nutr 115:454–459, 1985.
20. Humaloja T, Mykkänen HM: Intestinal absorption of ^{75}Se-labeled sodium selenite and selenomethionine in chicks: Effects of time, segment, selenium concentration and method of measurement. J Nutr 116:142–148, 1986.
21. Whanger PD, Pedersen ND, Hatfield J, Weswig PH: Absorption of selenite and selenomethionine from ligated digestive tract segments in rats. Proc Soc Exp Biol Med 153: 295–297, 1976.
22. Motsenbocker MA, Tappel AL: Effect of dietary selenium on plasma selenoprotein P, selenoprotein P1 and glutathione peroxidase in the rat. J Nutr 114:279–285, 1984.
23. Beilstein MA, Whanger PD: Distribution of selenium and glutathione peroxidase in blood fractions from humans, rhesus and squirrel monkeys, rats and sheep. J Nutr 113:2138–2146, 1983.
24. Beilstein MA, Whanger PD: Deposition of dietary organic and inorganic selenium in rat erythrocyte proteins. J Nutr 116:1701–1710, 1986.
25. Beilstein MA, Whanger PD: Chemical forms of selenium in rat tissues after administration of selenite or selenomethionine. J Nutr 116:1711–1719, 1986.
26. Deagen JT, Butler JA, Beilstein MA, Whanger PD: Effects of dietary selenite, selenocystine and selenomethionine on selenocystine lyase and glutathione peroxidase activities and on selenium levels in rat tissues. J Nutr 117:91–98, 1987.

27. Thomson CD, Robinson MF, Campbell DR, et al.: Effect of prolonged supplementation with daily supplements of selenomethionine and sodium selenite on glutathione peroxidase activity in blood of New Zealand residents. Am J Clin Nutr 36:24–31, 1982.
28. Thomson CD, Ong LK, Robinson MF: Effects of supplementation with high-selenium wheat bread on selenium, glutathione peroxidase and related enzymes in blood components of New Zealand residents. Am J Clin Nutr 41:1015–1022, 1985.
29. Levander OA, Alfthan G, Arvilommi H, et al.: Bioavailability of selenium to Finnish men as assessed by platelet glutathione peroxidase activity and other blood parameters. Am J Clin Nutr 37:887–897, 1983.
30. Luo X, Wei H, Yang C, et al.: Bioavailability of selenium to residents in a low-selenium area of China. Am J Clin Nutr 42:439–448, 1985.
31. Thomson CD, Robinson MF: Selenium in human health and disease with emphasis on those aspects peculiar to New Zealand. Am Clin Nutr 33:303–323, 1980.
32. Thomson CD, Robinson, MF: Urinary and fecal excretions and absorption of a large supplement of selenium: Superiority of selenate over selenite. Am J Clin Nutr 44:659–663, 1986.
33. Thomson CD, Burton CE, Robinson MF: On supplementing the selenium intake of New Zealanders. I. Short experiments with large doses of selenite and selenomethionine. Br J Nutr 39:579–587, 1978.
34. Stewart RDH, Griffiths NM, Thomson CD, Robinson MF: Quantitative selenium metabolism in normal New Zealand women. Br J Nutr 40:45–54, 1978.
35. Schroeder HA, Frost DV, Balassa JJ Essential trace metals in man: Selenium. J Chronic Dis 23:227–243, 1970.
36. Christensen MJ, Janghorbani M, Steinke FH, et al.: Simultaneous determination of absorption of selenium from poultry meat and selenite in young men: Application of a triple stable isotope method. Br J Nutr 50:43–50, 1983.
37. Kumpulainen J, Salmenperä L, Siimes MA, et al.: Selenium status of exclusively breastfed infants as influenced by maternal organic or inorganic selenium supplementation. Am J Clin Nutr 42:829–835, 1985.
38. Swanson CA, Reamer DC, Veillon C: Quantitative and qualitative aspects of selenium utilization in pregnant and nonpregnant women: An application of stable isotope methodology. Am J Clin Nutr 38:169–180, 1983.
39. Ip C: Selenium inhibition of chemical carcinogenesis. Fed Proc 44:2573–2578, 1985.
40. Milner JA: Effect of selenium on virally induced and transplantable tumor models. Fed Proc 44:2568–2572, 1985.
41. Shamberger RJ, Frost DV: Possible protective effect of selenium against human cancer. Can Med Assoc J 100:682, 1969.
42. Schrauzer GN, White DF, Schneider CJ: Cancer mortality correlation studies. III. Statistical associations with dietary selenium intakes. Bioinorg Chem 7:23, 1977.
43. Cowgill UM: The distribution of selenium and cancer mortality in the continental United States. Biol Trace Elem Res 5:345–361, 1983.
44. Clark LC: The epidemiology of selenium and cancer. Fed Proc 44:2584–2589, 1985.
45. Yu SY, Chu YJ, Gong DL, et al.: Regional variation of cancer mortality incidence and its relation to selenium levels in China. Biol Trace Elem Res 7:21, 1985.
46. Combs GF Jr, Clark LC: Can dietary selenium modify cancer risk? Nutr Rev 43:325–331, 1985.
47. Salonen JT, Salonen R, Lappetelainen R, et al.: Risk of cancer in relation to serum concentrations of selenium and vitamins A and E: Matched case-control analysis of prospective data. Br Med J 290:417–420, 1985.

48. Borek C, Ong A, Mason H, et al.: Selenium and vitamin E inhibit radiogenic and chemically induced transformation in vitro via different mechanisms. Proc Natl Acad Sci USA 83:1490–1494, 1986.
49. Fex G, Pettersson B, Åkesson B: Low plasma selenium as a risk factor for cancer death in middle-aged men. Nutr Cancer 1:221–229, 1987.
50. Menkes MS, Comstock GW, Vuilleumier JP, et al.: Serum beta-carotene, vitamins A and E, selenium, and the risk of lung cancer. N Engl J Med 315:1250–1254, 1986.
51. Beisel WR: Single nutrients and immunity. Am J Clin Nutr 35:417–468, 1982.
52. Lawson T, Birt DF: Enhancement of the repair of carcinogen-induced DNA damage in the hamster pancreas by dietary selenium. Chem Biol Interact 45:95–104, 1983.
53. Russell GR, Nader CJ, Patrick EJ: Induction of DNA repair by some selenium compounds. Cancer Lett 10:75–81, 1980.
54. Arciszewska LK, Martin SE, Milner JA: The antimutagenic effect of selenium on 7,12-dimethylbenz(a)anthracene and metabolites in the Ames Salmonella/microsome system. Biol Trace Elem Res 4:259–267, 1982.
55. Harbach PR, Swenberg JA: Effects of selenium on 1,2-dimethylhydrazine metabolism and DNA alkylation. Carcinogenesis 2:575–580, 1981.
56. Medina D, Osborn CJ: Differential effects of selenium on the growth of mouse mammary cells in vitro. Cancer Lett 13:333–344, 1981.
57. Medina D, Lane HW, Tracy CM: Selenium and mouse mammary tumorigenesis: An investigation of possible mechanisms. Cancer Res 43:2460–2464, 1983.
58. Balansky RM, Argirova RM: Sodium selenite inhibition of the reproduction of some oncogenic RNA-viruses. Experientia 37:1194–1195, 1981.
59. Schrauzer GN, McGuiness JE, Kuelin K: Effects of temporary selenium supplementation on the genesis of spontaneous mammary tumors in inbred female C3H/St mice. Carcinogenesis 1:199, 1980.
60. Burk, RF: Selenium and cancer: Meaning of serum selenium levels. J Nutr 116:1584–1586, 1986.
61. Clark LC, Combs GF Jr: Selenium compounds and the prevention of cancer: Research needs and public health implications. J Nutr 116:170–173, 1986.
62. Shamberger RJ, Willis CE, McCormak LJ: Selenium and heart disease. III. Blood selenium and heart mortality in 19 states. In: *Trace Substances in Environmental Health—XII*. Columbia: University of Missouri Press, 1979, p 59.
63. Salonen JT, Alfthan G, Huttunen JK, et al.: Association between cardiovascular death and myocardial infarction and serum selenium in a matched-pair longitudinal study. Lancet 2:175–179, 1982.
64. Xia Y, Hill KE, Burk RF: Effect of selenium deficiency on hydroperoxide-induced glutathione release from the isolated perfused rat heart. J Nutr 115:733–742, 1985.
65. Moore JA, Noiva R, Wells IC: Selenium concentrations in plasma of patients with arteriographically defined coronary atherosclerosis. Clin Chem 30:1171, 1984.
66. Wang YX, Bocher K, Recter H, et al.: Selenium and myocardial infarction: Glutathione peroxidase in platelets. Klin Wochenschr 59:817, 1981.
67. Kok FJ, De Bruijn AM, Vermeeren R, et al.: Serum selenium, vitamin antioxidants, and cardiovascular mortality: A 9-year follow-up study in the Netherlands. Am J Clin Nutr 45:462–468, 1987.
68. Masukawa T, Goto J, Iwata H: Impaired metabolism of arachidonate in selenium deficient animals. Experientia 39:405–406, 1983.
69. Schiavon R, Freeman GE, Guidi GC, et al.: Selenium enhances prostacyclin production

by cultured epithelial cells: Possible explanation for increased bleeding time in volunteers taking selenium as a supplement. Thromb Res 34:389-396, 1984.
70. Tolonen M, Halme M, Sarna S: Vitamin E and selenium supplementation in geriatric patients: A double-blind preliminary clinical trial. Biol Trace Elem Res 7:161-168, 1985.
71. Westermarck T, Santavuori P: Principles of antioxidant therapy in neuronal ceroid lipofuscinosis. Med Biol 62:148-151, 1984.
72. Venn R: The Sandoz clinical assessment - geriatric (SCAG) scale. Gerontology 29:185-198, 1983.
73. Fell GS, Stromberg P, Main A, et al.: Biochemical signs of human selenium depletion. Proc Nutr Soc 40:76A, 1981.
74. Aaseth J, Alexander J, Thomassen Y, et al.: Serum selenium levels in liver disease. Clin Biochem 15:281-283, 1982.
75. Valimaki MJ, Harju KJ, Ylikahri RH: Decreased serum selenium in alcoholics—a consequence of liver dysfunction. Clin Chim Acta 130:291-296, 1983.
76. Spooner RJ, Campbell RA, Rumley AG, Stromberg P: Glutathione peroxidase in human selenium deficiency. Proc Nutr Soc 39:37A, 1980.
77. Dworkin B, Newman LJ, Berezin S, et al.: Low blood selenium levels in patients with cystic fibrosis compared to controls and healthy adults. J Paren Ent Nutr 11:38-41, 1987.
78. Dutta SK, Miller PA, Greenberg LB, Levander OA: Selenium and acute alcoholism. Am J Clin Nutr 38:713-718, 1983.
79. Tanner AR, Bantock I, Hinks L, et al.: Depressed selenium and vitamin E levels in an alcoholic population. Dig Dis Sci 31:1307-1312, 1986.
80. Korpela H, Kumpulainen J, Luoma PV, et al.: Decreased serum selenium in alcoholics as related to liver structure and function. Am J Clin Nutr 42:147-151, 1985.
81. Van Rij AM, Thomson CR, McKenzie JM, Robinson MF: Selenium deficiency in total parenteral nutrition. Am J Clin Nutr 32:2076-2085, 1979.
82. Kien CL, Ganther HE: Manifestations of chronic selenium deficiency in a child receiving total parenteral nutrition. Am J Clin Nutr 37:319-328, 1983.
83. Baker SS, Lerman RH, Krey SH, et al.: Selenium deficiency with total parenteral nutrition: Reversal of biochemical and functional abnormalities by selenium supplementation: A case report. Am J Clin Nutr 38:769-774, 1983.
84. Lane HW, Dudrick S, Warren DC: Blood selenium levels and glutathione-peroxidase activities in university and chronic intravenous hyperalimentation subjects. Proc Soc Exp Biol Med 167:383-390, 1981.
85. Cohen HJ, Chovaniec ME, Mistretta D, Baker SS: Selenium repletion and glutathione peroxidase—differential effects on plasma and red blood cell enzyme activity. Am J Clin Nutr 41:735-747, 1985.
86. Fleming CR, McCall JT, O'Brien JF, et al.: Selenium status in patients receiving home parenteral nutrition. J Parent Ent Nutr 8:258-262, 1984.
87. Brown MR, Cohen HJ, Lyons JM, et al.: Proximal muscle weakness and selenium deficiency associated with long term parenteral nutrition. Am J Clin Nutr 43:549-554, 1986.
88. Amin S, Chen SY, Collipp PJ, et al.: Selenium in premature infants. Nutr Metab 24:331-340, 1980.
89. Cohen AM, Michaelson IC, Yanko L: Retinopathy in rats with disturbed carbohydrate metabolism following a high-sucrose diet. Am J Ophthalmol 73:863-869, 1972.
90. Papachristodoulou D, Heath H, Kang SS: The development of retinopathy in sucrose-fed and streptozotocin-diabetic rats. Diabetologia 12:367-374, 1976.

91. Thornber JM, Eckhert CD: Protection against sucrose-induced retinal capillary damage in the Wistar rat. J Nutr 114:1070–1075, 1984.
92. Michaelsson G, Edqvist LE: Erythrocyte glutathione peroxidase activity in acne vulgaris and the effect of selenium and vitamin E treatment. Acta Derm Venerol (Stockh) 64:9–14, 1984.
93. Thomson CD, Robinson MF, Campbell DR, et al.: Effect of prolonged supplementation with daily supplements of selenomethionine and sodium selenite on GSHPx activity in blood of New Zealand residents. Am J Clin Nutr 36:24–31, 1982.
94. Peterson FJ, Combs GF Jr, Holtzman JL, Mason RP: Effect of selenium and vitamin E deficiency on nitrofurantoin toxicity in the check. J Nutr 112:1741–1746, 1982.
95. Combs GF Jr, Peterson FJ: Protection against acute paraquat toxicity by dietary selenium in the chick. J Nutr 113:538–545, 1983.
96. Schrauzer GN, Ishmael D: Effects of selenium and of arsenic on the genesis of spontaneous mammary tumors in inbred C3H/St mice. Ann Clin Lab Sci 4:441, 1974.
97. Schrauzer GN, White DA, Schneider DJ: Inhibition of the genesis of spontaneous mammary tumors in C3H/St mice. Effects of selenium and of selenium antagonistic elements and their possible role in human breast cancer. Bioinorg Chem 6:265, 1976.
98. Schrauzer GN, Kuehn K, Hamm D: Effects of dietary selenium and of lead on the genesis of spontaneous mammary tumors in mice. Biol Trace Elem Res 3:185–196, 1981.
99. Jacobs MM, Griffin AC: Effects of selenium on chemical carcinogenesis. Comparative effects of antioxidants. Biol Trace Elem Res 1:1–13, 1979.
100. Newberry CL, Christian GD: Rapid colorimetric determination of microgram amounts of selenium in soils and biological samples, using millipore filters. J Assoc Off Anal Chem 48:322-326, 1965.
101. Shils ME, Levander OA: Selenium stability in TPN solutions. Am J Clin Nutr 35:829 (abstr), 1982.

17
Zinc

Introduction	286
Absorption, Metabolism, and Excretion	288
Dosage Forms	289
Clinical Studies	290
Nutritional Deficiency	290
The breast-fed infant	290
The formula-fed infant	291
The parenterally fed infant	291
Protein-calorie malnutrition	292
Growth retardation in children	292
Pregnancy	293
The elderly	295
Anorexia nervosa	296
Special diets	296
Acne and recurrent furunculosis	296
Pica	297
Conditioned Deficiency	297
Malabsorption	297
Renal disease	298
Alcoholism	299
Diabetes mellitus	299
Liver disease	300
Athletics and exercise	300
Genetic Disease	300
Acrodermatitis enteropathica	300
Wilson's disease	301
Sickle cell anemia	302
Beta-thalassemia	303
Down's syndrome	303
Other Clinical Studies and Potential Uses	303
The common cold	303
Leg ulcers	304
Surgery, wound healing, and burns	305
Arthritis and inflammatory diseases	305
Optic atrophy and disk pallor	306

Cigarette smoking 306
Cancer 307
Toxicity and Side Effects 307
Interactions 308
References 310

INTRODUCTION

Zinc is an essential nutrient because of its role as a constituent or cofactor in many Zn-dependent enzymes that participate in the metabolism of carbohydrate, lipids, protein, and nucleic acids. Some examples include carbonic anhydrase, hepatic alcohol dehydrogenase, retinal retinol (retinene) dehydrogenase, carboxypeptidase, alkaline phosphatase, deoxythymidine kinase, and delta-aminolevulinic acid dehydratase [1–5]. Other examples include nucleoside phosphorylase [6], intestinal conjugase [7], and bone collagenase [8].

Reduction in plasma Zn levels, which occurs rapidly when dietary zinc is withheld, has been shown in animal studies to impair and delay platelet aggregation in laboratory animals [9,10] and adult humans [11].

Although zinc itself is not considered an antioxidant or free radical scavenger, dietary Zn deficiency has been shown to stimulate the production of endogenous free radicals in rat lung microsomes by an NADPH- and cytochrome P-450-dependent system [12]. As a part of the zinc-copper metalloenzyme, cytosolic superoxide dismutase (SOD), zinc is vital to intracellular defense against superoxide radicals [13]. In addition, zinc plays a role in stabilizing cell membranes, both as a constituent, as in red cell membranes [14], and possibly by complexing with readily oxidizable moieties such as sulfhydryl groups in membrane proteins [15].

Adequate zinc nutriture is essential for normal bone collagenase activity and collagen turnover [8], and its dietary deficiency in the growing animal leads to skeletal deformities [8] and increased caries susceptibility of the teeth [16,17].

Zinc status is of major importance in maintenance of effective immune response, particularly T-cell-mediated response [1–5,18]. In experimental Zn deficiency, antibody-mediated responses to both T-cell-dependent and T-cell-independent antigens are reduced, and cytolytic T-cell responses, natural killer cell activity, and delayed-type hypersensitivity reactions are depressed [17]. Zinc deficiency causes thymus gland atrophy and depression of the activity of thymulin, a thymic hormone that requires Zn as a cofactor [18,19]. Using sheep red blood cells (SRBC) as antigen in Zn-adequate mice, DePasquale-Jardieu and Fraker [20] found that an ensuing 4-week period of Zn deprivation markedly reduced the secondary memory response to SRBC, and this was only partially restored after a subsequent 4 weeks of Zn repletion, suggesting that Zn deficiency may have destroyed a substantial portion of SRBC memory cells. Zinc deficiency markedly increases the severity of type A influenza virus infections in mice [21].

Subnomal plasma Zn levels (below 70 mcg/dl) in men and women volunteers, induced by inadequate dietary Zn for several weeks, produced a decrease in lymphocyte counts, impaired leukocyte chemotaxis, and clinical signs indicative of decreased resistance to infection (sore throat, aphthous stomatitis, seborrheic dermatitis, acne flare-up, sty, and furunculosis) [22,23]. Depressed cellular immunity and restoration by Zn supplementation have also been described in children with acrodermatitis enteropathica, a hereditary disease in which there is inefficient absorption of dietary zinc [24]. There is some indication that T-cell blast response to phytohemagglutinin can be increased even in healthy normal adults by administration of pharmacologic oral doses (150 mg Zn/d for 1 month) [25], or, in patients on TPN, by administering a relatively high parenteral daily dosage of Zn (12 mg/d for 20 days) in the intravenous infusate [26].

In young growing mice given various concentrations of Zn in their diets, although maximal carcass growth rate required only 5.4 mg Zn/kg of diet, optimal functional activity of natural killer cells from spleen was not reached until dietary Zn concentration was 40.4 mg Zn/kg of diet; maximal serum Zn levels were obtained with only 3.4 mg Zn/kg of diet, suggesting a hierarchy of Zn requirements, with optimal cell-mediated immunity requiring the highest dietary Zn intake [27]. Likewise, in Chilean infants recovering from malnutrition, Castillo-Duran et al. [28] report that Zn supplementation (2 mg/kg/d as Zn acetate) produced significant improvements in host defense (reduction in anergy, increased serum IgA) and growth, despite normal plasma Zn levels reached on the basal rehabilitation diet without Zn supplementation.

Adequate Zn intakes are needed in the young male for gonadal maturation [1] and in middle-aged men for normal testicular function and maintenance of normal serum testosterone levels [29]. In young men, mild Zn deficiency was associated with decline in basal metabolic rate and decrease in circulating thyroxine (T-4) levels [10].

The RDA for infants is 3–5 mg Zn/d; for children aged 1–10 years, 10 mg/d; for adolescents and adults, 15 mg/d; for pregnant women, 20 mg/d; and for lactating women, 25 mg/d [31]. These allowances are somewhat arbitrary and perhaps excessive; the mean intakes for most age and sex groups in these categories in North America are below the RDA values without apparent adverse effects on health [32,33]. The best food sources among commonly eaten foods are beef, pork, veal, and dark poultry meats. Oysters are an excellent source, although not as often eaten. Fish and white poultry meat are less satisfactory sources than red meats. For vegetarians, peas, beans, and nuts are the best dietary sources, although not as good as meat [33].

Clinical signs of zinc deficiency stemming from inadequate dietary Zn intake are rare in western countries unless there is some underlying disease process that interferes with Zn absorption or utilization [33]. Such conditions are discussed in the Clinical Studies section of this chapter.

Various methods have been used to assess Zn nutriture in the laboratory. Zinc determinations in red cells and hair have been employed, but have the disadvantage that they do not reflect recent changes in body Zn stores [34]. Serum alkaline phosphatase has been proposed as a useful indicator, but Prasad [34] suggests assay of neutrophil alkaline phosphatase activity as a sensitive indicator of the zinc status; the method has been well verified in extensive studies of experimental Zn depletion in human volunteers in his laboratory [35]. The most common measurement used to assess current status is the plasma Zn concentration, but it must be made certain that the plasma is unhemolyzed and that conditions that cause redistribution of plasma Zn to other tissues, such as infections, acute stress, and myocardial infarction, are ruled out [34]. Serum Zn concentration has often been used, but tends to give higher Zn values than plasma Zn determinations, probably because of transfer of Zn from erythrocytes and other cellular elements to serum during the clot retraction period [36]. In fasting plasma samples prepared by adding 0.15 ml of 30% sodium citrate per 15 ml of whole blood, Smith and coworkers found a mean of 79 mcg Zn/dl, range 62–100 mcg/dl, in ten healthy adults [36]. Readings below 70 mcg/dl are considered to indicate risk of deficiency [36]. Variations of as much as 19 mcg/dl have been observed over a 24-hour period [37], so that time of blood sampling should be standardized, e.g., before breakfast after overnight fasting.

During weight loss caused by starvation or wasting diseases, the Zn released from degraded protein tissues maintains plasma Zn at normal levels in spite of depletion of the total body Zn pool [38]. Infection and stress of injury induce segregation of plasma Zn into the liver, so that under these conditions, a low plasma Zn level is not necessarily indicative of Zn deficiency [34,39].

ABSORPTION, METABOLISM, AND EXCRETION

Food zinc is largely bound to proteins and therefore probably released below the common duct, to be absorbed in the ileum. Free ionic zinc is probably absorbed mostly in the duodenum at the lower pH levels prevailing there. The absorption of zinc is under regulation to some degree by the Zn status of the body, being greater when there is zinc deficiency. Metallothionein, a sulfur-rich metal-binding protein present in the intestinal mucosa and believed inducible by zinc, may be involved in this regulatory mechanism. From the enterocyte, zinc is transferred to the portal circulation, where it is bound mostly to albumin, but also to transferrin and globulins; only 2–3% of circulating zinc is found in ultrafiltrable form, either bound to amino acids or present as free ionic zinc. Circulating zinc delivered to the body cells is used for synthesis of the Zn metalloenzymes and other metalloproteins such as hepatic Zn metallothionein [2,40,41].

There is a substantial enteropancreatic circulation of zinc, as the pancreas secretes Zn-containing proteolytic carboxypeptidases into the intestinal lumen; it is assumed that a large part of this endogenous Zn is reabsorbed [40,42].

Total body zinc in adults is estimated at 1.5–2.5 grams [1,43,44]. Tissues high in Zn concentration include the liver, pancreas, spleen, lung, retina, prostate, skeletal muscle, and bone [1]. Because of their mass, skeletal muscle and bone contain most of the body's zinc (60–62% and 20–28% respectively), while liver contains 2–4% of the total [45].

The principal routes of Zn excretion are dermal, urinary, and fecal. At ordinary Zn intakes, dermal losses (sweat, hair and nail growth, skin shedding) are 0.5–0.7 mg/d [22,46]. During Zn depletion, sweat losses gradually decline [46]. The extensive human studies of Spencer and her associates [47] show the predominance of the fecal route during usual dietary intakes and the complex interplay between nutritional state, Zn intake, and the relative magnitudes of urinary and fecal excretion. In healthy men consuming 14.1–15.1 mg/d of dietary Zn, about 2% of intake was excreted in the urine and 85–93% in the feces. Fecal Zn includes unabsorbed exogenous Zn as well as endogenous Zn from pancreatic and intestinal wall secretions [40]. When pharmacologic dosages of Zn are given, urinary excretion remains at about 2% of the dose, nearly all of the excess intake going out in the stool [47]. During weight loss on a reducing diet, the degradation of Zn-containing tissues releases Zn to the circulation, with an attendant marked increase in urinary Zn excretion [47].

The absorption of zinc (ionic) administered in solution on an empty stomach ranges from 41–79%, while the zinc present in foods or that given as a supplement with meals is absorbed in the range of 10–40%. The zinc of foods high in phytate and fiber, e.g., whole grains and legumes, is less bioavailable than the zinc in meats and other low-phytate, low-fiber foods [40].

DOSAGE FORMS

A number of supplement dosage forms are sold over the counter. These include amino acid-chelated zinc in tablets providing 15 and 22 mg elemental Zn per tablet; timed-release chelated zinc, 50 mg per tablet; zinc gluconate,

10-, 25-, 50-, and 100-mg Zn per tablet; and zinc orotate, 75 mg Zn per tablet. Lozenges containing zinc gluconate or chelated zinc are also available, at 25 mg elemental Zn per lozenge. Zinc acetate dihydrate and zinc sulfate heptahydrate are available in pharmacies.

CLINICAL STUDIES
Nutritional Deficiency

Although nutritional Zn deficiency as severe as that observed by Prasad in the Middle East [3] is rarely seen in western countries, numerous instances of marginal or poor Zn status have been reported from North America and Europe. Frequently some secondary or conditioning factor, as well as nutritional deficiency, is involved. Larger-than-nutritional amounts of supplemental Zn may be required to expedite recovery.

The breast-fed infant. The breast-fed full-term infant orinarily does not experience any difficulties with Zn nutrition in spite of the fact that after the first month of lactation the Zn concentration of breast milk falls to a level much below that of cow's milk [48]. The Zn in breast milk has been demonstrated to be much more bioavailable to the infant than the Zn of cow's milk or formulas based on cow's milk [49]. Krebs and Hambidge [50] have calculated that the Zn concentration of breast milk is adequate for the infant up to 5–6 months of age, and may be adequate for a longer period if the mother's Zn intake is at the RDA level of 25 mg/d. As compared with a group of nonsupplemented lactating women ingesting a mean of 10.7 mg Zn/d, a supplemented group whose intake was raised to a mean of 25 mg/d had a mean increase in their milk Zn concentration of about 10% relative to the control group in the first 6 months of lactation and about 50% during months 7–9 [48].

Infants of very low birth weight may not get enough Zn from breast milk, even during the first few months of lactation, to provide for their rapid growth rate. The immature gastrointestinal tract appears to be an inefficient absorber of ionic zinc [51]. Moore et al. [52] describe an infant with an acrodermatitis-like rash at 20 weeks of age who required an oral supplement of 15 mg Zn/d to clear up the rash and restore a normal Zn status. Aggett et al. [53] describe a case of acute Zn deficiency in a 2-month-old preterm male infant who had been exclusively breast-fed; plasma Zn was only 30.7 mcg/dl. Laboratory tests excluded acrodermatitis enteropathica. Treatment was 30 mg oral $ZnSO_4 \cdot 7\ H_2O$ (13.5 mg elemental Zn) twice daily, which induced complete remission. By age 12 months the boy no longer needed supplemental Zn. Zimmerman et al. [54] have reported similar instances of acrodermatitis in breast-fed premature infants; low Zn contents in breast milk, unresponsive to maternal Zn supplementation, appeared responsible. Symptoms were cleared with oral zinc sulfate (5.5 mg Zn) administered to the infant t.i.d. [54].

Goldberg and Sheehy [55] report a seizure syndrome occurring around the fifth day of life in otherwise normal full-term breast-fed babies, called "fifth day fits"; the observed incidence in Melbourne, Australia, has been about 2 per 1,000 live births. Zn concentration in CSF fluid is usually below 6.5 mcg/dl versus a normal value of about 24 mcg/dl. The authors suggest delayed maturation of Zn-binding ligands in the intestinal mucosa. The condition resolves spontaneously on continued breast-feeding without Zn therapy.

The formula-fed infant. Krieger et al. [56] describe a transient neonatal Zn deficiency that developed during the first month of life in a full-term female infant fed a Zn-adequate milk-based formula. Diaper rash and irritability developed at 2 weeks, progressing to anorexia, lethargy, blepharitis, monilial diaper rash, and oral thrush at 5½ weeks, when she was hospitalized. During the following 8 days, while under treatment with Mycostatin® and antibiotics, clinical condition worsened, with development of diarrhea, perioral rash, and additional neurologic symptoms. Plasma Zn was 57 mcg/dl (normal 95 ± 7.1). Zn therapy at 45 mg/d orally for 9 days cleared all symptoms. There was a recurrence 3 weeks later, again corrected by Zn supplementation. However, discontinuance of Zn therapy 3 months later did not bring on a recurrence, and plasma Zn levels, growth, and development remained normal during 4 years of follow-up. The initial development of Zn deficiency symptoms on a Zn-adequate milk-based formula, together with the response to therapeutic Zn, was suggestive of acrodermatitis enteropathica, but the later clearing of the condition without further Zn supplementation indicated only a transient inadequacy of Zn absorption.

Thorp et al. [57] report that 14 of 39 premature infants, most of whom were given a commercially available pemature infant formula containing zinc, developed low plasma Zn levels (45 mg/dl or less) by approximately 6 weeks of age; one infant of the 39 received human milk and about one-half of the group received some parenteral nutrition (median, 12 days), but the Zn content of the parenteral solutions was far below the 100–300 mcg/kg/d recommended by the American Medical Association [58]. True clinical Zn deficiency was observed in 2 of the 14 who had low plasma Zn levels, manifested by increased gastric residuals, poor suck, and decreased growth, but skin and gastrointestinal signs were absent. The authors consider plasma Zn levels below 60 mcg/dl in these infants to be abnormal and recommend that the Zn status of premature infants be monitored for the first 6–8 months with supplementation given when necessary. In the case of the two Zn-deficient infants described, they gave Zn acetate orally to supply 2 mg Zn/d/kg body weight, which cleared symptoms within 4 days.

The parenterally fed infant. Lockitch et al. [59] find that premature infants requiring long-term parenteral nutrition (birth weight 2,000 g or less) appear to maintain plasma Zn levels within the normal range only when receiving Zn intake in parenteral fluids close to the top of the American Medical Association's recommendation for premature infants of 100–300 mcg/kg/d [58].

Rothbaum et al. [60] describe development of Zn deficiency with an acrodermatitis-like skin rash and low serum Zn and alkaline phosphatase levels in seven infants with chronic diarrhea while receiving central venous alimentation for nutritional support. With parenteral Zn supplementation at 200–300 mcg/kg/d, the rash healed and alkaline phosphatase levels rose to normal. Naveh et al. [61] have documented the decline of serum Zn levels in children with chronic and protracted diarrhea, ascribed to high losses of Zn-containing intestinal secretions in the watery stools.

Palma et al. [62] describe development of Zn deficiency in three premature infants, manifested as an acrodermatitis skin rash, 6 or 7 weeks after going on TPN following abdominal surgery (first day of life in two infants, ninth day in the third). The parenteral fluid supplied 40 mcg Zn/kg/d. These infants had experienced losses of fluid from a gastrostomy, nasogastric drainage, and ileostomy. Increase of Zn in the TPN infusates to provide 200–400 mcg/kg/d cleared the skin rashes and returned serum Zn levels to normal.

The decline in immune defenses brought on by zinc deficiency can result in infections and sepsis presenting a major hazard to life [56,60].

Protein-calorie malnutrition. Bizarre feeding regimens for infants may bring on deficiencies of zinc as well as other nutrients. For example, Canfield et al. [63] describe two cases of protein-calorie malnutrition (PCM) with low plasma Zn levels, one in an 8-month-old boy resulting from doctrinaire feeding of a vegetarian diet, the other a 6-month-old girl who had been fed a water-flour-sugar mixture because of parental ignorance. Change to a complete formula plus Zn supplements brought about recovery in both instances, but with persisting deficits in height and weight for age.

Growth retardation in children. An increased prevalence of below-average Zn levels in hair and plasma has been found in children of short stature for their age, often associated with anorexia and hypogeusia [64–67]. Zinc supplementation has been shown to stimulate growth and normalize plasma growth hormone, testosterone, and somatomedin C levels in children whose short stature was linked to Zn deficiency [68–70]. Collipp and coworkers [68,69], for children aged 7–14 years, used dosages of 50 or 100 mg elemental Zn/d as $ZnSO_4 \cdot 7\ H_2O$, administered at mealtimes to reduce gastric upset; dosage was continued for 1 to 2 years. Walravens et al. [70] used a smaller dosage, 5 mg Zn twice daily, given as $ZnSO_4 \cdot 7\ H_2O$ in cherry-flavored syrup; whether parents were instructed to give the syrup at mealtimes or otherwise is not stated. Absorption of zinc administered before or between meals is much greater than that of zinc given with food or immediately after meals [40]. In the study of

Walravens and associates, involving 40 low-income Spanish-American children with growth retardation, both linear growth [70] and food intake [71] were increased by Zn supplementation, with the increases primarily occurring among the boys.

An epidemiologic study of 106 preschool Canadian (Ontario) children by Vanderkooy and Gibson [72] suggests that, at similar dietary Zn intakes, the composition of the diet may influence Zn status. Suboptimal Zn nutriture (hair Zn below 70 mcg/g and/or height-for-age below 15th percentile) was associated with lower intakes of readily bioavailable Zn from flesh foods and higher intakes of milk.

Walravens, in his review of zinc nutrition in infants and children [73], points out that recognition of Zn deficiency may be difficult, particularly in marginal cases in which plasma and hair Zn levels may be within accepted limits, and recommends in case of doubt an empirical trial of Zn supplementation at 1 mg/kg/d. Although zinc sulfate is often used, zinc acetate and zinc gluconate seem to be better tolerated [74]. The American Academy of Pediatrics recommends a solution of 2.8 mg/ml of Zn acetate dihydrate (about 1 mg elemental Zn) administered in two or three divided doses per day at least 1 hour before meals, to a total daily dosage of 1 mg Zn per kg of body weight per day [75]; the solution can be flavored.

Pregnancy. Hurley and Baly [76] have reviewed the extensive evidence obtained from animal studies that zinc deficiency during pregnancy causes a variety of congenital defects, including malformations of skeleton, nervous system, and lungs, probably because Zn deficiency inhibits DNA synthesis and therefore cell division. Jameson [77] has reviewed the associations between low Zn status in pregnant women and incidence of comparable congenital defects in the human infant, including skeletal malformations, spina bifida, anencephaly, and achondroplasia. Cherry et al. [78] noted an association between low maternal plasma Zn levels in pregnant teenagers and the incidence of undescended testes and metatarsus varus in their infants.

A large-scale, placebo-controlled study of Zn supplementation (30 mg/d as gluconate) in pregnant adolescents showed a 30% reduction in prematurity rates in normal weight women and elimination of the need for assisted respiration in their preterm infants [79]. Underweight Zn-supplemented women in this study experienced longer mean gestation periods and heavier birthweights than underweight controls.

Meadows et al. [80], in a study involving more than 100 pregnant women, found that leukocyte Zn concentrations were significantly lower in mothers giving birth to term babies who were small for gestational age than in mothers with normal term babies, suggesting that maternal tissue Zn depletion is associated with fetal growth retardation.

In a study group of 450 pregnant women, Mukherjee et al. [81] noted a significant association between the total occurrence of fetomaternal complications and plasma Zn and albumin levels in the lowest quartile for these values. Complications assessed included maternal infections, excessive maternal bleeding, fetal distress, prematurity or stillbirth, spontaneous abortion, toxemia, and maternal tissue fragility at delivery. An association has been observed between low maternal plasma or serum Zn levels and spontaneous abortion [77,82]. In their studies of pregnant teenagers, Cherry et al. [78] found that among 62 women having plasma Zn levels below 50 mcg/dl, 35% developed hypertension/toxemia, while among 153 with levels above 50 mcg/dl, 14% became toxemic. Manci et al. [83] analyzed the dry weight concentrations of Zn, Ca, Cu, and Fe in 20 human placentae from term gestations complicated by hypertension, compared with 30 placentae from uncomplicated term gestations, and found significantly lower concentrations of Zn and Ca in the former, while mean Cu and Fe levels were not significantly different between the two groups.

Limited trials of Zn supplementation have been carried out by Jameson [77], comparing maternal and fetal complications between supplemented and unsupplemented women. One study was done in the third trimester and another in the second and third trimesters, using in both cases a daily supplement of 45 mg Zn (200 mg $ZnSO_4 \cdot 7\ H_2O$). Maternal and fetal complications were reduced to a moderate but significant extent.

Jameson [77] recommends consideration of Zn supplementation at 25–45 mg Zn/d for any pregnant patient with a plasma Zn level early in pregnancy below 65 mcg/dl. In evaluating the significance of serum and plasma Zn levels, it should be borne in mind that values in pregnant women average 20–30% lower than those in nonpregnant women [84,85], largely because of expansion of circulating volume, and that serum Zn readings are somewhat higher than those of plasma Zn [36]. Hair Zn is unreliable as an indicator of Zn deficiency, as paradoxical increases in hair Zn level can occur in severe deficiency, possibly because of slow hair growth [77]. Bergmann et al. [86] noted elevated hair Zn concentrations in mothers of infants with spina bifida, and Vir et al. [87] found that of 60 pregnancies followed, two ended in spontaneous abortion, and these two subjects had serum Zn in the lower range and hair Zn in the higher range of group values.

The pregnant patient should be counseled on good dietary sources of zinc. If she customarily eats little or no meat, counseling should be done with special care, and Zn supplementation should be considered, but in view of remaining uncertainties regarding the value of Zn supplementation in pregnancy [88] the possibility of adverse nutrient interactions if too much Zn is taken (see below), *supplements larger than the RDA for pregnant women of 20 mg Zn/d should be used with caution.*

Two commonly used micronutrient supplements, iron and folic acid, can interfere with absorption of Zn supplements if taken at the same time [81,82,89]. Breskin et al. [82] found that in first-trimester pregnant women receiving 15–25 mg/d of supplemental Zn, administration of 30–60 mg/d of Fe depressed serum Zn to levels as low as those of women taking no Zn supplement. Most prenatal vitamin-mineral supplements contain 60–65 mg Fe. Ordinary "one-a-day" multivitamin-mineral supplements containing 18 mg Fe did not seem to interfere with Zn absorption [82]. Mukherjee et al. [81] caution that folic acid depresses absorption, suggesting that the mechanism may be by formation of an insoluble, poorly absorbed Zn salt in the intestinal lumen, although ordinarily the molar ratio of Zn to folate is very high in the diet or in supplements. In contrast, the extensive studies of Keating et al. [90] in both laboratory rats and male human volunteers showed no inhibition of Zn absorption over a wide range of Zn:folate ratios. No such interaction has been reported between food Zn, which is mostly complexed with protein, and food folate, which is present in polyglutamyl form. It is probably best to give supplemental Zn at a different time of day than a prenatal or "one-a-day" multivitamin-mineral preparation, or than therapeutic iron.

The elderly. A number of investigators have noted a decline in mean serum or plasma Zn levels of the elderly compared with the norms of young adults [91–93]. Low income and poor dietary habits are responsible in part, resulting in low Zn intakes [93,94]. Mean red-cell Zn levels also show a decline with age [91–95]. Elderly individuals show a lower absorption of dietary Zn than young people [96] and a lower plasma Zn response to a test dose of zinc acetate given on an empty stomach [97]. Although the elderly may adapt by reducing endogenous Zn losses and remain in Zn balance [97], it is a fact that clinical symptoms resulting from or associated with Zn deficiency, such as impaired immune response, poor wound healing, and diminished taste acuity occur more frequently in elderly than in young adults [98]. Bunker et al. [99] noted frequent negative Zn balances in housebound elderly subjects in England, whereas healthy active people appeared to have an adequate Zn status.

Stiedemann and Harrill [100] found that post-titer antibody readings after administration of flu vaccine to 36 women aged 67–96 years were positively correlated with serum Zn levels and Zn intakes. Duchateau et al. [101] administered 220 mg $ZnSO_4 \cdot 7\,H_2O$ (50 mg Zn) twice daily for 1 month to 15 healthy subjects over 70 years of age, delayed cutaneous hypersensitivity responses, and immunoglobulin IgG antibody response to tetanus vaccine. Zinc supplementation may also improve protein synthesis in Zn-deficient old people. A trial of 21 days' supplementation with 50 mg Zn/d in 14 institutionalized elderly with plasma Zn levels below 70 mcg/dl resulted not only in an increase in plasma Zn but also in an increase in mean plasma albumin from

3.35 to 3.63 g/dl; upon cessation of Zn supplementation, the plasma albumin returned to pretrial levels in 21 days [102]. Bogden et al. [103], studying Zn nutriture and immune function in 100 noninstitutionalized American elderly, noted that Zn intake was below the 15 mg RDA for more than 90% of the subjects, and that incidence of anergy to a panel of seven skin test antigens was 41%. Trials on Zn supplementation are proceeding [103].

Supplementation with Zn, as well as other essential nutrients, should be considered if the patient is not eating enough to maintain normal weight for height, sex and age.

Anorexia nervosa. A few investigators have reported promising results in empirical trials of oral Zn supplementation in anorexia nervosa [104–107]. Five anorectic young women studied by Safai-Kutti and Kutti [107] in Sweden gained a median of 0.8 kg per month over observation times of 5–17 months while being treated with 45–90 mg Zn/d as $ZnSO_4 \cdot 7 H_2O$. No double-blind trials have been reported as of this writing.

Special diets. Although well-designed vegetarian or lacto-ovo-vegetarian diets can meet Zn needs, their lack of Zn-rich red meat and their increased fiber and phytate levels militate against adequate Zn intake and absorption. Measured by plasma Zn response to an oral load of 50 mg Zn on an empty stomach, a lacto-ovo-vegetarian diet did not maintain as good Zn nutrition in young women as a nonvegetarian diet, although dietary Zn intakes were the same [108].

Kramer et al. [109] analyzed the Zn contents of seven weight-reduction diets composed of common foods. Most of the diets provided about 1,000 kcal/d, one about 600 kcal/d. The diets tended to be adequate or nearly adequate in Zn if the protein source was mainly beef and if the RDA for protein was exceeded. However, if the principal protein sources were fish, poultry, milks, eggs, or cheese, the Zn intake could be low despite adequate protein intake. Zinc should be included in the micronutrient supplements prescribed for patients on weight-reducing diets.

Acne and recurrent furunculosis. Michaelson et al. [110] reported striking reductions of acne papules, pustules, and infiltrates in 38 male and 26 female acne patients treated with supplemental Zn for 12 weeks. The dosage used was 200 mg $ZnSO_4 \cdot 7H_2O$ (45 mg Zn) three times a day. Orris et al. [111], however, obtained no improvement in a trial with 22 acne patients. Zinc deficiency per se probably does not cause acne, but since it impairs immune response, could worsen it once established.

Brody [112] obtained excellent therapeutic results with oral Zn treatment (45 mg Zn t.i.d.) of recurrent furunculosis. In 15 patients aged 24–50 years, he found low serum Zn levels ranging from 49 to 59 mcg/dl (normal 70–115). The patients had had recurrent boils for 3–10 years. In seven patients (four males, three females), the boils were incised and antibiotics prescribed. During a 3-month follow-up, new boils continued to appear; serum Zn remained low. The other eight patients (four males, four females) were prescribed oral Zn as the only treatment. Serum Zn rose to normal in one month, active lesions regressed, and no new boils appeared during 3 months' follow-up. It was not clear why serum Zn was low in these patients; possibly some people simply need more dietary Zn than average to attain normal serum Zn levels.

In recurrent infections of any kind, determination of serum or plasma Zn levels is advisable, and if low levels are found, a trial of oral Zn therapy should be made.

Pica. Pica may be defined as a pathological craving either for specific foods or for nonfood substances, e.g., geophagia [113]. Historically, this craving has been associated with metal nutrient deficiency, notably of iron, but low plasma Zn levels are also observed in a high percentage of individuals with pica. If geophagia is the form of the disorder taken, both Zn and Fe may be tightly bound by the soil or clay eaten and worsen the plasma deficiencies. Controlled trials of Zn and Fe supplementation are needed [113].

Conditioned Deficiency

Malabsorption. McClain [114] has recently reviewed the subject of Zn metabolism in various malabsorption syndromes, including Crohn's disease (regional enteritis), ulcerative colitis, celiac sprue, short bowel syndrome, jejunoileal bypass, chronic diarrhea, cystic fibrosis, acute pancreatitis, and pancreatic insufficiency. Reduced plasma Zn levels (below 70 mcg/dl) are often found, as well as reduced plasma Zn response to an oral $ZnSO_4$ load taken on an empty stomach [114].

Crohn's disease patients with Zn deficiency may have acute symptoms, such as acrodermatitis, or more subtle disturbances, such as impaired immune response or defective night vision [114]. Zinc depletion can be due not only to poor absorption, but to anorexia, protein-losing enteropathy, diarrhea, enterocutaneous fistular drainage, reaction with dietary fats to form insoluble Zn soaps, and loss of enteropancreatic cycling [114].

The plasma Zn level is a useful indicator of Zn status in the stable patient, but may be within the normal range in a patient who is losing lean tissue mass and supplying endogenous Zn to the circulation from degraded Zn-containing tissue enzymes [40]. Flare-up of the condition or infection causes release of the monokine interleukin-1 (leukocyte endogenous mediator), which depresses plasma Zn by segregating it into the liver [34,39,114]; thus, the total body pool of Zn could be of normal magnitude in the presence of low plasma Zn in the acutely ill patient. Response to Zn supplementation may sometimes be the only practical means of diagnosing Zn deficiency.

Supplementation regimens have generally involved long-term administration of 50–150 mg oral Zn/d (220–660 mg/d of $ZnSO_4 \cdot 7\ H_2O$) as capsules in two or three daily doses [115,116]. In a Crohn's disease patient whose main clinical symptom was impaired dark adaptation, the condition was remedied by administration of 10 mg Zn as Zn gluconate four times daily, before meals and h.s., for 1 year [117]. Improvement was believed due to restoration of normal activity of Zn-dependent retinol dehydrogenase in the retina. Dark adaptation deteriorated when Zn supplementation was discontinued [117].

Patients being treated by total parenteral nutrition may develop acute Zn deficiency as they convalesce, with provision of all nutrients in adequate amounts except zinc. The synthesis of Zn-proteins in regained lean tissues drains the circulation of Zn, producing very low plasma Zn readings. Provision of 3 mg Zn/d in the parenteral fluid, which would be adequate for a stable patient, was inadequate in two patients studied by Allen et al. [26], and parenteral Zn had to be increased to 12 mg/d to bring about recovery.

Defective Zn absorption is found in celiac sprue (gluten enteropathy) and dermatitis herpetiformis [118,119]. Absorption usually improves to normal levels when there is good adherence to a gluten-free diet [118]. Zinc supplementation should be considered during recovery, as the rate and degree of restoration of mucosal integrity could be influenced by the Zn deficiency that developed before diet therapy [119]. Some clinical benefits in terms of increased plasma Zn levels and greater activity of jejunal maltase, sucrase, and lactase have been claimed for Zn supplementation (220 mg $ZnSO_4 \cdot 7H_2O$ t.i.d. for 8 weeks) in so-called nonresponsive adult celiac disease, in which there is failure to regenerate the villi in the jejunal mucosa on a gluten-free diet [120].

Renal disease. Low plasma Zn levels are common in chronic renal failure, as well as low concentrations of Zn in leukocytes and hair [4]. In part, a low Zn status in advanced kidney failure may be due to anorexia and low food intake; hemodialysis patients with higher intakes of protein, especially meat, had higher plasma Zn levels than patients eating less protein [121].

Zinc supplementation is effective in overcoming many of the clinical and biochemical abnormalities of dialyzed renal patients, which include hypogeusia, impotence, and elevated plasma ammonia and ribonuclease activity [4]. Mahajan et al. [122] carried out a double-blind, placebo-controlled trial of Zn supplementation in 24 hemodialysis patients. Patients were randomly assigned to placebo capsules or Zn acetate capsules (25 mg Zn) to be taken orally twice daily 1–2 hours before meals. After 6 months, the Zn-supplemented, but not the placebo, group demonstrated significant increases in mean plasma, leukocyte, and hair Zn concentrations and decreases in plasma ammonia and ribonuclease activity. Abnormalities of taste and sexual function improved significantly in patients receiving Zn but not in those receiving placebo. These investigators also found that Zn supplementation reverses hyperprolactinemia, another abnormality seen in male hemodialysis patients [123].

Reimold [124] reported markedly lower mean plasma Zn and hair Zn in 105 pediatric patients with nephrotic syndrome or nephritis-nephrosis than in healthy control children (plasma Zn 52 mcg/dl versus 85 mcg/dl in controls).

Alcoholism. Low plasma zinc levels are frequently observed in chronic alcoholics. Deficiency may be in part due to an effect of ethanol on the kidney to increase Zn excretion and in part to low dietary Zn intake [4]. The alcoholic is less efficient in taking up $ZnSO_4$ administered on an empty stomach [125], but appeared to absorb Zn acetate taken with food as well as nonalcoholic controls [126].

Goldiner et al. [127] investigated the effects of $ZnSO_4$ administration on the impotence and/or hypogonadism of 25 male outpatients with alcohol-induced hepatic cirrhosis. Patients were stable and had ingested no alcohol during the previous 6 months. They were assigned randomly in double-blind fashion to placebo or 50 mg Zn/d (220 mg $ZnSO_4 \cdot 7\ H_2O$) and took the medication for 4–8 months. Serum Zn levels rose in the Zn-supplemented group, but there was no change in sexual function or serum testosterone and urinary gonadotropin levels. However, impaired dark adaptation in stable Zn-deficient alcoholic cirrhotics was corrected by Zn supplementation [128].

Recovering alcoholics should have Zn included at the RDA level in their oral micronutrient supplements as a general precautionary measure.

Low cord plasma Zn concentrations have been observed in neonates with the fetal alcohol syndrome, and it has been suggested that the developmental anomalies are due to Zn deprivation of the fetus [129]. Animal studies have shown that alcohol ingestion inhibits Zn transfer across the placenta, inducing fetal growth retardation [130,131], but Zn supplementation of pregnant rats while daily alcohol ingestion continued did not overcome the defect in placental transport of zinc.

Diabetes mellitus. Low plasma Zn levels occur more frequently among diabetics than in the general population [92,132]. Kinlaw et al. [132] found that 5 of 20 patients with type II diabetes had serum Zn levels below 70 mcg/dl, and all of the diabetic patients exhibited hyperzincuria. Heise et al. [133], comparing 18 insulin-dependent diabetic women with 15 nondiabetic controls, found that the Zn concentration in the urine of the diabetic was 2.7 times that of the controls; plasma Zn levels, however, were not reduced in these diabetic women. The increase in urinary Zn excretion appears to be due to hyperglycemia, as glucose infusion in normal dogs increased their urinary output of Zn; the investigators speculate that Zn repletion of diabetics could improve insulin sensitivity in patients with type II diabetes or reduce the severity of the disease [132]. Niewoehner et al. [134] found that in patients with type II diabetes mellitus and Zn deficiency, there was an improvement in lymphocyte response to phytohemagglutinin but no consistent change in natural killer cell activity following Zn supplementation, indicating that at least part of the impairment in immune response in diabetes may be due to low Zn status. Obviously, additional controlled studies of Zn supplementation in diabetes are needed.

Liver disease. Keeling and Thompson [135] have shown that mean leukocyte Zn concentration is significantly lower than that of healthy controls in patients with chronic liver disease (alcoholic cirrhosis, primary biliary cirrhosis, and active chronic hepatitis); muscle Zn was also below normal in patients with alcoholic cirrhosis. Patients with low leukocyte Zn showed impaired dark adaptation. Other measures of Zn status were evaluated, but appeared less useful, i.e., red cell and plasma Zn levels and urinary Zn excretion [135]. Liver cirrhosis patients show significantly lower uptakes of Zn from a single-dose supplement than healthy controls [136].

Reding et al. [137] have reported clinical improvement with Zn supplementation (200 mg Zn acetate dihydrate t.i.d. with meals for 7 days) in patients with hepatic encephalopathy; an unsupplemented control group showed no improvement. Zinc deficiency has been observed to cause elevation of plasma ammonia levels in both animals and man, probably because of an associated decline in the activity of hepatic ornithine carbamoyltransferase, an enzyme required for urea synthesis; the neurotoxicity of ammonia contributes to the development of encephalopathy and coma in patients with advanced liver disease [138].

Schölmerich et al. [136] found a Zn-histidine 1:2 complex to be more effective than $ZnSO_4$ as a supplement for cirrhosis patients; administration 1 hour before a meal was an order of magnitude more effective than administration with or after a meal.

Athletics and exercise. Haralambie [139] observed reduced plasma Zn levels in athletes in training, probably reflecting increased sweat losses. A modest Zn supplement may be required by athletes or those who engage in regular strenuous exercise in order to maintain a normal plasma Zn concentration. Confirmatory clinical trials are needed.

Genetic Disease

Acrodermatitis enteropathica. Acrodermatitis enteropathica (AE) is an autosomal recessive heritable disease of Zn malabsorption occurring usually in infants of Italian, Armenian, or Iranian lineage [138]. The disease remains inapparent while the infant is receiving breast milk, which contains Zn-binding ligands that facilitate Zn absorption, but emerges a few weeks after weaning. Infants receiving formula instead of breast milk develop symptoms at 1–2 months of age. Classical findings, besides low plasma Zn levels, comprise diarrhea, vomiting, alopecia, and skin lesions of erythematous, vesicular-bullous, or eczematous nature of the extremities and the oral, anal, and genital areas. Other signs include growth retardation, paronychia, nail deformation, and emotional lability. The eyes may be affected, with photophobia, blepharitis, conjunctivitis, and corneal opacities [138,140]. Milder variants of AE have been described [141].

Pharmacologic doses of zinc salts allow sufficient Zn absorption to raise plasma Zn to normal levels and clear all AE signs and symptoms. The recommended dosage is 1–2 mg elemental Zn per kg bodyweight per day. After institution of Zn therapy, mood improves first, followed by resolution of the gastrointestinal problems and skin lesions. Catch-up growth will take place in young childen [140].

Brenton et al. [142] reported two normal pregnancies in a woman with AE who was maintained on a dosage of 300 mg zinc sulfate per day, increased to 450 mg/d in the last month of pregnancy to maintain normal plasma Zn levels. The two infants had no congenital abnormalities and appeared to be developing normally.

Wilson's disease. Wilson's disease is an autosomal, recessively inherited genetic disorder involving excessive accumulation of copper in the liver, brain, cornea, and other organs. As the disorder progresses, cirrhosis develops, as well as neurologic symptoms and Kayser-Fleischer rings in the cornea. Among possible biochemical causes is failure to excrete normal amounts of copper in the bile, but the primary biochemical defect(s) remains unknown. The traditional treatment has been oral administration of penicillamine, which chelates copper so that it may be excreted in the urine. However, penicillamine causes acute sensitivity reactions (skin eruptions, fever, eosinophilia, leukopenia, thrombocytopenia, and lympadenopathy) in 30% of patients within the first 2 weeks of therapy, controllable in most by withdrawing therapy for a few days and restarting at a lower dose, but about 10% of patients cannot tolerate the drug at all [143].

Investigators in the Netherlands and the United States have shown that pharmacologic Zn doses provide a relatively nontoxic alternative to penicillamine. High-dose Zn causes inhibition of copper absorption, probably by inducing synthesis of intestinal metallothionein at high levels in the mucosal cells, forming a barrier to absorption of copper ion. Under this treatment, excretion of copper in the stool increases. In patients previously "decoppered" with penicillamine, a state of Zn deficiency may exist, as the drug complexes with Zn as well as copper. This deficiency state must be overcome before inhibition of copper absorption occurs, so that there may be a lag time of 2–7 weeks before negative copper balance is secured [143].

The American investigators recommend a dosage schedule of 25 mg of elemental Zn every 4 hours during the day for four doses, and one 50-mg dose at bedtime. Zinc acetate dihydrate is preferred over zinc sulfate. Food and beverages other than water are avoided for 1 hour before and after Zn administration [143].

The Netherlands group has had good results with zinc sulfate ($ZnSO_4 \cdot 7 H_2O$) in powders to be dissolved in water and taken orally 30 minutes before meals [144,145]. Dosages were 100–400 mg of zinc sulfate t.i.d. corresponding to 68–272 mg Zn/d. They found considerable differences among patients in the dosage required and recommend adjustment of dose on the basis of plasma response to an oral ^{64}Cu load test, which is also used in subsequent monitoring [146]. Frequent determinations of plasma Cu, ceruloplasmin, and Zn are also recommended, so that dosage may be modified as needed. Clinical improvements were noted with respect to tremor, rigidity, dystonia, dysarthria, and Kayser-Fleischer rings [144,145]. Liver biopsies showed substantial reductions of Cu content [144,145].

Improvements will probably be made in the manner of Zn administration in Wilson's disease. For example, Van den Hamer et al. [147] determined that the antagonistic effect of zinc on copper absorption remains strong for 3–5 days after Zn supplementation is stopped, suggesting that in Zn-saturated patients it may be possible to allow control of Cu absorption with less frequent Zn dosage. In animals, a depot parenteral Zn preparation, given weekly or biweekly, increased urinary Cu excretion and lowered Cu concentrations in liver, kidney, and spleen; such a dosage form could be useful in treating Wilson's disease patients who have poor compliance or difficulty with oral therapy [148]. Once a Wilson's disease patient has been effectively decoppered by between-meals Zn treatment, it is conceivable that he could be maintained by administration of Zn supplements with meals, when they would be less likely to cause gastrointestinal side effects.

Sickle cell anemia. Zinc deficiency of moderate degree, associated with hyperzincuria, has frequently been reported in sickle cell anemia (SCA) patients, along with clinical manifestations characteristic of Zn deficiency, namely growth retardation, hypogonadism in males, hyperammonemia, abnormal dark adaptation, delay in wound healing, and impaired cell-mediated immune response. Oral supplementation of SCA patients with Zn in pharmacologic doses has normalized Zn levels in plasma, erythrocytes, and neurophils; normalized plasma ammonia levels; brought about increased growth and development of secondary sex characteristics; increased the activity of Zn-dependent enzymes; reversed dark-adaptation abnormality; and corrected anergy [138,149,150].

Brewer [151] suggests that Zn supplementation be initiated in children with SCA at a dosage of 1 mg Zn per kg bodyweight per day in two or three divided doses, avoiding food 1 hour before and after each dose. In adult man, pharmacologic doses (150 mg Zn/d), in five or six divided doses between meals, reduced irreversible sickling of red cells by about one-third [151]. Subramanian and Prasad [152] have presented the case history of a teenage boy followed for 4 years, 2 years on placebo and 2 years on 30 mg Zn/d (as Zn acetate b.i.d.).

During the placebo period, he experienced three painful crises, two of which required hospital admission. On Zn supplementation he has not needed hospitalization [152].

Hypocupremia may develop during long-term administration of Zn in pharmacologic dosages. This is remediable with a daily supplement of 1 or 2 mg copper [153].

Beta-thalassemia. Homozygous beta-thalassemia patients have much low er plasma, erythrocyte, and hair Zn levels than healthy controls, along with hyperzincuria [154]. Arcasoy et al. [155] have reported favorable preliminary results of Zn treatment of 11 homozygous thalassemic patients. Initial therapy involved administration of 200 mg $ZnSO_4 \cdot 7H_2O$ (45 mg Zn) daily for 3–5 years. The usual transfusion therapy continued. Plasma and red cell Zn levels rose, but the effect on growth retardation was minimal. On a trial basis the dosage was increased to 300–400 mg/d of zinc sulfate in four subjects, who then showed height increases comparable with those of normal children.

Down's syndrome. For reasons not understood, patients with Down's syndrome often show reduced Zn levels in plasma and leukocytes, together with immune response defects associated with Zn deficiency, e.g., defective white cell chemotaxis and impaired cellular immunity, manifested clinically as increased susceptibility to infections [156]. Bjorksten et al. [157], in a study of 12 Down's patients ranging in age from 8 to 22 years, found that administration of 200 mg $ZnSO_4 \cdot 7 H_2O$ (45 mg Zn) t.i.d. for 2 months doubles the serum Zn concentration, restored leukocyte chemotaxis in eight of nine patients in whom it had initially been defective, improved delayed skin reactivity, and partially restored the lymphoblast transformation response to phytohemagglutinin. The dosage selected was probably higher than needed, as mean serum Zn after 2 months of supplementation was 50% higher than in normal control children. In addition, circulating copper levels were somewhat depressed. In follow-up studies, the investigators were trying lower Zn doses [157].

Other Clinical Studies and Potential Uses

The common cold. Based upon literature reports of inhibition of rhinoviruses and herpes simplex viruses in vitro by unphysiologically high concentrations of Zn ion [158–163], as well as serendipitous observations in a 3-year-old girl who sucked instead of swallowed her Zn gluconate tablets, Eby et al. [164] undertook a double-blind placebo-controlled study of the effect of Zn gluconate on duration of the common cold. In all, 146 volunteers were enrolled, suffering from colds of various durations, of which 83 were randomly assigned to Zn gluconate tablets (23 mg Zn) and 63 to identically appearing calcium lactate tablets. A topical effect on the mucous membranes was desired, rather than a systemic effect, as it would not be possible or safe to raise plasma Zn levels to those found effective against viruses in vitro. Subjects were therefore instructed to allow the tablets to dissolve slowly in the mouth. The initial loading dose was two tablets, followed by one tablet every 2 waking hours, not exceeding 12 tablets per day (275 mg Zn/d) for adults and nine tablets a day for youths. Children under 60 pounds (27 kg) received one-half tablet every 2 waking hours, not exceeding six tablets per day. Duration of the experiment was 7 days.

At the end of 7 days, 90% of the Zn-supplemented group reported themselves to be free of symptoms, while only 49% of the placebo group were symptom-free. A closer analysis was made of 65 subjects (37 Zn, 28 placebo) who had reported having cold symptoms for 3 days or less before joining the experiment. After 7 days, 86% of these Zn-treated subjects were asymptomatic, compared with 46% of the placebo subjects. In the Zn-treated group, 11% became asymptomatic within 12 hours and 22% within 24 hours. Based on the relative rates of decline of symptoms in the two groups, the Zn treatment appeared to shorten the average duration of colds by about 7 days. Subsequent uncontrolled observations indicated that incipient colds of only a few hours in duration could often be aborted within several hours [164].

Side effects were mainly related to taste, i.e., unpalatability or distortion of taste. There were a few complaints of mouth irritation and nauseous feelings. Vomiting occurred in two subjects after they ingested the loading dose of two tablets (46 mg Zn). The investigators tried Zn ascorbate and Zn aspartate as alternative Zn salts, obtaining results similar to those obtained with gluconate. Zinc orotate was substantially less effective, probably because of low solubility. Zinc sulfate and zinc chloride were not tried because of known caustic effects on the mucous membranes [164]. Zinc gluconate lozenges are now listed in the catalogs of suppliers of vitamin and mineral supplements.

As of this writing, confirmatory trials of the efficacy of Zn lozenges in aborting the common cold have not been reported. *In view of the high doses employed, use of Zn lozenges should be terminated as soon as cold symptoms subside and should not in any case be continued for longer than 7 days.*

Leg ulcers. Stasis leg ulcers have been treated with oral Zn sulfate in double-blind trials by several investigators, one-half of whom obtained increases in healing rate [165–167] and one-half of whom did not [168–170], as compared with patients receiving placebos. It seems clear that Zn supplementation improves healing rate in patients who are initially Zn-deficient but not those whose Zn status is adequate [167]. Thus, Hallböök and Lanner [167], in a double-blind trial involving 27 patients, found that leg ulcers in patients with a serum Zn level higher than 110 mcg/dl initially or after receiving oral $ZnSO_4·7 H_2O$ (200 mg t.i.d.), healed significantly faster than the leg ulcers in patients with lower serum Zn levels. They administered $ZnSO_4$ or placebo as an effervescent drink with meals; tablets or capsules of $ZnSO_4$ tended to cause gastric irritation. No hematologic or biochemical evidence of toxicity was noted during 4 months of therapy at 600 mg $ZnSO_4 \cdot 7$ H_2O or 135 mg Zn/d. They recommend that Zn treatment be tried for at least 7–8 weeks to see if it is having an effect on healing.

Surgery, wound healing, and burns. Van Rij [171] has critically reviewed the subject of zinc supplements in surgery. As in the case of leg ulcers, Zn supplementation improves wound healing only in patients with pre-existing Zn deficiency. It is important to assure an adequate Zn nutrition in the patient, preferably before surgery, to achieve a normal rate of wound healing and to diminish the likelihood of sepsis due to impaired immune response. Van Rij [171] lists risk factors for Zn deficiency that should be assessed before and after surgery. Previous diet, drugs taken, alcoholism, malabsorptive disorders, and renal disease are examples of factors that may have led to low Zn status preceding hospitalization. For correction and/or prevention of Zn deficiency, oral Zn supplements have traditionally been given as $ZnSO_4 \cdot 7$ H_2O, 220 mg (50 mg Zn) t.i.d.; taking the supplement with food reduces absorption somewhat, but also reduces gastric upset. Van Rij suggests that Zn acetate dihydrate t.i.d. (15 mg Zn t.i.d.) is also effective and less likely to cause GI disturbance [171]. With patients on intravenous alimentation, the recommended level of Zn infusion for stable adult patients is 2.5–4 mg/d [172]. However, when Zn losses are excessive and in need of treatment for deficiency, larger supplements may be needed, e.g., up to 20 mg Zn daily, administered IV over several hours [26,171]. The sooner Zn deficits are corrected and the system "primed," the better the patient is prepared for the trauma of surgery [171]. Obviously this generalization applies to other essential nutrients as well as zinc.

Arthritis and inflammatory diseases. Svenson et al. [173] studied Zn concentrations in the plasma and blood cells of patients with various inflammatory conditions (rheumatoid arthritis, spondarthritis, scleroderma), compared with values in healthy control subjects. Zn levels in patients were significantly and markedly depressed in erythrocytes, platelets, and granulocytes of patients versus controls, and significantly but less markedly depressed in plasma. Upon corticosteroid treatment, Zn levels rose, but not to normal values. Zn supplementation was not tried in this study.

Simkin [174] investigated the effects of Zn supplementation in 24 patients with chronic refractory rheumatoid arthritis. Twelve patients each were randomly allocated to placebo or 220 mg $ZnSO_4 \cdot 7$ H_2O (50 mg Zn) t.i.d. for 12 weeks; previous drug therapies were continued. This double-blind trial was followed by an open 12-week period when all patients received Zn. During the double-blind phase, Zn-treated patients showed a rise of mean serum Zn from 84 to 116 mcg/dl and fared better than placebo-controlled patients with regard to joint swelling, morning stiffness, walking time, and the patient's own impressions. These indices as well as joint tenderness also improved with Zn treatment in both groups during the second 12-week period [174].

Clemmensen et al. [175] conducted a similar double-blind trial for 6 weeks, followed by a 24-week open trial, in 24 patients who had suffered from psoriatic arthritis for several years. Zn was supplied as 220 mg $ZnSO_4 \cdot 7\ H_2O$ in tablets taken t.i.d. to a total of 660 mg (150 mg/d of Zn). The treatment was well tolerated except for slight dyspepsia of 1–2 weeks' duration in five patients on Zn treatment and two on placebo. The initial 6-week Zn treatment produced some remission with respect to overall condition, morning stiffness, joint pains, and functional capacity of the joints, as well as a significant reduction in joint pains. During the extended open trial, improvements were maintained, but functional capacity of the joints did not increase. Mean daily intake of analgesics was reduced.

These encouraging observations deserve clinical follow-up by others.

The desirability of exploration of Zn therapy in inflammatory diseases is further supported by the laboratory studies of Kretzschmar et al. [176] in mice with systemic lupus erythematosus (SLE). In this disease there are antinuclear antibodies and antibodies directed against DNA. Zn was administered as a depot subcutaneous injection of Zn pamoate (55 mg Zn/kg bodyweight once weekly) to maintain a sustained elevation (about twice normal) of plasma Zn. After 24 weeks, there was a significant reduction in glomerular lesions in the kidneys of Zn-treated SLE mice as compared with untreated SLE mice. The authors recommend follow-up in human SLE and other inflammatory diseases [176].

Optic atrophy and disk pallor. Leopold [177] and Stuvtevant [178] have reviewed clinical evidence relating optic neuropathy to Zn deficiency. The tissues of the eye and optic nerve are normally high in Zn. During Zn deficiency in acrodermatitis enteropathica, there have been observations of photophobia, gaze aversion, depressed visual acuity, centrocecal scotomata, and optic disk pallor. These reviewers recommend that Zn status be investigated in patients who have optic disk pallor, usually thought to be idiopathic or drug-induced. Also, ophthalmologic examination would be a reasonable precaution in patients who are at risk of Zn deficiency for such reasons as malabsorption, chronic diarrhea, or other Zn-depleting disorders [177,178].

Cigarette smoking. Chowdhury and Chandra [179] report observations of altered immune function in cigarette smokers, ages 23–55 years, who had smoked at least 20 cigarettes per day for 6 years or more, compared with healthy controls. Leukocyte, lymphocyte, T-cell, and natural killer (NK) cell counts were normal in smokers, but B-cell count was low and suppressor T-cell count was high, relative to controls. Supplemental oral Zn at 100 mg/d for 4 weeks, given to both groups, reduced the suppressor T-cell numbers to control levels in the smokers and increased the ratio of helper to suppressor cells. Zn increased NK-cell activity and serum Zn levels in both groups. The authors suggest that moderate Zn dosage can correct some smoking-induced alterations of cellular immune function.

Cancer. Limited studies suggest that good Zn nutriture is important to prevention of cancer. In part this is expected, in view of the importance of zinc to immune function. Newberne and coworkers [180–182] have shown that moderate Zn deficiency in rats enhances the incidence of chemically induced esophageal cancer and that addition of ethanol increases the incidence still more. In human lung cancer patients, Allen et al. [183] observed low serum zinc and hyperzincuria associated with a depressed T-cell phytohemagglutinin (PHA) response; the PHA response was normalized by 6 weeks of Zn supplementation.

Tobey et al. [184] found that Zn ion in cultures of Chinese hamster ovary cells, at a concentration only slightly above that in human plasma, provided a 4.5-fold enhancement of survival after treatment with the alkylating agent, melphalan, suggesting that Zn might be useful as a protective agent for normal cells during cancer therapy with alkylating agents.

TOXICITY AND SIDE EFFECTS

The acute and chronic toxicity of Zn is low. Ingestion of 12,000 mg of $ZnSO_4 \cdot 7 H_2O$ (2,720 mg Zn) over a 2-day period caused drowsiness, lethargy, and an increase in serum lipase and amylase levels [185]. Chronic ingestion by Wilson's disease patients of 300–1,200 mg $ZnSO_4 \cdot 7 H_2O$ per day, dissolved in water, for 3 years caused no gastric discomfort and no other side effects; two Wilson's disease patients have received zinc sulfate for 25 years or more without adverse effects [144,145]. Brewer et al. [143] noted no side effects in administering Zn acetate dihydrate solutions (150 mg Zn/d) to Wilson's disease patients orally in five or six divided doses between meals and at bedtime over several weeks' observations. Tablets or capsules of zinc sulfate tend to cause gastric irritation [167]. A Zn-histidine complex has been proposed by German investigators as a particularly well absorbed and well tolerated form of oral Zn [186].

Infants and children with acrodermatitis enteropathica have tolerated chronic oral zinc sulfate and zinc acetate dosage very well at 1–2 mg Zn/kg/d [140]. A woman with this genetic disease successfully bore two healthy full-term infants while taking 300 mg/d of $ZnSO_4 \cdot 7 H_2O$ through most of each pregnancy, increasing to 450 mg/d during the final month [120].

Not many long-term studies have been carried out with high-dose Zn administration in healthy adults without genetic disease. However, Aamondt et al. [187], as part of a study of zinc metabolism, administered 100 mg Zn/d as zinc sulfate to 50 men and women, aged 27–72 years, for as long as 440 days without adverse effects. There is a tendency for plasma Zn levels to stabilize during chronic high-dose administration. For example, Babcock and associates [188] found in their Zn metabolism studies that the mean plasma Zn mass increase above preload levels in adults was only 37% in face of an 11-fold increase in Zn intake. Freeland-Graves et al. [189], administering 15, 50, and 100 mg Zn/d to young women, noted that plasma Zn levels rose until about the fourth week of Zn administration and then declined toward initial presupplementation values in the next 4 weeks in the case of the 15-mg and 50-mg groups, with a smaller decline in the 100-mg group.

Administration of 50–100 mg Zn/d as zinc sulfate to adolescent and preadolescent children, who were of short stature but otherwise healthy, for as long as 2 years was without adverse effects [68,69].

No hematologic or biochemical abnormalities were noted after 4 months' administration of 600 mg $ZnSO_4 \cdot 7\ H_2O$ (135 mg Zn) per day as an effervescent solution to leg ulcer patients [167]. Occasional nausea and vomiting were noted in 7-day trials on Zn gluconate lozenges as therapy for the common cold; dosages were as high as 250 mg Zn/d for adults and 125 mg/d for children [164].

In parenteral alimentation, Palma et al. [62] administered 250–400 mcg/kg bodyweight/d for 10–14 days to Zn-deficient infants weighing 1,300–2,400 g without adverse effect. Rothbaum et al. [60] treated Zn-deficient infants and children of ages 5–32 months for 15 weeks at 200–300 mcg Zn/kg/d in parenteral fluids, obtaining good clinical results without adverse findings.

Van Rij [171], in discussing hyperalimentation of the critically ill adult patient, notes that infusions as high as 40–80 mg Zn/d, administered over several hours, have been used to make up Zn deficits. Acute injections of 5-mg doses have been given in a period of several minutes in normal adult subjects without detrimental effects [171], *but a daily 10-mg dose infused over a 1-hour period was reported after 5 days to cause tachycardia, hypothermia, profuse sweating, and blurred vision* [190]. One death has been reported due to acute Zn toxicity in a patient on TPN because of an error in dose computation, resulting in hypotension, jaundice, anemia, and renal failure [191].

INTERACTIONS

Long-term administration of Zn in pharmacologic doses can result in hypocupremia, which is treatable with 1–2 mg Cu per day [153]. This is because high-dose Zn depresses copper absorption by a poorly understood mechanism involving induction by Zn of high levels of metallothionein in the intestinal mucosa, followed by sequestration of copper as a stable Cu-metallothionein complex in the mucosal cells [192]. This effect of pharmacologic Zn is used to advantage in decoppering Wilson's disease patients, as discussed earlier in this chapter, but may be a cause for concern in individuals taking Zn supplements for general health reasons. Fischer et al. [193], in a trial of Zn supplementation in 26 healthy men, determined that 50 mg Zn/d as the gluconate, taken in two doses, morning and night, caused within 6 weeks a small but significant depression in erythrocyte Cu,Zn-superoxide dismutase activity, although plasma copper and ferroxidase activities were not affected. With the ordinary dietary range of Zn intakes of 9–20 mg/d, the retention and excretion of copper does not seem to be affected [194]. However, elderly subjects receiving 23.26 mg Zn/d showed lower retentions of copper than when they received 7.8 mg/d [195].

Inasmuch as copper deficiency tends to increase plasma cholesterol levels, a number of studies have been carried out to see whether the mild deficiency induced by long-term pharmacologic Zn dosage may affect blood lipid levels. In general, these studies have not shown any effect on total cholesterol levels. Some, but not all, studies have indicated that high-dose Zn causes a decline in HDL-cholesterol (HDLC) levels. Hooper et al. [197] noted a 25% decline in mean HDLC level in adult men taking 440 mg of zinc sulfate daily for 5 weeks. The same group carried out an epidemiologic study of the relations between exercise, serum HDLC levels, and use of Zn supplements in 270 healthy men and women over age 60, finding that supplemental Zn ingestion appeared to block exercise-induced increases in HDLC and that (in 22 subjects habitually taking Zn supplements) stopping the use of Zn supplements for 8 weeks increased HDLC levels slightly but significantly [198]. In the same report, however, they note that they found no change in HDLC levels in trained young runners who took 50 mg Zn/d for 8 weeks [198].

Freeland-Graves et al. [189] observed no changes in HDLC levels in young women taking 15, 50, or 100 mg Zn/d as Zn acetate for 8 weeks, except for a transient 8% drop in the 100-mg group at 4 weeks when plasma Zn levels peaked before starting to decline toward presupplementation levels. Pachotikarn et al. [199] report that in 23 young men who ingested 50 mg Zn/d as Zn gluconate for 6 weeks, there was a small but statistically nonsignificant increase in HDLC levels and a decrease in total cholesterol, together with a statistically significant reduction in diastolic blood pressure; plasma copper levels were unchanged. A 6-week treatment with 100 mg Zn/d orally was observed to reduce elevated serum cholesterol in cigarette smokers [180].

In the light of knowledge to date, and pending further studies, it would appear prudent for healthy persons taking vitamin-mineral supplements to limit

Zn supplementation to RDA levels and to accompany this with 2 mg/d of copper to diminish the likelihood of developing a reduced copper status over the long term.

Ionic iron present in solution with zinc ion at molar ratios of 2 or 3 to 1 significantly inhibits the absorption of zinc in human subjects, whereas heme iron in the same molar excess does not have an inhibitory effect [200]. Furthermore, when Zn was given in "organic" form (as oysters), inorganic Fe did not inhibit Zn absorption [200]. Valberg et al. [201] and Sandström et al. [202] confirmed the antagonism of inorganic Fe and Zn ions given in solution on an empty stomach, but found that when these supplements were given with a meal, iron did not adversely affect Zn absorption. In infants, administration of 30 mg Fe/d as ferrous fumarate given 30 minutes before a meal did not affect serum Zn or Cu levels in comparison with controls not receiving iron treatment [203]. However, use of prenatal multivitamin-mineral supplements containing 60–65 mg Fe per daily dose appeared to cause some inhibition of Zn absorption in first-trimester pregnant women [82], suggesting that Zn supplements should be taken at another time of day than therapeutic Fe supplements.

Folic acid has appeared in some studies [81] to interfere somewhat with Zn absorption, but not in others [90]. Pending further investigations of this potential interference, it may be best to give folic acid, like iron, at another time of day than Zn supplements.

Calcium, which has been reported to interfere with Zn absorption in laboratory animals, was not found to do so in human subjects receiving a constant Zn intake of 14.5 mg/d and calcium intakes varying from 200 to 2,000 mg/d [204]. Conversely, a high Zn intake (140 mg/d) depressed calcium absorption significantly in adult men when calcium intake was low (200 mg/d), but not when calcium intake was at recommended levels (800 mg/d) [205].

A number of drugs, most with metal-chelating properties, may depress plasma Zn levels: ethambutol, diiodohydroxyquin, iodochlorhydroxyquin, disulfiram, oxyquinolines, penicillamine, iproniazid, nialamide, and isocarboxazid [177]. Chelators cause hyperzincuria [2]. Chlorothiazides also cause hyperzincuria, and Prasad advocates monitoring for Zn deficiency in hypertensive patients receiving long-term therapy with chlorothiazide [2]. Glucagon is also known to cause hyperzincuria [2]. Oral contraceptives have been reported to induce mild hypozincemia [206]. The recent review by King [207] reaches the conclusion that Zn nutriture is probably not significantly compromised by oral contraceptive use and that extra dietary Zn is not needed.

REFERENCES

1. Prasad AS: Clinical and biochemical manifestations of zinc deficiency in human subjects. J Am Coll Nutr 4:65–72, 1985.

2. Prasad AS: Clinical, biochemical and nutritional spectrum of zinc deficiency in human subjects: An update. Nutr Rev 41:197–208, 1983.
3. Prasad AS: Discovery and importance of zinc in human nutrition. Fed Proc 43:2829–2834, 1984.
4. Prasad AS: Clinical and biochemical spectrum of zinc deficiency in human subjects. In: *Clinical, Biochemical, and Nutritional Aspects of Trace Elements*, AS Prasad (ed). New York: Alan R. Liss, 1982, pp 3–62.
5. Forbes RM: Use of laboratory animals to define physiological functions and bioavailability of zinc. Fed Proc 43:2835–2839, 1984.
6. Anon.: Nucleoside phosphorylase: A zinc metalloenzyme and a marker of zinc deficiency. Nutr Rev 42:279–281, 1984.
7. Tamura T, Shane B, Baer MT, et al.: Absorption of mono- and polyglutamyl folates in zinc-depleted man. Am J Clin Nutr 31:1984–1987, 1978.
8. Starcher BC, Hill CH, Madaras JC: Effect of zinc deficiency on bone collagenase and collagen turnover. J Nutr 110:2095–2102, 1980.
9. Gordon PR, O'Dell BL: Zinc deficiency and impaired platelet aggregation in guinea pigs. J Nutr 113:239–245, 1983.
10. Gordon PR, Browning JD, O'Dell BL: Platelet arachidonate metabolism and platelet function in zinc-deficient rats. J Nutr 113:766–772, 1983.
11. Gordon PR, Woodruff CW, Anderson HL, O'Dell BL: Effect of acute zinc deprivation on plasma zinc and platelet aggregation in adult males. Am J Clin Nutr 35:113–119, 1982.
12. Bray TM, Kubow S, Bettger WJ: Effect of dietary zinc on endogenous free radical production in rat lung microsomes. J Nutr 116:1054–1060, 1986.
13. McCord JM, Keele BB Jr, Fridovich I: An enzyme-based theory of obligate anaerobiosis: The physiological function of superoxide dismutase. Proc Natl Acad Sci USA 68:1027, 1971.
14. Chvapil M, Montgomery D, Ludwig JC, Zukoski CF: Zinc in erythrocyte ghosts. Proc Soc Exp Biol Med 162:480–484, 1979.
15. Chvapil, M: New aspects in the biological role of zinc: A stabilizer of macromolecules and biological membranes. Life Sci 13:1041–1049, 1973.
16. Cerklewski FL: Effect of suboptimal zinc nutrition during gestation and lactation on rat molar tooth composition and dental caries. J Nutr 111:1780–1783, 1981.
17. Fang MM, Lei KY, Kilgore LT: Effects of zinc deficiency on dental caries in rats. J Nutr 110:1032–1036, 1980.
18. Fraker PJ, Gershwin ME, Good RA, Prasad AS: Interrelationships between zinc and immune function. Fed Proc 45:1474–1479, 1986.
19. Prasad AS, Daridenne M, Abdallah J, et al.: Serum thymulin and zinc deficiency in humans. Am J Clin Nutr 45:873A, 1987.
20. DePasquale-Jardieu P, Fraker PJ: Interference in the development of a secondary immune response in mice by zinc deprivation: Persistence of effects. J Nutr 114:1762–1769, 1984.
21. Gabliks J, Nauss K, Newberne P: Increased severity of influenza A virus infection in zinc-deficient mice. Nutr Rep Int 31:911–917, 1985.
22. Hess FM, King JC, Margen S: Effect of low zinc intake and oral contraceptive agents on nitrogen utilization and clinical findings in young women. J Nutr 107:2219–2227, 1977.
23. Baer MT, King JC, Tamura T, et al.: Nitrogen utilization, enzyme activity, glucose intolerance and leukocyte chemotaxis in human experimental zinc depletion. Am J Clin Nutr 41:1220–1235, 1985.
24. Clinical nutrition case: Zinc therapy of depressed cellular immunity in acrodermatitis enteropathica. Nutr Rev 39:168–170, 1981.

25. Duchateau J, Delespesse G, Vereecke P: Influence of oral zinc supplementation on the lymphocyte response to mitogens of normal subjects. Am J Clin Nutr 34:88–93, 1981.
26. Allen JI, Kay NE, McClain CJ: Severe zinc deficiency in humans: Association with a reversible T-lymphocyte dysfunction. Ann Intern Med 95:154–157, 1981.
27. Bunk MJ, Yung JP, Galvin JE, Blaner WS: Divergent zinc needs for growth and for natural killer cell activity in weanling mice: Evidence for a hierarchy of zinc requirements. Am J Clin Nutr 45:845A, 1987.
28. Castillo-Duran C, Heresi G, Fisberg M, Uauy R: Controlled trial of zinc supplementation during recovery from malnutrition: Effects on growth and immune function. Am J Clin Nutr 45:602–608, 1987.
29. Abbasi AA, Prasad AS, Rabbani P, DuMouchelle E: Experimental zinc deficiency in man. Effect on testicular function. J Lab Clin Med 96:544–550, 1980.
30. Wada L, King JC: Effect of low zinc intakes on basal metabolic rate, thyroid hormones and protein utilization in adult men. J Nutr 116:1045–1053, 1986.
31. National Research Council: *Recommended Dietary Allowances,* 9th ed. Washington, DC: National Academy of Sciences, 1980.
32. Smith JC Jr, Morris ER, Ellis R: Zinc: Requirements, bioavailabilities and recommended dietary allowances. In: *Zinc Deficiency in Human Subjects,* AS Prasad, AO Çavdar, GJ Brewer, PJ Aggett (eds). New York: Alan R. Liss, 1983, pp 147–169.
33. Sandstead HH: Requirement of zinc in human deficiency. J Am Coll Nutr 4:73–82, 1985.
34. Prasad AS: Laboratory diagnosis of zinc deficiency. J Am Coll Nutr 4:591–598, 1985.
35. Rabbani PI, Prasad AS, Tsai R, et al.: Dietary model for production of experimental zinc deficiency in man. Am J Clin Nutr 45:1514–1525, 1987.
36. Smith JC, Holbrook JT, Danford DE: Analysis and evaluation of zinc and copper in human plasma and serum. J Am Coll Nutr 4:627–638, 1985.
37. Markowitz ME, Rosen JF, Mizruchi M: Circadian variations in serum zinc (Zn) concentrations: Correlations with blood ionized calcium, serum total calcium and phosphate in humans. Am J Clin Nutr 41:689–696, 1985.
38. Gordon EF, Gordon RC, Passal DB: Zinc metabolism: Basic, clinical and behavioral aspects. J Pediatr 99:341–349, 1981.
39. Beisel WR: The role of zinc in neutrophil function. In: *Clinical, Biochemical, and Nutritional Aspects of Trace Elements,* AS Prasad (ed). New York: Alan R. Liss, 1982, pp 203–210.
40. Solomons NW: Biological availability of zinc in humans. Am J Clin Nutr 35:1048–1075, 1982.
41. Cousins RJ: Mechanism of zinc absorption. In: *Clinical, Biochemical and Nutritional Aspects of Trace Elements,* AS Prasad (ed). New York: Alan R. Liss, 1982, pp 117–128.
42. Anon.: On the entero-pancreatic circulation of endogenous zinc. Nutr Rev 39:163–163, 1981.
43. Wiegand E, Kirchgessner M: Analysis of a dynamical factorial model of trace element utilization. Z Tierphysiol 39:325–341, 1977.
44. Henkin RI: Zinc in humans, In; *Zinc*. Subcommittee on Zinc, Committee on Medical and Biological Effects of Pollutants, National Research Council. Baltimore: University Park Press, 1979, pp 123–172.
45. Aamondt RL, Rumble WF, Babcock AK, et al.: Effects of oral zinc loading on zinc metabolism in humans. I. Experimental studies. Metabolism 31:326–334, 1982.
46. Milne DB, Canfield WK, Mehalko JR, Sandstead HH: Effect of dietary zinc on whole body surface loss of zinc: Impact on estimation of zinc retention by balance method. Am J Clin Nutr 38:181–186, 1983.

47. Spencer H, Kramer L, Osis D: Zinc balances in humans. In: *Clinical, Biochemical, and Nutritional Aspects of Trace Elements,* AS Prasad (ed). New York: Alan R. Liss, 1982, pp 103–115.
48. Krebs NF, Hambidge KM, Jacobs MA, Rasbach JO: The effects of a dietary zinc supplement during lactation on longitudinal changes in maternal zinc status and milk zinc concentrations. Am J Clin Nutr 41:560–570, 1985.
49. Sandström B, Cederblad Å, Lönnerdal B: Zinc absorption from human milk, cow's milk, and infant formulas. Am J Dis Child 137:726–729, 1983.
50. Krebs NF, Hambidge KM: Zinc requirements and zinc intakes of breast-fed infants. Am J Clin Nutr 43:288–292, 1986.
51. Peirce P, Hambidge M, Fennessey P, et al.: Evaluation of zinc (Zn) absorption in the preterm infant using a stable isotope technique. Fed Proc 46:748A, 1987.
52. Moore MEC, Moran JR, Greene HL: Zinc supplementation in lactating women: Evidence for mammary control of zinc secretion. J Pediatr 105:600–602, 1984.
53. Aggett PJ, Atherton DJ, More J, et al.: Symptomatic zinc deficiency in a breast-fed preterm infant. Arch Dis Child 55:547–550, 1980.
54. Zimmerman AW, Hambidge KM, Lepow ML, et al.: Acrodermatitis in breast-fed premature infants: Evidence for a defect of mammary zinc secretion. Pediatrics 69:176–183, 1982.
55. Goldberg HJ, Sheehy JM: Fifth day fits: An acute zinc deficiency syndrome? Arch Dis Child 57:633–635, 1982.
56. Krieger I, Alpern BE, Cunnane SC: Transient neonatal zinc deficiency. Am J Clin Nutr 43:955–958, 1986.
57. Thorp JW, Boeckx RL, Robbins S, et al.: A prospective study of infant zinc nutrition during intensive care. Am J Clin Nutr 34:1056–1060, 1981.
58. Shils ME, Burke AW, Greene HL, et al.: Guidelines for essential trace element preparations for parenteral use. A Statement by an expert panel. JAMA 241:2051, 1979.
59. Lockitch G, Pendray MR, Godolphin WJ, Quigley G: Serial changes in selected serum constituents in low birth weight infants on peripheral parenteral nutrition with different zinc and copper supplements. Am J Clin Nutr 42:24–30, 1985.
60. Rothbaum RJ, Maur PR, Farrell MK: Serum alkaline phosphatase and zinc undernutrition in infants with chronic diarrhea. Am J Clin Nutr 35:595–598, 1982.
61. Naveh Y, Lightman A, Zinder O: Effect of diarrhea on serum zinc concentrations in infants and children. J Pediatr 101:730–732, 1982.
62. Palma PA, Conley SB, Crandell SS, Denson SE: Zinc deficiency following surgery in zinc-supplemented infants. Pediatrics 69:801–803, 1982.
63. Canfield WK, Menge R, Walravens PA, Hambidge KM: Plasma zinc values in children recovering from protein-energy malnutrition. J Pediatr 97:87–89, 1980.
64. Hambidge KM, Hambidge C, Jacobs M, Baum JD: Low levels of zinc in hair, anorexia, poor growth and hypogeusia in children. Pediatr Res 6:868–874, 1972.
65. Solomons NW, Rosenfeld RL, Jacob RA, Sandstead HH: Growth retardation and zinc nutrition. Pediatr Res 10:923–927, 1976.
66. Hambidge KM, Walravens PA, Brown RM, et al.: Zinc nutrition of preschool children in the Head Start program. Am J Clin Nutr 29:734, 1976.
67. Buzina R, Jusic M, Sapunar J, Milanovic N: Zinc nutrition and taste acuity in school children with impaired growth. Am J Clin Nutr 33:2262–2267, 1980.
68. Collipp PJ, Castro-Magana M, Petrovic M, et al.: Zinc deficiency: Improvement in growth and growth hormone levels with oral zinc therapy. Ann Nutr Metab 26:287–290, 1982.

69. Ghavami-Maibodi SZ, Collipp PJ, Castro-Magana M, et al.: Effect of oral zinc supplements on growth hormonal levels, and zinc in healthy short children. Ann Nutr Metab 27:214–219, 1983.
70. Walravens PA, Krebs NF, Hambidge KM: Linear growth of low income preschool children receiving a zinc supplement. Am J Clin Nutr 38:195–201, 1983.
71. Krebs NF, Hambidge KM, Walravens PA: Increased food intake of young children receiving a zinc supplement. Am J Dis Child 138:270–273, 1984.
72. Vanderkooy PDS, Gibson RS: Food concentration patterns of Canadian preschool children in relation to zinc and growth status. Am J Clin Nutr 45:609–616, 1987.
73. Walravens PA: Zinc nutrition in infants and children. In: *Clinical, Biochemical, and Nutritional Aspects of Trace Elements,* AS Prasad (ed). New York: Alan R. Liss, 1982, pp 129–144.
74. Prasad AS: Clinical, biochemical and pharmacological role of zinc. Annu Rev Pharmacol Toxicol 20:393, 1979.
75. Committee on Nutrition, American Academy of Pediatrics: Zinc. Pediatrics 62:408, 1978.
76. Hurley LS, Baly DL: The effects of zinc deficiency during pregnancy. In: *Clinical, Biochemical, and Nutritional Aspects of Trace Elements,* AS Prasad (ed). New York: Alan R. Liss, 1982, pp 145–159.
77. Jameson S: Zinc nutrition and human pregnancy. In: *Zinc Deficiency in Human Subjects,* AS Prasad, AO Çavdar, GJ Brewer, PJ Aggett (eds). New York: Alan R. Liss, 1983, pp 53–69.
78. Cherry FF, Bennett EA, Bazzano GS, et al.: Plasma zinc in hypertension/toxemia and other reproductive variables in adolescent pregnancy. Am J Clin Nutr 34:2367–2375, 1981.
79. Cherry F, Sandstead H, Bazzano G, et al.: Zinc nutriture in adolescent pregnancy: Response to zinc supplementation. Fed Proc 46:748A, 1987.
80. Meadows NJ, Ruse W, Smith MF, et al.: Zinc and small babies. Lancet 2:1135–1136, 1981.
81. Mukherjee MD, Sandstead HH, Ratnaparkhi MV, et al.: Maternal zinc, iron, folic acid, and protein nutriture and outcome of human pregnancy. Am J Clin Nutr 40:496–507, 1984.
82. Breskin MW, Worthington-Roberts BS, Knopp RH, et al.: First trimester serum concentrations in human pregnancy. Am J Clin Nutr 38:943–953, 1983.
83. Manci EA, Cooley NR, Temple K, Blackburn WR: Trace metals in human term placentae in hypertensive disorders. Fed Proc 46:524A, 1987.
84. Swanson CA, King JC: Zinc utilization in pregnant and nonpregnant women fed controlled diets providing the zinc RDA. J Nutr 112:697–707, 1982.
85. Swanson CA, Turnlund JR, King JC: Effect of dietary zinc sources and pregnancy on zinc utilization in adult women fed controlled diets. J Nutr 113:2557–2567, 1983.
86. Bergmann KE, Makosch G, Tews KH: Abnormalities of hair zinc concentration in mothers of newborn infants with spina bifida. Am J Clin Nutr 33:2145–2150, 1980.
87. Vir SC, Love AHG, Thompson W: Zinc concentration in hair and serum of pregnant women in Belfast. Am J Clin Nutr 34:2800–2807, 1981.
88. Swanson CA, King JC: Zinc and pregnancy outcome. Am J Clin Nutr 46:763–771, 1987.
89. Krebs NF, Hambidge KM, Sibley L, English J: Acute effects of iron therapy on zinc status during pregnancy. Fed Proc 46:747A, 1987.
90. Keating JN, Wada L, Stokstad ELR, King JC: Folic acid: Effect on zinc absorption in humans and the rat. Am J Clin Nutr 46:835–839, 1987.
91. Lindeman RD, Clark ML, Colmore JP: Influence of age and sex on plasma and red-cell zinc concentrations. J Gerontol 26:358–363, 1971.
92. Chooi MK, Todd JK, Boyd ND: Influence of age and sex on plasma zinc levels in normal and diabetic individuals. Nutr Metab 20:135–142, 1976.

93. Wagner PA, Krista ML, Bailey LB, et al.: Zinc status of elderly black Americans from urban low-income households. Am J Clin Nutr 33:1771–1777, 1980.
94. Holden JM, Wolf WR, Mertz W: Zinc and copper in self-selected diets. J Am Diet Assoc 75:23, 1979.
95. McWilliams PL, Agarwal RP, Henkin RI: Zinc concentration in erythrocyte membranes in normal volunteers and in patients with taste and smell dysfunction. Biol Trace Elem Res 5:1–8, 1983.
96. Turnlund JR, Durkin N, Costa F, Margen S: Stable isotope studies of zinc absorption and retention in young and elderly men. J Nutr 116:1239–1247, 1986.
97. Bales CW, Steinman LC, Freeland-Graves JH, et al.: The effect of age on plasma zinc uptake and taste acuity. Am J Clin Nutr 44:664–669, 1986.
98. Sandstead HH, Henriksen LK, Greger JL, et al.: Zinc nutrition in the elderly in relation to taste acuity, immune response, and wound healing. Am J Clin Nutr 36:1046–1059, 1982.
99. Bunker VW, Hinks LJ, Stansfield MF, et al.: Metabolic balance studies for zinc and copper in housebound elderly people and the relationship between zinc balance and leukocyte zinc concentrations. Am J Clin Nutr 46:353–359, 1987.
100. Stiedemann M, Harrill I: Relation of immunocompetence to selected nutrients in elderly women. Nutr Rep Int 21:931–942, 1980.
101. Duchateau J, Delespesse G, Vereecke P: Influence of oral zinc supplementation on the immune response of old people. Am J Med 70:1001–1004, 1981.
102. Parry WH, Flint DM, Wahlqvist ML, et al.: The effect of zinc supplementation in aged persons on serum albumin and binding capacity. Proc Nutr Soc 40:40A, 1981.
103. Bogden JD, Oleske JM, Munves EM, et al.: Zinc and immunocompetence in the elderly: Baseline data on zinc nutriture and immunity in unsupplemented subjects. Am J Clin Nutr 46:101–109, 1987.
104. Esca SA, Brenner W, Mock K, Gschnait F: Kwashiorkor-like zinc deficiency syndrome in anorexia nervosa. Acta Derm Venereol (Stockh) 59:361–364, 1979.
105. Safai-Kutti S, Kutti J: Zinc and anorexia nervosa. Ann Intern Med 100:317–318, 1984.
106. Bryce-Smith D, Simpson RID: Case of anorexia nervosa responding to zinc sulphate. Lancet 2:350, 1984.
107. Safai-Kutti S, Kutti J: Zinc supplementation in anorexia nervosa. Am J Clin Nutr 44:581–582, 1986.
108. Freeland-Graves JH, Ebangit ML, Hendrikson PJ: Alterations in zinc absorption and salivary sediment zinc after a lacto-ovo-vegetarian diet. Am J Clin Nutr 33:1757–1766, 1980.
109. Kramer L, Spencer H, Osis D: Zinc and mineral content of weight reducing diets. Am J Clin Nutr 34:1372–1378, 1981.
110. Michaelsson G, Juhlin L, Vahlquist A: Effects of oral zinc and vitamin A in acne. Arch Dermatol 113:31–36, 1977.
111. Orris L, Shalita AR, Sibulkin D, et al.: Oral zinc therapy of acne. Arch Dermatol 114:1018–1020, 1978.
112. Brody I: Treatment of recurrent furunculosis with oral zinc. Lancet 2:1358, 1977.
113. Danford D: Pica and zinc. In: *Zinc Deficiency in Human Subjects,* AS Prasad, AO Çavdar, GJ Brewer, PJ Aggett (eds). New York: Alan R. Liss, 1983, pp 185–195.
114. McClain CJ: Zinc metabolism in malabsorption syndromes. J Am Coll Nutr 4:49–64, 1984.
115. McClain C, Soutor C, Zieve L: Zinc deficiency: A complication of Crohn's disease. Gastroenterology 78:272–279, 1980.
116. Clinical nutrition case: Zinc deficiency in Crohn's disease. Nutr Rev 40:109–112, 1982.
117. McClain CJ, Su LC, Gilbert H, Cameron D: Zinc-deficiency-induced retinal dysfunction in Crohn's disease. Dig Dis Sci 28:85–87, 1983.

118. Crofton RW, Glover SC, Ewen SWB, et al.: Zinc absorption in celiac disease and dermatitis herpetiformis: A test of small intestinal function. Am J Clin Nutr 38:706–712, 1983.
119. Naveh Y, Lightman A, Zinder O: A prospective study of serum zinc concentration in children with celiac disease. J Pediatr 102:734–736, 1983.
120. Jones PE, Peters TJ: Oral zinc supplements in non-responsive coeliac syndrome: Effect of jejunal morphology, enterocyte production, and brush border disaccharidase activities. Gut 22:194–198, 1981.
121. Blendis LM, Ampil M, Wilson DR, et al.: The importance of dietary protein in the zinc deficiency of uremia. Am J Clin Nutr 34:2658–2661, 1981.
122. Mahajan SK, Prasad AS, Rabbani P, et al.: Zinc deficiency: A reversible complication of uremia. Am J Clin Nutr 36:1177–1183, 1982.
123. Mahajan SK, Hamburger RJ, Flamenbaum W, et al.: Effect of zinc supplementation on hyperprolactinaemia in uremic men. Lancet 2:750–751, 1985.
124. Reimold EW: Changes in zinc metabolism during the course of the nephrotic syndrome. Am J Dis Child 134:46–50, 1980.
125. Sullivan JF, Jetton MM, Burch RE: A zinc tolerance test. J Lab Clin Med 93:485–492, 1979.
126. Dinsmore WW, Callender ME, McMaster D, Love AHG: The absorption of zinc from a standardized meal in alcoholics and in normal volunteers. Am J Clin Nutr 42:688–693, 1985.
127. Goldiner WH, Hamilton BP, Hyman PD, Russell RM: Effect of the administration of zinc sulfate on hypogonadism and impotence in patients with chronic stable hepatic cirrhosis. J Am Coll Nutr 2:157–162, 1983.
128. Morrison SA, Russell RM, Carney EA, Oaks EV: Zinc deficiency: A cause of abnormal dark adaptation in cirrhotics. Am J Clin Nutr 31:276–281, 1978.
129. Flynn A, Miller SI, Martier SS, et al.: Zinc status of pregnant alcoholic women: A determinant of fetal outcome. Lancet 1:572–574, 1981
130. Ghishan FK, Patwardham R, Greene HL: Fetal alcohol syndrome: Inhibition of placental transport of zinc as a potential mechanism for fetal growth retardation. J Lab Clin Med 100:45, 1982.
131. Zidenberg-Cherr S, Rosenbaum J, Keen CL: Reduced placental transfer of Zn during organogenesis: A mechanism underlying fetal alcohol syndrome. Fed Proc 46:598A, 1987.
132. Kinlaw WB, Levine AS, Morley JE, et al.: Abnormal zinc metabolism in type II diabetes mellitus. Am J Med 75:273–277, 1983.
133. Heise C, King JC, Costa F, Fitzmiller J: Renal function, glycemic control and urinary zinc excretion in insulin-dependent diabetic women. Fed Proc 46:1007A, 1987.
134. Niewoehner CB, Allen JI, Boosalis M, et al.: The role of zinc supplementation in type II diabetes mellitus. Am J Med 81:63–68, 1986.
135. Keeling, PWN, Thompson RPH: Tissue zinc status in patients with chronic liver disease In: *Zinc Deficiency in Human Subjects,* AS Prasad, AO Çavdar, GJ Brewer, PJ Aggett (eds). New York: Alan R. Liss, 1983, pp 235–254.
136. Schölmerich J. Krauss E, Wietholtz H, et al.: Bioavailability of zinc from zinc-histidine complexes. II. Studies on patients with liver cirrhosis and the influence of the time of application. Am J Clin Nutr 45:1487–1491, 1987.
137. Reding P, Duchateau J, Bataille C: Oral zinc supplementation improves hepatic encephalopathy. Lancet 2:493–495, 1984.
138. Prasad AS: Zinc deficiency in human subjects. In: *Zinc Deficiency in Human Subjects,* AS Prasad, AO Çavdar, GJ Brewer, PJ Aggett (eds). New York: Alan R. Liss, 1983, pp 1–33.

139. Haralambie G: Serum zinc in athletes in training. Int J Sports Med 2:135–138, 1981.
140. Walravens PA: Zinc nutrition in infants and children: In: *Clinical, Biochemical, and Nutritional Aspects of Trace Elements,* AS Prasad (ed). New York: Alan R. Liss, 1982, pp. 129–144.
141. Krieger I, Evans GW, Zelkowitz PS: Zinc dependency as a cause of chronic diarrhea in variant acrodermatitis enteropathica. Pediatrics 69:773–777, 1982.
142. Brenton DP, Jackson MJ, Young A: Two pregnancies in a patient with acrodermatitis enteropathica treated with zinc sulphate. Lancet 2:500–502, 1981.
143. Brewer GJ, Hill GM, Prasad AS, et al.: Oral zinc therapy for Wilson's disease. Ann Intern Med 99:314–320, 1983.
144. Hoogenraad, TU, Van den Hamer CJA: 3 years of continuous oral zinc therapy in 4 patients with Wilson's disease. Acta Neurol Scand 67:356–364, 1983.
145. Hoogenraad TU, Van den Hamer CJA, Van Hattum J: Effective treatment of Wilson's disease with oral zinc sulphate: Two case reports. Br Med J 289:273–276, 1984.
146. Van den Hamer CJA, Hoogenraad TU: ^{64}Cu loading tests for monitoring zinc therapy in Wilson's disease. Trace Elem Med 1:84–87, 1984.
147. Van den Hamer CJA, Hoogenraad TU, Klompjan ERK: Persistence of the antagonistic influence of zinc on copper absorption after cessation of zinc supplementation for more than 5 days. Trace Elem Med 1:88–90, 1984.
148. Hill GM, Brewer GJ, Hogikyan ND, Stellini MA: The effect of depot parenteral zinc on copper metabolism in the rat. J Nutr 114:2283–2291, 1984.
149. Prasad AS: Clinical disorders of zinc deficiency. In: *Clinical Applications of Recent Advances in Zinc Metabolism,* AS Prasad, IE Dreosti, BS Hetzel (eds). New York: Alan R. Liss, 1982, pp 89–119.
150. Brewer GJ, Hill GM, Dick RD, et al.: Interactions of trace elements: Clinical significance. J Am Coll Nutr 4:33–38, 1985.
151. Brewer GJ: Zinc supplementation treatment of growth retardation and hypogonadism in sickle cell patients (editorial). Am J Hematol 10:195–198, 1981.
152. Subramanian L, Prasad AS: Zinc deficiency in a patient with sickle cell disease. Nutr Rev 41:217–219, 1983.
153. Prasad AS, Brewer GJ, Shoomaker EB, Rabbani P: Hypocupremia induced by zinc therapy in adults. JAMA 240:2166–2168, 1978.
154. Dogru U, Arcasoy A, Çavdar AO: Zinc levels of plasma, erythrocyte, hair and urine in homozygote beta-thalassemia. Acta Haematol 62:41–44, 1979.
155. Arcasoy A, Çavdar AO, Ertug H, Gurpinar F: Zinc treatment in homozygous beta-thalassemia (a preliminary study). In: *Zinc Deficiency in Human Subjects,* AS Prasad, AO Çavdar, GJ Brewer, PJ Aggett (eds). New York: Alan R. Liss, 1983, pp 107–116.
156. Anon.: Correction of impaired immunity in Down's syndrome by zinc. Nutr Rev 38:365–367, 1980.
157. Björksten B, Back O, Gustavson KH, et al.: Zinc and immune function in Down's syndrome. Acta Paediatr Scand 69:183–187, 1980.
158. Butterworth BE, Grunert RR, Korant BD, et al.: Replication of rhinoviruses. Arch Virol 51:169–189, 1976.
159. Butterworth BE, Korant BD: Characterization of the large picornaviral polypeptides produced in the presence of zinc ion. J Virol 14:282–291, 1974.
160. Fridlender B, Chejanovsky N, Becker Y: Selective inhibition of herpes simplex virus type I DNA polymerase by zinc ions. Virology 84:551–554, 1978.
161. Gupta P, Rapp F: Effect of zinc ions on synthesis of herpes simplex type 2-induced polypeptides. Proc Soc Exp Biol Med 152:455–458, 1976.

162. Korant BD, Butterworth BE: Inhibition by zinc of rhinovirus protein cleavage: Interaction of zinc with capsid polypeptides. J Virol 18:298–306, 1976.
163. Korant BD, Kauer JC, Butterworth BE: Zinc ions inhibit replications of rhinoviruses. Nature 248:588–590, 1974.
164. Eby GA, Davis DR, Halcomb WW: Reduction in duration of common colds by zinc gluconate lozenges in a double-blind study. Antimicrob Agents Chemother 25:20–24, 1984.
165. Husain SL: Oral zinc sulphate in leg ulcers. Lancet 1:1069–1071, 1969.
166. Serjeant GR, Galloway RE, Gueri MC: Oral zinc sulphate in sickle cell ulcers. Lancet 2:891–893, 1970.
167. Halböök T, Lanner E: Serum zinc and healing of leg ulcers. Lancet 2:780–782, 1972.
168. Greaves MW, Ives FA: Double blind trial of zinc sulphate in the treatment of chronic venous leg ulceration. Br J Dermatol 87:632–633, 1972.
169. Clayton RJ: Double blind trial of oral zinc sulphate in patients with leg ulcers. Br J Clin Pract 26:368–370, 1972.
170. Myers MB, Cherry G: Zinc and the healing of chronic leg ulcers. Am J Surg 120:77–81, 1970.
171. Van Rij AM: Zinc supplements in surgery. In: *Clinical, Biochemical, and Nutritional Aspects of Trace Elements,* AS Prasad (ed). New York: Alan R. Liss, 1982, pp 259–276.
172. Expert Panel: Guidelines for essential trace element preparations for parenteral use. JAMA 241:2052–2054, 1979.
173. Svenson K, Hallgren R, Johansson E, Lindh U: Reduced zinc in peripheral blood cells from patients with inflammatory connective tissue diseases. Inflammation 9:189–199, 1985.
174. Simkin PA: Oral zinc sulphate in rheumatoid arthritis. Lancet 2:539–542, 1976.
175. Clemmensen OJ, Siggaard-Andersen J, Worm AM, et al.: Psoriatic arthritis treated with oral zinc sulphate. Br J Dermatol 103:411–415, 1980.
176. Kretzschmar P, Brewer GJ, Walker SE: Depot-zinc therapy of systemic lupus erythematosus in B/W mice. Proc Soc Exp Biol Med 168:301–305, 1981.
177. Leopold IH: Zinc deficiency and visual impairment. Am J Ophthalmol 85:871–875, 1978.
178. Sturtevant FM: Zinc deficiency, acrodermatitis, enteropathica, optic atrophy, subacute myelo-optic neuropathy, and 5,7-dihalo-8-quinolinols. Pediatrics 65:610–613, 1980.
179. Chowdhury BA, Chandra BK: Alteration of immune functions in cigarette smokers is corrected by zinc therapy. J Am Coll Nutr 6:443A, 1987.
180. Fong, LY, Sivak A, Newberne PM: Zinc deficiency and methylbenzylnitrosamine-induced esophegeal cancer in rats. J Natl Cancer Inst 61:145–150, 1978.
181. Gabrial GN, Schrager TF, Newberne PM: Zinc deficiency, alcohol and a retinoid: Association with esophageal cancer in rats. J Natl Cancer Inst 68:785–789, 1982.
182. Mobarhan S, Russell RM, Newberne PM, Ahmed SB: The effect of zinc deficiency and alcohol feeding on esophageal epithelium of rats. Nutr Rep Int 29:639–645, 1984.
183. Allen JI, Bell E, Boosalis MG, et al.: Association between urinary zinc excretion and lymphocyte dysfunction in patients with lung cancer. Am J Med 79:209–215, 1985.
184. Tobey RA, Enger MD, Griffith JK, Hildebrand CE: Zinc-induced resistance to alkylating agent toxicity. Cancer Res 42:2980–2984, 1982.
185. Murphy JV: Intoxication following ingestion of elemental zinc. JAMA 212:2119, 1970.
186. Schölmerich J, Freudemann A, Köttgen E, et al.: Bioavailability of zinc from zinc-histidine complexes. I. Comparison with zinc sulfate in healthy men. Am J Clin Nutr 45:1480–1486, 1987.
187. Aamondt RL, Rumble WF, Babcock AK, et al.: Effects of oral zinc loading on zinc metabolism in humans—I. Experimental studies. Metabolism 31:326–334, 1982.
188. Babcock AK, Henkin RI, Aamondt RL, et al.: Effects of oral zinc loading on zinc metabolism in humans. II: In vivo kinetics. Metabolism 31:335–347, 1982.

189. Freeland-Graves JH, Friedman BJ, Han WH, et al.: Effect of zinc supplementation on plasma high-density lipoprotein cholesterol and zinc. Am J Clin Nutr 35:988–992, 1982.
190. Bos LP, Vam Vloten WA, Smit AFD, Nube M: Zinc deficiency with skin lesions as seen in acrodermatitis enteropathica and intoxication with zinc during parenteral nutrition. Neth J Med 20:263, 1977.
191. Brocks A, Reid H, Glazer G: Acute intravenous zinc poisoning. Br Med J 1:1390, 1977.
192. Fischer PWF, Giroux A, L'Abbé MR: Effects of zinc on mucosal copper binding and on the kinetics of copper absorption. J Nutr 113:462–469, 1983.
193. Fischer PWF, Giroux A, L'Abbé MR: Effect of zinc supplementation on copper status in adult man. Am J Clin Nutr 40:743–746, 1984.
194. Colin MA, Taper LJ, Ritchey SJ: Effect of dietary zinc and protein levels on the utilization of zinc and copper by adult females. J Nutr 113:1480-1488, 1983.
195. Burke DM, DeMicco FJ, Taper LJ, Ritchey SJ: Copper and zinc utilization in elderly adults. J Gerontol 36:558–563, 1981.
196. Klevay LM, Inman L, Johnson LK, et al.: Increased cholesterol in plasma in a young man during experimental copper depletion. Metabolism 33:1112–1118, 1984.
197. Hooper PL, Visconti L, Garry PJ, Johnson GE: Zinc lowers high-density lipoprotein-cholesterol levels. JAMA 244:1960–1961, 1980.
198. Goodwin JS, Hunt WC, Hooper P, Garry PJ: Relationship between zinc intake, physical activity, and blood levels of high-density lipoprotein cholesterol in a healthy elderly population. Metabolism 34:519–523, 1985.
199. Pachotikarn C, Madeiros DM, Windham F: Effect of oral zinc supplementation upon plasma lipids, blood pressure, and other variables in young adult white males. Nutr Rep Int 32:373–382, 1985.
200. Solomons NW, Jacob RA: Studies on the bioavailability of zinc in humans: Effects of heme and nonheme iron on the absorption of zinc. Am J Clin Nutr 34:475–482, 1981.
201. Valberg LS, Flanagan PR, Chamberlain MJ: Effects of iron, tin, and copper on zinc absorption in humans. Am J Clin Nutr 40:536–541, 1984.
202. Sandström B, Davidsson L, Cederblad A, Lönnerdal B: Oral iron, dietary ligands and zinc absorption. J Nutr 115:411–414, 1985.
203. Yip R, Reeves JD, Lönnerdal B, et al.: Does iron supplementation compromise zinc nutrition in healthy infants? Am J Clin Nutr 42:683–687, 1985.
204. Spencer H, Kramer L, Osis D: Zinc balances in humans during different intakes of calcium and phosphorus. In: *Nutritional Bioavailability of Zinc,* GE Inglett (ed). Washington, DC: ACS Symposium Series 210, 1983, pp 223–232.
205. Spencer H, Rubio N, Kramer L, et al.: Effect of zinc supplements on the intestinal absorption of calcium. J Am Coll Nutr 6:47–51, 1987.
206. Prasad AS, Oberleas D, Lei KY, et al.: Effect of oral contraceptive agents on nutrients. Am J Clin Nutr 28:377–384, 1975.
207. King JC: Do women using oral contraceptive agents require extra zinc? J Nutr 117:217–219, 1987.

Appendix

FOOD AND NUTRITION BOARD, NATIONAL ACADEMY OF SCIENCES–NATIONAL RESEARCH COUNCIL
RECOMMENDED DAILY DIETARY ALLOWANCES,[a] REVISED 1980
Designed for the maintenance of good nutrition of practically all healthy people in the U.S.A.

	Age (years)	Weight (kg)	Weight (lb)	Height (cm)	Height (in)	Protein (g)	Fat-Soluble Vitamins Vitamin A (μg RE)[b]	Vitamin D (μg)[c]	Vitamin E (mg α-TE)[d]	Water-Soluble Vitamins Vitamin C (mg)	Thiamin (mg)	Riboflavin (mg)	Niacin (mg NE)[e]	Vitamin B-6 (mg)	Folacin[f] (μg)	Vitamin B-12 (μg)	Minerals* Calcium (mg)	Phosphorus (mg)	Magnesium (mg)	Iron (mg)	Zinc (mg)	Iodine (μg)
Infants	0.0–0.5	6	13	60	24	kg × 2.2	420	10	3	35	0.3	0.4	6	0.3	30	0.5[g]	360	240	50	10	3	40
	0.5–1.0	9	20	71	28	kg × 2.0	400	10	4	35	0.5	0.6	8	0.6	45	1.5	540	360	70	15	5	50
Children	1–3	13	29	90	35	23	400	10	5	45	0.7	0.8	9	0.9	100	2.0	800	800	150	15	10	70
	4–6	20	44	112	44	30	500	10	6	45	0.9	1.0	11	1.3	200	2.5	800	800	200	10	10	90
	7–10	28	62	132	52	34	700	10	7	45	1.2	1.4	16	1.6	300	3.0	800	800	250	10	10	120
Males	11–14	45	99	157	62	45	1000	10	8	50	1.4	1.6	18	1.8	400	3.0	1200	1200	350	18	15	150
	15–18	66	145	176	69	56	1000	10	10	60	1.4	1.7	18	2.0	400	3.0	1200	1200	400	18	15	150
	19–22	70	154	177	70	56	1000	7.5	10	60	1.5	1.7	19	2.2	400	3.0	800	800	350	10	15	150
	23–50	70	154	178	70	56	1000	5	10	60	1.4	1.6	18	2.2	400	3.0	800	800	350	10	15	150
	51+	70	154	178	70	56	1000	5	10	60	1.2	1.4	16	2.2	400	3.0	800	800	350	10	15	150
Females	11–14	46	101	157	62	46	800	10	8	50	1.1	1.3	15	1.8	400	3.0	1200	1200	300	18	15	150
	15–18	55	120	163	64	46	800	10	8	60	1.1	1.3	14	2.0	400	3.0	1200	1200	300	18	15	150
	19–22	55	120	163	64	44	800	7.5	8	60	1.1	1.3	14	2.0	400	3.0	800	800	300	18	15	150
	23–50	55	120	163	64	44	800	5	8	60	1.0	1.2	13	2.0	400	3.0	800	800	300	18	15	150
	51+	55	120	163	64	44	800	5	8	60	1.0	1.2	13	2.0	400	3.0	800	800	300	10	15	150
Pregnant						+30	+200	+5	+2	+20	+0.4	+0.3	+2	+0.6	+400	+1.0	+400	+400	+150	h	+5	+25
Lactating						+20	+400	+5	+3	+40	+0.5	+0.5	+5	+0.5	+100	+1.0	+400	+400	+150	h	+10	+50

[a] The allowances are intended to provide for individual variations among most normal persons as they live in the United States under usual environmental stresses. Diets should be based on a variety of common foods in order to provide other nutrients for which human requirements have been less well defined.

[b] Retinol equivalents. 1 retinol equivalent = 1 μg retinol or 6 μg β carotene.

[c] As cholecalciferol. 10 μg cholecalciferol = 400 IU of vitamin D.

[d] α-Tocopherol equivalents. 1 mg d-α tocopherol = α-TE.

[e] 1 NE (niacin equivalent) is equal to 1 mg of niacin or 60 mg of dietary tryptophan.

[f] The folacin allowances refer to dietary sources as determined by *Lactobacillus casei* assay after treatment with enzymes (conjugases) to make polyglutamyl forms of the vitamin available to the test organism.

[g] The recommended dietary allowance for vitamin B-12 in infants is based on average concentration of the vitamin in human milk. The allowances after weaning are based on energy intake (as recommended by the American Academy of Pediatrics) and consideration of other factors, such as intestinal absorption.

[h] The increased requirement during pregnancy cannot be met by the iron content of habitual American diets nor by the existing iron stores of many women; therefore the use of 30–60 mg of supplemental iron is recommended. Iron needs during lactation are not substantially different from those of nonpregnant women, but continued supplementation of the mother for 2–3 months after parturition is advisable in order to replenish stores depleted by pregnancy.

(Note: The above table is reproduced from the publication by the Food and Nutrition Board: *Recommended Dietary Allowances*, 9th ed. Washington, DC: National Academy of Sciences, 1980.)

*RDAs have not been issued for other minerals. The suggested "safe and adequate" adult intakes of chromium, copper, selenium are, respectively, 50–200 mcg/d, 2–3 mg/d, and 50–200 mcg/d.

Index

Abetalipoproteinemia
  vitamin E, 71
  vitamin K, 99
Abruptio placentae
  folic acid and, 165
  vitamin A, 22
  vitamin C, 218
Acetyl-CoA carboxylase, 193
Acetylcoenzyme A, 103
Aciduria
  organic, riboflavin and, 124–125
  xanthuremia, and vitamin B-6, 152
Acne vulgaris
  selenium, 277–278
  vitamin A/Retin A, 12–13
  zinc, 296–297
Acrodermatitis enteropathica, 287, 292, 300–301
Acute phase reaction, copper, 262
Acyl carrier protein, 189, 190
Acyl CoA-dehydrogenase, 125
Adenosylcobalamin deficiency, 184
S-Adenosyl-methionine, 167
Adhesions, surgical, vitamin E, 75–76
Adriamycin®. *See* Doxorubicin
Adult respiratory distress syndrome, vitamin E, 74
Aging
  free radical theory, 219–220
  photoaging and vitamin A acid, 13–14
    tretinoin (all-*trans*-retinoic acid, Retin A®), 13–14
  vitamin D and bone loss, 39–43
  see also Elderly
Alagille syndrome, 70
Alcohol
  vitamin A, 23
  vitamin C, 231
Alcoholics/alcoholism
  folic acid, 167
  niacin and niacinamide, 131, 134

pantothenic acid, 191
riboflavin, 123
selenium, 276
thiamin, 106–108
vitamin A and cirrhosis, 5, 10–11
  dark adaptation, 10
vitamin B-6, 150
vitamin C, 218, 221–222
vitamin D, 46
vitamin E, 62, 77
zinc, 299
  fetal alcohol syndrome, 299
  and RBP, 10–11, 21
Alkaline phosphatase and vitamin D, 34, 39, 41, 51
Allergies and vitamin C, 224
Allithiamin, 112
Amblyopia, tobacco, vitamin B-12 and, 182
Amine oxidases and copper, 255
Amiodarone and vitamin B-6, 155
Amitryptyline and vitamin B-6, 155
Amphetamines and vitamin B-6, 155
Amyotrophic lateral sclerosis and thiamin, 112
Anaphylaxis
  thiamin, 105, 113
  vitamin K, 96, 100
Anemia
  macrocytic, 167
  megaloblastic, 112, 171, 180, 182–185
  pernicious, and vitamin B-12, 182–183
    cf. folic acid deficiency, 171, 183
  riboflavin and, in pregnancy/lactation, 121
  sickle-cell
    riboflavin, 123
    vitamin B-6, 151
    vitamin C, 224
    vitamin E, 74
    zinc in, 302–303
  sideroblastic, 150

vitamin B-6, 151
Anorexia nervosa and zinc, 292, 296, 298
Antacids and folate, 171
Antibiotics, cephalosporin, and vitamin K, 101
Antiinflammatory action, vitamin E, 74
Antimalarials and folate, 171
Antimutagen
   beta-carotene, 6
   vitamin C, 225-227, 232-233
   vitamin E, 79
Antioxidant functions
   copper, 255-256
   selenium, 270, 273, 276
   vitamin A, 6, 12, 17
   vitamin C, 202-203, 205, 211-212, 220-221
   vitamin E, 60, 61, 62, 68, 71, 75-78, 80
alpha-1-Antitrypsin deficiency, cholestasis, 98
Apgar scores, vitamin B-6, 142
Aquacobalamin (vitamin B-12b), 179
Aquamephyton®, 96
Aquasol A®, 9, 15
Arachidonate and selenium, 275
Arachidonic acid, 72
Arrhythmias, copper and, 261-262
Arsenic and selenium, 279
Arteriosclerosis, folic acid, 168
Arthritis
   vitamin E, 61, 74
   zinc, 305-306
Ascorbic acid. *See* Vitamin C *entries*
Ascorbicap®, 226
Aspirin/salicylates, 72-73, 132, 133
   vitamin C, 231
   vitamin E, 73
   vitamin K, 101
Asthma
   vitamin B-6, 150-151
   vitamin C, 224
Atherosclerosis
   selenium, 274-275
   vitamin A, 20
   vitamin C, 214, 217
   vitamin E, 73, 76
Athletic training. *See* Exercise/athletic training

Autism and vitamin B-6, 153
Autoimmune disorders, vitamin E, 76-77
Avidin-induced deficiency, biotin, 194, 195, 197, 199
   raw egg white, 194, 195, 197
6-Azauridine and vitamin B-6, 155

Barbiturates and folate, 171
Bedsores, vitamin C, 222-223
Benign breast disease, vitamin E, 68-69
Benzoyl peroxideand tretinoin, 13
Beriberi and thiamin, 104, 107
Berocca PN®, 9
Biliary atresia
   riboflavin, 121
   vitamin E, 70
Bioflavonoids, 209
Biotin, 193-199
   absorption, metabolism, and excretion, 194-195
   alopecia, 194, 195
   avidin-induced deficiency, 194, 195, 197, 199
   biotinidase deficiency, 197
   carboxylase deficiency, multiple, 196
   clinical studies, 195-198
   -dependent enzymes, 193-194
   dermatitis, exfoliative, 194
   diabetes mellitus, 198
   dosage forms, 195
   hyperlipidemia, 197-198
   interactions, 199
      insulin, 198
   Leiner's disease, 196
   seborrheic dermatitis, 196
   structure, 194
   sudden infant death syndrome, 198
   total parenteral nutrition, 194, 195
   toxicity and side effects, 198
Biotinidase deficiency, 197
Blood, fecal occult, and vitamin C, 233
Bone loss, vitamin D, 39-43
Branched-chain alpha-ketoacid dehydrogenase deficiency, 111
Breast cancer
   vitamin A, 17, 19
   vitamin D, 49
   vitamin E, 79
Breast disease, benign, vitamin E, 68-69

Breast-fed infants
 riboflavin, 120
 vitamin D, 38–39
 vitamin K, 97, 98
 zinc, 290–291
Breast implants, vitamin E, 75
Bronchopulmonary dysplasia
 vitamin A, 9
 vitamin E, 67
Burning feet syndrome, pantothenic acid, 191
Burns, 262
 vitamin C, 223, 224
 vitamin E, 80
 zinc, 305
Bursitis, 152
Butylated hydroxytoluene and vitamin A, 22

Cadmium
 and selenium, 279
 and vitamin C, 232
Caffeine and vitamin A, 23
Calcifediol (25-hydroxyvitamin D), 32–41, 44, 48, 50
Calciferol®, 36
Calcitonin and vitamin D, 32
Calcitriol (1,25-dihydroxyvitamin D), 32–45, 47–52
Calcium
 chelation, vitamin C and, 229–230
 and vitamin D, 36, 39–42, 44–45, 49
 and zinc, 310
Calcium oxalate stone formation
 vitamin B-6 and, 146–148
 vitamin C and, 228–229
Calderol®, 36
Cancer, 16–19
 beta-carotene (provitamin A), 16–18
 folic acid, 169–170
 selenium, 273–274, 278, 279
 vitamin A, 16–19
 vitamin B-6, 151–152
 vitamin C, 224–227
 vitamin E, 78–80
 zinc, 307
 see also specific sites and types
Candida albicans and selenium, 270
Carboxylase deficiency, multiple, 196
gamma-Carboxylation, vitamin K, 94

Carcinoma in situ, uterocervical
 folic acid, 170
 vitamin A, 17
Cardiopulmonary bypass, vitamin E, 75–76
Cardiovascular disease
 selenium, 274–275
 vitamin C, 211–215
 vitamin E, 72–73
beta-Carotene (provitamin A), 5, 6, 8, 19, 166
 antimutagen, 6
 cancer, 16–19
 oxygen, singlet, quenching, 6, 16, 17
 pharmacokinetics, 7
 side effects, 21–22
 structural formula, 4
Carotenes, carotenoids, 3–6
Carpal tunnel syndrome, vitamin B-6, 149
 related disorders, 152
Cataracts
 riboflavin, 122–123
 vitamin C, 221
Celiac sprue, 297, 298; *see also* Malabsorption
Cephalosporin antibiotics and vitamin K, 101
Ceruloplasmin (ferroxidase I), 232, 255–259, 262, 264, 302
Cervical carcinoma in situ
 folic acid, 170
 vitamin A, 17
Cervical dysplasia
 folic acid, 170
 vitamin C, 226–227
Cheilosis, 119
Chinese restaurant syndrome and vitamin B-6, 152
Chlorpromazine and riboflavin, 125
Chlorthiazides and zinc, 310
Cholecalciferol (vitamin D-3), 31, 32, 34, 35, 45
Cholestasis
 alpha-1-antitrypsin deficiency, 98
 chronic, and vitamin E, 70, 71
 neonatal, vitamin A, 9
Cholesterol
 HDL
  chromium, 251
  copper, 260

vitamin B-6, 148
vitamin C, 212, 214, 219
vitamin E, 72
zinc, 309
LDL, 61, 260
copper, 60
receptors, 215
vitamin C, 212, 215
vitamin E, 61
niacin and niacinamide (vitamin B-3), 132, 133
serum, vitamin A, 20
vitamin C and, 212–214, 215, 219
vitamin E, 61, 72
see also Hypercholesterolemia
Cholesterol 7-alpha-hydroxylase, cytochrome P-450 dependent, 212
Cholestyramine and folate, 171
Choline acetylation, 191
Choline acetyltransferase, 256, 260
Chromium, 247–253, 277
absorption, metabolism, excretion, 248–249
clinical studies, 250–253
diabetes, 251, 252
dosage forms, 250
elderly, 252
exercise, 252–253
glucose tolerance factor, 247, 249, 250
hyperglycemia, glucose intolerance, and insulin resistance, 248–251, 253
interactions, 253
lipids, blood, 251–253
malnutrition, protein–calorie, 248
pregnancy, 252
total parenteral nutrition, 247, 248
toxicity and side effects, 253
Chronic disease and selenium, 276
Chylomicrons, 61, 95
Cigarette smoking. See Smokers
Cleft palate, folic acid, 165
Cobalt and vitamin C, 232
Cochlear deafness, vitamin D, 47
Coenzyme A, 189–191
Colchicine and vitamin B-12, 186
Cold sores, vitamin C, 209
Colestipol, 132–133
and folate, 171

Colitis
chronic granulomatous, 191
ulcerative, 99, 191
Collagen
thiamin and wound healing, 113
vitamin C, 222
Collagenase, vitamin A, 19–20
Common cold
vitamin C, 208–209
zinc, 303–304, 308
Complement, vitamin C, 210
Congenital dependencies, vitamin B-6, 152
Congenital methemoglobinemia, riboflavin and ascorbic acid, 124
Contraceptives. See Oral contraceptives
Conversion disorders, 110
Copper, 255–264
absorption, metabolism, and excretion, 257–258
acute phase reaction, 262
clinical studies, 259–263
deficiency, 256–257
dosage forms, 258
elderly, 260
enzymes, 255–256
familial benign copper deficiency, 262
glucose tolerance, abnormal, 260–261
diabetes, 261
heartbeat irregularities (arrhythmias), 261–262
hypercholesterolemia, 256, 260
hypertension, 263
Indian childhood cirrhosis, 258
infants, 259
interactions, 263–264
lactation, 259–260
Menkes' syndrome, 258, 262
pregnancy, 259–260
and selenium, 279
toxicity, 263
and vitamin C, 232
Wilson's disease. See Wilson's disease
and zinc, 308–310
Copper-reduction glucose test and vitamin C, 233
Coumarin and vitamin K, 100
Crohn's disease, 71, 99
vitamin A, 11–12

vitamin C, 223
zinc, 297, 298
Cyanocobalamin. *See* Vitamin B-12 (cyanocobalamin)
Cycloserine, vitamin B-6, 154
Cystathioninuria, vitamin B-6, 152
Cystic fibrosis
  vitamin A, 11
  vitamin D, 45
  vitamin E, 70
  vitamin K, 99
Cystic mastitis, vitamin E, 68–69
Cytochrome oxidase, copper, 256
Cytochrome P450-dependent system, 212, 286
Cytosolic superoxide dismutase
  copper, 256–257
  zinc, 256, 286, 309

Darier's disease, vitamin A, 15
Dark adaptation, vitamin A, 10
7-Dehydrocholesterol, 34–35
Dementia, folic acid, 167
Depression
  vitamin B-3 (niacin and niacinamide), 136
  vitamin B-6, 153
  vitamin B-12, 185
  vitamin E, 70
De Quervain's disease, 152
Dermatitis
  exfoliative, biotin, 194
  herpetiformis, 297, 298
Dermatoses, ichthyosiform, vitamin A and, 14
Dexpanthenol, paralytic ileus, 191
DHT®, 36
Diabetes mellitus
  biotin, 198
  chromium, 251, 252
  copper, 261
  mononuclear and polymorphonuclear leukocytes and vitamin C, 216
  niacin and niacinamide (vitamin B-3), 134–135
  pantothenic acid, 191
  retinopathy, and selenium, 277
  riboflavin, 122
  thiamin (vitamin B-1), 108, 112
  vitamin B-6, 146

vitamin C, 213, 216–217
vitamin E, 80
zinc, 299–300
Dialysis, renal
  riboflavin, 124
  vitamin A contraindicated, 21
  vitamin C toxicity, 229
  vitamin D, 44
  zinc, 98
Diamine oxidase, vitamin B-6, 144
Diet, vegetarian. *See* Vegetarian diet
Digitalis, 50
Dihydrofolate reductase, 164
  deficiency, 169
Dihydrotachysterol, 36, 37, 44, 50
Diiodohydroxyquin and zinc, 310
1,2-Dimethylbenzanthracene (DMBA), 79, 225
1,2-Dimethylhydrazine (DMH), 225
Disulfiram and zinc, 310
DMBA, 79, 225
DNA synthesis, folic acid, 163, 165
L-DOPA and vitamin B-6, 154
Dopamine-beta-hydroxylase, copper, 255
Down's syndrome
  vitamin A, 9–10
  zinc in, 303
Doxorubicin (Adriamycin®), micronutrient interactions, 79–80, 227, 263
Dupuytren's contracture, 152
  vitamin E, 80

Edema of feet and ankles, 152
E-Ferol syndrome, 81
Egg white, raw, and avidin, 194, 195, 197
Elbows, painful, 152
Elderly
  chromium, 252
  copper, 260
  folic acid, 166–167
  riboflavin, 122
  selenium, 275–276, 278
  thiamin (vitamin B-1), 108–109
  vitamin B-6, 145
  vitamin B-12, 184
  vitamin C, 218–221
  vitamin C deficiency and vitamin K, 98–99
  vitamin E, 76

zinc, 295–296
   see also Aging
Encephalopathy, subacute necrotizing, thiamin, 111–112
Endothelial lipoprotein lipase, 256
Enterocolitis, necrotizing
   vitamin A, 8
   vitamin E, 64, 66, 81
Enteropathy, gluten, 297, 298
Ephynal, 64, 66, 67, 70, 71
Epilepsy, 51
Ergocalciferol (vitamin D-2), 31, 32, 34, 36, 40
Erythrocyte
   glutathione reductase (EGR), 120, 121, 123
   transketolase, thiamin (vitamin B-1), 108–109, 111, 112
Erythropoietic protoporphyria, vitamin A, 15
Ethambutol and zinc, 310
Ethinylestradiol, vitamin C, 231
Exercise/athletic training
   chromium, 252–253
   niacin and niacinamide (vitamin B-3), 135
   riboflavin, 123–124
   thiamin (vitamin B-1), 111
   vitamin A, 20
   vitamin E, 78
   zinc, 300

Familial benign copper deficiency, 262
Fatigue, chronic, vitamin B-12, 185
Fatty acids, omega-6, 62
Febrile lymphadenopathy, recurrent, thiamin (vitamin B-1), 112
Fecal occult blood and vitamin C, 233
Feet
   burning, 191
   edema, 152
Ferroxidase II, copper, 255
Fetal alcohol syndrome, zinc, 299
Fibrinolysis, niacin and niacinamide (vitamin B-3), 135
Fibrinolytic activity, vitamin C, 211–212
Fibroplasia, retrolental, vitamin E, 64–66
Fibrositis, vitamin E, 80
Finger, trigger, 152
Fish oil, enhanced requirement for vitamin E, 60

Flavin adenine dinucleotide (FAD), 117–119, 124
Fluphenazine, vitamin C, 231
Folate compounds, 161–162
Folic acid, 161–172, 185–186, 221
   absorption, metabolism, and excretion, 164
   alcoholism, 167
   cancer, 169–170
   clinical studies, 164–170
   dietary sources, 163
   DNA synthesis, 163, 165
   dosage forms, 164
   elderly, 166–167
   gastrointestinal disease, 168
   genetic disorders, 169
   hemolytic anemias, 169
   infants, 166, 168
   infertility, 166
   interactions, 171–172
   neurologic/neuromuscular disorders, 167–168
   neutrophil hypersegmentation, 163
   postmenopausal homocystinemia, 168–169
   pregnancy, 163–166, 168
   structural formula, 162
   toxicity, 171
   and zinc, 295, 310
Follicle stimulating hormone, 68
Folyl polyglutamates, 164
N-Formiminoglutamic acid (FIGLU), 162
Free radical(s)
   theory of aging, 219–220
   vitamin A/beta-carotene, 6
   vitamin C, 202–203, 220
   vitamin E, 60–61, 65, 75, 77
   zinc, 286
   see also Antioxidant functions; Oxygen
Frostbite, vitamin C, 224
Fructose-1,6-diphosphatase deficiency, folic acid, 169
Furunculosis, recurrent, zinc, 296–297

Galactose intolerance and cataracts, 122–123
Gallstones, 212
Gastrectomy, vitamin B-12, 183–184
Gastrointestinal abnormalities/disease
   folic acid, 168

vitamin B-12, 180
Genetic dependency, vitamin K, 99
Genetic disorders
  biotin, 196–197
  folic acid, 169
  niacin, 135
  riboflavin, 124–125
  thiamin (vitamin B-1), 111–112
  vitamin B-6, 152
  vitamin B-12, 184–185
  zinc, 300–303
Geophagia, 297
Glaucoma
  thiamin, 110
  vitamin A, 12
  vitamin C, 12
  vitamin E, 12
Glucocorticoids, 51–52
  vitamin A and, interactions, 23
Glucose intolerance, chromium, 248–251, 253
Glucose-6-phosphate dehydrogenase (G6PD) deficiency, 230
  vitamin E, 73–74
Glucose test, copper-reduction, and vitamin C, 233
Glucose tolerance, abnormal, copper, 260–261
  diabetes, 261
Glucose tolerance factor, chromium, 247, 249, 250
Glutathione peroxidase, Se-dependent (GSHPx), 269–278
Glutathione reductase, 269
  erythrocyte, 120, 121, 123
Gluten enteropathy, 297, 298
Growth hormone, zinc deficiency, 292
Growth retardation in children, zinc, 292–293

Hair zinc, unreliable indicator, 294
Haloperidol, vitamin C, 231
Hartnup's disease, niacin and niacinamide, 135
Heartbeat irregularities, copper, 261–262
Hemodialysis. *See* Dialysis, renal
Hemolytic anemias, folic acid, 169
Hemolytic disorders, vitamin D, 48
Hemorrhagic disease of the newborn, 97

Heparin, vitamin D, interaction with, 52
Hepatic failure, nonalcoholic fulminant, thiamin (vitamin B-1), 108
Hepatitis, infectious, vitamin A, 8
Herpes labialis, vitamin C, 209
Histamine, vitamin C, 208–209, 224
Homocysteinemia, postmenopausal, folic acid, 168–169
Homocystinuria, vitamin B-6, 152
Hydantoin and folate, 171
Hydralazine, 149, 154
Hydroxocobalamin (vitamin B-12a), 179, 182, 184, 186
Hydroxylation, vitamin D, 32–34, 36, 46, 48
Hydroxyl radical
  vitamin C, 202, 203, 220
  vitamin E, 61
  see also Free radical(s)
Hyperbilirubinemia, phototherapy
  riboflavin, 120–121
  vitamin E, 67
Hypercalcemia and vitamin D, 40, 41, 45, 47, 49–50, 52
  digitalis, 50
Hypercalciuria and vitamin D, 41, 45, 49
Hypercholesterolemia
  copper, 256, 260
    HDL- and LDL-cholesterol, 260
  familial, niacin and niacinamide (vitamin B-3), 133
  vitamin C and, 212–214
Hyperemesis gravidarum, vitamin B-6, 143–144
Hyperglycemia, glucose intolerance, and insulin resistance, chromium, 248–251, 253
Hyperkinesis, vitamin B-6, 152, 153
Hyperlipidemia, biotin, 197–198
Hyperoxaluria, 229
  vitamin B-6, 146–148
Hyperphosphatemia, vitamin D, 44
Hyperprolactinemia, zinc deficiency, 298
Hypertension
  copper, 263
  and vitamin B-12, 186
  vitamin C, 215–216
  vitamin D, 48
Hyperthyroidism, riboflavin and, 124

Hypertriglyceridemia, vitamin C, 215
Hypervitaminosis A, 20–22
Hypocalcemia, vitamin D, 37–38, 43, 47
Hypocupremia, zinc, 308
Hypogeusia, zinc deficiency, 292, 298
Hypoparathyroidism, vitamin D, 46
Hypophosphatemia
    vitamin D, 38
    X-linked rickets, 46–47
Hypothrombinemia of infancy, vitamin K, 95–99, 101
Hypotonia in infants, folic acid, 168
Hysterical conversion, 110

Ichthyosiform dermatoses, vitamin A, 14
Ileus, paralytic, 191
Immunoregulatory functions
    selenium, 270, 273
    vitamin A, 5, 17
    vitamin C, 210–211, 220–221, 225
    vitamin D, 33–34
    vitamin E, 76, 79
    zinc, 286–287, 295
Indanedione and vitamin K, 100
Indian childhood cirrhosis, copper, 258
Infants
    breast-fed
        riboflavin, 120
        vitamin D, 38–39
        vitamin K, 97, 98
        zinc, 290–291
    copper, 259
    folic acid, 166, 168
    hypothrombinemia, vitamin K, 95–99, 101
    hypotonia and folic acid, 168
    preterm
        intravenous vitamin E dangerous, 64–66, 81
        riboflavin, 120
        vitamin A, 8–9
        seizures, vitamin B-6, 152
        vitamin E, malabsorption, 70–71
Infections
    vitamin A, 8
    vitamin C and immune function, 210–211, 220

Infectious hepatitis, vitamin A, 8
Infertility
    folic acid, 166
    vitamin C, 224
    vitamin E, 80
Inflammatory disease
    vitamin E, 74
    zinc, 305–306
Insulin
    and biotin, 198
    resistance, chromium, 248–251, 253
    vitamin C, 216, 217
    gamma-Interferon, 33
Interstitial keratitis, vitamin E, 80
Intraventricular hemorrhage, infants, vitamin E, 64, 65, 67, 81
Intrinsic factor, 181–183
Iodochlorhydroxyquin and zinc, 310
Iproniazid
    and vitamin B-6, 154
    and zinc, 310
Iron, 284
    and folate, 172
    and vitamin C, 232
        iron overload, 230
    and zinc, 295, 297, 310
Irritability, vitamin B-6, 152
Ischemic heart disease, vitamin C, 211
Isoalloxazine, 117
Isocarboxazid
    and vitamin B-6, 154
    and zinc, 310
Isoniazid, 149, 154, 155
    niacin and niacinamide (vitamin B-3), 136
Isonicotinylhydrazide and vitamin B-6, 154

Joints, stiffness in, 152

Kaschin-Beck disease, 270
Kayser-Fleischer rings, 301
Keratitis, interstitial, vitamin E, 80
Keshan disease, 270
Ketogenic diet, optic neuropathy during, thiamin (vitamin B-1), 109–110
Knees, painful, 152
Konakion®, 96
Korsakoff's psychosis, thiamin, 106

Lactation
    copper, 259–260
    riboflavin, 120–121
    vitamin B-6, 141–143
Lead
    and selenium, 279
    and vitamin C, 232
    and vitamin D, children, 52
Leg cramps/restless legs
    folic acid, 167, 168
    vitamin E, 77
Leg ulcers, zinc, 304–305
Leigh's disease, thiamin (vitamin B-1) in, 111–112
Leiner's disease, biotin, 196
Leukemia, myeloid, vitamin D, 48–49
Leukocytes, mononuclear and polymorphonuclear, diabetes, and vitamin C, 216
Lipid peroxidation, vitamin E (alpha-tocopherol), 75–77
Lipids, blood, chromium, 251–253; see also Cholesterol
Lipoprotein lipase, endothelial, 256
Lipoproteins, 95; see also under Cholesterol
Liver
    disease
        Indian childhood cirrhosis, copper, 258
        nonalcoholic fulminant hepatic failure, 108
        and zinc, 300
    vitamin A storage in, 5–7
    vitamin D, 34
Lumiflavin, 120
Lumisterol, 35
Lupus erythematosus, systemic, 306
    vitamin E, 76–77
Luteinizing hormone, 68
Lysyl oxidase, copper, 255

Macrocytic anemia, 167
Macrocytosis (MCV elevation), 181, 183
Magnesium, postmenopausal bone loss, 42
Malabsorption
    thiamin (vitamin B-1), 108, 109
    vitamin A, 11
    vitamin B-12, 182–184
    vitamin D, 45–46
    vitamin E, 62, 70–72
    vitamin K, 99
    zinc, 297–298
Mammary dysplasia, vitamin E, 68–69
Manic-depression and vitamin C, 232–233
Mannosyl retinyl phosphate, 5
MAO inhibitors and vitamin B-6, 154
Maple syrup urine disease, thiamin, 111
Megaloblastic anemia, 112, 171, 180, 182–185
Melanoma
    vitamin A, 18
    vitamin D, 49
Menadiol, 96, 100, 101
Menadione, 93, 95–97, 99–101
Menaquinone, 93; see also Vitamin K
Ménètrier's disease, folic acid, 168
Menkes' syndrome, copper, 258, 262
Mephyton®, 96, 99
Mercury
    and selenium, 279
    and vitamin C, 232
Metallothionein, 258, 263, 288–289, 301, 309
Metformin and vitamin B-12, 186
Methemoglobinemia, congenital, riboflavin, 124
Methionine synthetase, 181
Methotrexate, 164
3-Methycrotonyl-CoA carboxylase, 194
    deficiency, 196–197
Methylcobalamin, 180, 185
    deficiency, 184
Methylene tetrahydrofolate reductase deficiency, folic acid, 169
Methylmalonyl coenzyme A, 181
2-Methyl-1,4-naphthoquinone, 93
N-Methyl-N′-nitro-N-nitrosoguanidine (MNNG), 225
2-Methyl-3-phytyl-1,4-naphthoquinone, 93
5-Methyltetrahydrofolic acid, 162, 164, 167, 180, 186
Methyl trap phenomenon, 163
Mononuclear leukocytes, diabetes, and vitamin C, 216
Monosodium glutamate and vitamin B-6, 152
Morning sickness, vitamin B-6, 143–144
Muscular dystrophy, vitamin E, 80
Mycobacterium tuberculosis and vitamin D, 34

Myocardial infarction
    niacin and niacinamide (vitamin B-3), 135–136
    selenium, 274

NAD, 129–131, 136
NADH, vitamin C, 203
NADP, 129–131
NADPH, 129, 286
Natural killer cells, zinc and, 306–307
Necrotizing enterocolitis
    vitamin A, 8
    vitamin E, 64, 66, 81
Neomycin and vitamin B-12, 186
Neonatal cholestasis, vitamin A and, 9
Nephritis, vitamin E, 80
Nephrotic syndrome, vitamin D, 43
Neural tube defects, folic acid, 165
Neurologic/neuromuscular disorders
    folic acid, 167–168
    thiamin, 107–108
    vitamin E, 70, 71, 72
Neuronal ceroid lipofuscinosis, 275
Neurotic tension, thiamin, 110–111
Neutrophil(s)
    alkaline phosphatase as indicator of zinc status, 288, 292
    hypersegmentation, 180, 181, 183
        folic acid, 163
        vitamin C, 211
Niacin and niacinamide (vitamin B-3), 129–137
    absorption, metabolism, and excretion, 131
    alcoholism, 131, 134
    clinical studies, 132–136
    deficiency, symptoms, 130–131
    depression, 136
    diabetes mellitus, 134–135
    dietary sources, 131
    dosage forms, 131
    exercise and athletic training, 135
    fibrinolysis, 135
        myocardial infarction, 135–136
    Hartnup's disease, 135
    interactions, 136
    nicotinic acid in hyperlipidemia, 132–134
    pyridine-3-carboxylic acid and pyridine-3-carboxylic acid amide, 129
    schizophrenia, 136
    structures, 130
    toxicity, 136
Nialamide
    and vitamin B-6, 154
    and zinc, 310
Nicotinamide adenine dinucleotide (NAD), 129–131, 136
Nicotinamide adenine dinucleotide phosphate (NADP), 129–131
Nicotinic acid in hyperlipidemia, 132–134
Nitrocobalamin (vitamin B-12c), 179
Nitrofurantoin and selenium, 279
Nitrosamine synthesis inhibition, vitamin C, 224–225, 232
Nitrous oxide and vitamin B-12, 186
NK cells, zinc, 306–307
Nocturnal paralysis, 152

Omega-6 fatty acids, 62
Optic atrophy
    and disk pallor, zinc, 306
    in glaucoma, thiamin (vitamin B-1), 110
    vitamin C, 110
Optic neuropathy during ketogenic diet, thiamin (vitamin B-1), 109–110
Oral contraceptives
    folic acid, 170, 171
    pantothenic acid, 191
    riboflavin, 119, 125
    vitamin A, 23
    vitamin B-6, 144–145
    vitamin C, 231
    zinc, 310
Oral lichen planus, vitamin A, 15
Organic acidurias, 124–125
Osteitis fibrosa, vitamin D, 43
Osteoarthritis, vitamin E, 74
Osteomalacia, vitamin D, 34, 40, 43, 47
Osteopenia, vitamin D, 40
Osteoporosis
    folic acid, 169
    postmenopausal, vitamin D, 40–43
Osteosclerosis, vitamin D, 43
Oxalosis
    vitamin B-6, 146–148
    vitamin C, 228–229
Oxygen
    administration, vitamin E, 64–65, 67, 75

singlet, quenching, beta-carotene (provitamin A), 6, 16, 17
see also Free radical(s)
Oxyquinolines and zinc, 310

Pancreatitis
    vitamin B-12, 184
    vitamin E, 71, 72
Pantotheine, 190
Pantothenic acid, 189–191
    absorption, metabolism, excretion, 190
    clinical studies, 191
    dosage forms, 190
    interactions, 191
    structure, 190
    toxicity, 191
Paralysis, nocturnal, 152
Paralytic ileus, dexpanthenol, 191
Paraquat, 137, 279
Parathyroid hormone and vitamin D, 32, 34,
    36, 39–41, 43, 44, 46–48, 51, 52
Parkinson's disease, 155
    vitamin E, 61
Pellagra, 130, 131, 134
D-Penicillamine, 258, 301
    and vitamin B-6, 155
    and zinc, 310
Periarthritis, shoulder, 152
Periodontal disease, vitamin C, 222
Peritoneal adhesion reduction, vitamin E, 75
Pernicious anemia, vitamin B-12, 182–183
    cf. folic acid deficiency, 171, 183
Peyronie's disease, vitamin E, 80
Phenalazine and vitamin B-6, 154
Phenformin and vitamin B-12, 186
Phenobarbital and vitamin A, 22
Phenothiazines and riboflavin, 125
Phenytoin, 51
    thiamin, 113
Phosphate and vitamin D, 34
4'-Phosphopantetheine, 189
Photosensitivity, vitamin A, 15–16
Phototherapy of hyperbilirubinemia
    riboflavin, 120–121
    vitamin E, 67
Phrynoderma, vitamin E, 77
Phylloquinone, 93, 95–98, 100, 101
Phytonodione, 93
Pica, zinc, 297

Pityriasis rubra pilaris, vitamin A, 14–15
Placental separation, premature, vitamin C,
    217–218
Plasma clotting factors, vitamin K, 93–95
Plasma lecithin:choline acetyltransferase,
    256, 260
Platelet aggregation inhibition
    vitamin B-6, 153
    vitamin C, 212
    vitamin E, 73, 77, 81
Polychlorinated biphenyls, vitamin C, 232
Polymorphonuclear leukocytes, diabetes, and
    vitamin C, 216
Polyposis coli, familial, vitamin C, 226
Porphyria, vitamin E, 76–77
Postmenopausal bone loss, 42
Postmenopausal homocystinemia, folic acid,
    168–169
Postmenopausal osteoporosis, vitamin D and,
    40–43
Potassium chloride and vitamin B-12, 186
Pregnancy
    chromium, 252
    copper, 259–260
    folic acid, 163–166, 168
    riboflavin, 121–122
    thiamin (vitamin B-1), 109
    vitamin B-6, 141–143
        and toxemia, 144
    vitamin C, 217–218
    vitamin D, 51
    vitamin E, 68
    zinc, 293–295
Premenstrual syndrome
    vitamin B-6, 144
    vitamin E, 69–70
Pressure sores, vitamin C, 222–223
Primidone and folate, 171
Probucol, 133, 134
Procarbazine and vitamin B-6, 154
Propionyl-CoA carboxylase, 193–194, 197
Prostacyclin, 73
    selenium, 275
Prostaglandins, vitamin E, 69–70, 73, 76
Protein and postmenopausal bone loss, 42
Prothrombin, vitamin K, 93, 95
Psoriasis, 152
    vitamin A, 14

vitamin D, 48
Psoriatic arthritis, zinc in, 306
Psychiatric disorders
    niacin and niacinamide, 136
    vitamin B-12, 185
    and vitamin C, 232–233
Psychosis, organic, vitamin B-12, 185
Pteroyl glutamate. *See* Folic acid
Pulmonary tuberculosis, vitamin D, interaction with, 52
Purpura, vitamin E, 80
Pyridine-3-carboxylic acid and pyridine-3-carboxylic acid amide, niacin and niacinamide (vitamin B-3), 129
Pyridoxal, 139, 140; *see also* Vitamin B-6 (pyridoxine, pyridoxol)
Pyridoxal 5′-phosphate (PLP), 139–154
Pyridoxamine, 139, 140; *see also* Vitamin B-6 (pyridoxine, pyridoxol)
Pyridoxamine 5′-phosphate, 139
Pyridoxine · HCl, 140–151, 153–155
Pyrimethamine and folate, 171
Pyruvate carboxylase, 193
Pyruvate decarboxylase deficiency, thiamin (vitamin B-1) in, 111

Radiation sickness, vitamin B-6, 151
Raynaud's disease, 77
Renal disease
    vitamin C and hyperoxalemia risk, 215
    zinc, 298–299
    see also Dialysis, renal
Renal failure
    vitamin A, 21–22
    vitamin B-6, 148–149
        dialysis, 148–149
        HDL-cholesterol, 148
    vitamin E, 77–78
Renal function, vitamin D, 49
Renal hypophosphatemic rickets, X-linked, 46–47
Renal osteodystrophy, vitamin D, 43–45
Respiratory distress syndrome, vitamin E, 67, 74
Restless legs, folic acid, 167, 168
Retina, 3
    Wald cycle, 4
Retinoic acid, 3, 6, 14, 15, 19, 20
    cancer and, 19

Retinol, 3–5, 6, 7, 16–18
    cancer and, 16–19
Retinol-binding protein, 6–7, 9, 10
    zinc and, in alcoholism, 10–11, 21
Retinol dehydrogenase, zinc-dependent, 298
Retinopathy
    diabetic, selenium, 277
    of prematurity, vitamin E, 64–66
Retrolental fibroplasia, vitamin E, 64–66
Rheumatism, 152
Rheumatoid arthritis, zinc in, 305–306
Riboflavin (vitamin B-2), 117–125, 149, 151, 152
    absorption, metabolism, and excretion, 119–120
    in alcoholism, 123
    biliary atresia, 121
    cataracts, 122–123
    clinical studies, 120–125
    congenital methemoglobinemia, 124
    diabetes mellitus, 122
    dietary sources, 119
    dosage forms, 120
    in elderly, 122
    exercise/athletic training, 123–124
    hemodialysis, 124
    hyperthyroidism, 124
    interactions, 125
    oral contraceptives and, 119, 129
    organic acidurias, 124–125
    phototherapy of hyperbilirubinemia, 120–121
    pregnancy and lactation, 121–122
    preterm infant, 120
    sickle cell anemia, 123
    socioeconomic status, 121, 122
    structure of, 118
    toxicity, 125
Riboflavin 5′-phosphate (FMN, flavin mononucleotide), 117, 118, 119
Rickets
    renal hypophosphatemic X-linked, 46–47
    vitamin D, 34
        genetic forms, 46–47
        nutritional, 39, 46, 47
        of prematurity, 37–38
Rocaltrol®, 36, 45

Roniacol® (nicotinyl alcohol), 134
R protein, 181, 182, 184

Salicylates. *See* Aspirin/salicylates
Schizophrenia
  folic acid, 167
  niacin and niacinamide (vitamin B-3), 136
  vitamin B-6, 153
Scleroderma, vitamin E, 80
Scurvy, 204
  rebound, following high-dose vitamin C, 229
Seborrheic dermatitis, biotin, 196
Selenitrace®, 272
Selenium, 77, 78, 269–279
  absorption, metabolism, and excretion, 270–271
  acne vulgaris, 277–278
  alcoholism, 276
  cancer, 273–274, 278, 279
  cardiovascular disease, 274–275
  chronic disease, 276
  clinical studies, 273–278
  deficiency, 270
  diabetic retinopathy, 277
  dietary sources, 271
  dosage forms, 272
  elderly, 275–276, 278
  glutathione peroxidase, Se-dependent (GSHPx), 269–278
  interactions, 279
  malnutrition, 276
  parenteral nutrition, home and total, 276–278
  selenomethionine, 271, 274, 279
  toxicity, 278–279
  and vitamin C, 232
  vitamin E, 270, 275, 276, 279
Selenium-yeast, 273–274, 278
Selenocysteine, 269, 271
Selenomethionine, 271, 274, 279
Senile vaginitis, vitamin E, 80
Shock lung syndrome, vitamin E, 74
Short-bowel syndrome, 71
Shoulders, painful, 152
Sickle-cell anemia
  riboflavin, 123
  vitamin B-6, 151
  vitamin C, 224

vitamin E, 74
zinc in, 302–303
Side effects. *See* Toxicity/side effects
Sideroblastic anemia, 150
Skin synthesis, vitamin D, 34
Small intestine bacterial overgrowth, vitamin B-12, 184
Smokers
  folic acid, 170
  tobacco amblyopia, vitamin B-12, 182
  vitamin A/beta-carotene, 7–8, 17
  vitamin C, 206
  vitamin E, 80
  zinc, 306–307
Sodium nitroprusside and vitamin B-12, 186
Sodium selenate, 272, 278, 279
Sodium selenite, 272, 278, 279
Solatene®, 8, 16
Somatomedin C, zinc deficiency, 292
Spina bifida, folic acid, 165
Sprue, celiac, 297, 298; *see also* Malabsorption
Subacute necrotizing encephalopathy, thiamin (vitamin B-1) in, 111–112
Sudden infant death syndrome
  biotin, 198
  thiamin (vitamin B-1), 112–113
Sulfasalazine and folate, 171
Sunlight, vitamin D, 32, 35, 46, 51
Superoxide dismutase, cytosolic
  copper, 256–257
  zinc, 256, 286, 309
Surgery
  vitamin C, 223
  vitamin E, 75–76
    adhesions, inhibition, 76
    cardiopulmonary bypass, 76
    contracture inhibition, 75–76
    skin flap survival, 76
  zinc, 305
Synkayvite®, 96
Systemic lupus erythematosus, 306
  vitamin E, 76–77

Tachysterol, 35, 37
T cells
  vitamin C, 220
  zinc, 286, 306–307
Tendinous xanthomas, 133

Tennis elbow, 152
Teratogenicity, vitamin A, 8, 11, 12,15, 21
Testicular maturation, zinc, 287, 293
Testosterone, zinc deficiency, 292, 299
beta-Thalassemia, zinc in, 303
Thiamin (vitamin B-1), 103–113, 123
   absorption, metabolism, and excretion, 105
   alcoholism, 106–108
      Korsakoff's psychosis, 106
      Wernicke-Korsakoff syndrome, 104, 106, 107, 111
      Wernicke's encephalopathy, 106
   amyotrophic lateral sclerosis, 112
   beriberi, 104, 107
   clinical studies, 106–113
   collagen formation and wound healing, 113
   diabetes mellitus, 108, 112
   dietary sources, 104–105
   dosage forms, 105
   elderly, 108–109
   erythrocyte transketolase, 108–109, 111, 112
   exercise and athletic training, 111
   febrile lymphadenopathy, recurrent, 112
   genetic disorders, 111–112
   hepatic failure, nonalcoholic fulminant, 108
   interactions, 113
   malabsorption, 108, 109
   neurologic disorders, 107–108
   neurotic tension, 110–111
   optic atrophy in glaucoma, 110
   optic neuropathy during ketogenic diet, 109–110
   pregnancy, 109
   structure, 103, 104
   sudden infant death syndrome, 112–113
   toxicity, 113
      anaphylaxis, 105, 113
Thiamin alkyl disulfides, 106–107
Thiamin hydrochloride, 103–113, 134
Thiamin tetrahydrofurfuryl disulfide, 112
Thromboemboli, folic acid, 168
Thrombophlebitis, vitamin E toxicity, 80–81
Thromboxane A_2 inhibition, vitamin E, 72–73

Thymulin, zinc, 286
Thyroxine (T4), zinc, 287
Tobacco amblyopia, vitamin B-12, 182; *see also* Smokers
alpha-Tocopherol. *See* Vitamin E *entries*
d-alpha-Tocopherol acetate, 11, 12
d-beta-Tocopherol, 59, 60
d-delta-Tocopherol, 59, 60
d-gamma-Tocopherol, 59, 60, 63
all-rac-alpha-Tocopheryl acetate, 63, 74
alpha-Tocopheryl acetate, 64, 67,70, 71, 72, 75, 76, 79, 80
Toxicity/side effects
   beta-carotene (provitamin A), 21–22
   copper, 263
   folic acid, 171
   niacin and niacinamide (vitamin B-3), 136
   selenium, 278–279
   thiamin, 113
   vitamin A, 8, 20–22
   vitamin B-6, 153–154
   vitamin B-12, 185
   vitamin C, 227–231
   vitamin D, 49–51
   vitamin E (alpha-tocopherol), 80–81
   vitamin K, 96, 98, 100
   zinc, 304, 307–308
Transcobalamin II and III, 181
Transcobalamin II deficiency, hereditary, 184–185
Transketolase, erythrocyte, thiamin (vitamin B-1), 108–109, 111, 112
Trauma, vitamin C, 223
Tretinoin (all-*trans*-retinoic acid, Retin A®), 8, 13
   acne vulgaris, 12–13
   aging and photoaging, 13–14
   benzoyl peroxide and, 13
Triamterene and folate, 171
Trigger finger, 152
Trimethoprim and folate, 171
Tryptophan, 131, 135, 136, 145
Tuberculosis, pulmonary, vitamin D, interaction with, 52
Turner's syndrome, 251
Tyrosinase, copper, 255

Ulcerative colitis, 99, 191
Urine acidification, vitamin C, 223–224

Uterine dysplasia, vitamin C, 226–227
Uterocervical dysplasia and carcinoma in situ, vitamin A, 17

Vaginitis, senile, vitamin E, 80
Vanadium and vitamin C, 232
Vegetarian diet
    vitamin A, 6
    vitamin B-12, 180
    zinc, 292, 296
Vitamin A, 3–23, 165, 256, 275
    absorption, metabolism, and excretion, 6–8
        smokers, 7–8, 17
    acne vulgaris, 12–13
    aging and photoaging, 13–14
    alcoholism and cirrhosis, 5, 10–11
        dark adaptation, 10
    cancer, 16–19
    cholestasis, neonatal, 9
    clinical studies, 8–20
    Crohn's disease, 11–12
    cystic fibrosis, 11
    Darier's disease, 15
    deficiency, symptoms, 5
    dietary sources, 5
    dosage forms, 8
    Down's syndrome, 9–10
    functions, 4–5
    glaucoma, 12
    ichthyosiform dermatoses, 14
    infants, preterm, 8–9
    infections and stress, 8
    interactions, 22–23
    liver, storage in, 5–7
    malabsorption, 11
    oral lichen planus, 15
    photosensitivity, 15–16
        erythropoietic protoporphyria, 15
    pityriasis rubra pilaris, 14–15
    psoriasis, 14
    renal dialysis, contraindicated in, 21
    structural formula, 4
    teratogenicity, 8, 11, 12, 15, 21
    toxicity, 8, 20–22
    wound healing, 19
Vitamin B-1. *See* Thiamin (vitamin B-1)
Vitamin B-2. *See* Riboflavin (vitamin B-2)
Vitamin B-3. *See* Niacin and niacinamide (vitamin B-3)
Vitamin B-6 (pyridoxine, pyridoxol), 123, 139–155
    absorption, metabolism, and excretion, 141
    alcoholism, 150
    asthma, 150–151
    autism, 153
    cancer, 151–152
    carpal tunnel syndrome, 149
        related disorders, 152
    clinical studies, 141–153
    congenital dependencies, 152
    depression, 153
    diabetes mellitus, 146
    dietary sources, 140
    dosage forms, 141
    elderly, 145
    hyperemesis gravidarum, 143–144
    hyperkinesis, 152, 153
        diamine oxidase, 144
    infant seizures, 152
    interactions, 154–155
    irritability, 152
    mental retardation, 153
    monosodium glutamate (Chinese restaurant syndrome), 152
    oral contraceptives, 144–145
    oxalosis, hyperoxaluria, and oxalateurolithiasis, 146–148
    platelet aggregation inhibition, 153
    pregnancy and lactation, 141–143
    pregnancy, toxemia of, 144
    premenstrual syndrome, 144
        Apgar scores, 142
    pyridoxal 5′-phosphate (PLP), 139–154
    pyridoxamine 5′-phosphate, 139
    radiation sickness, 151
    renal failure, 148–149
    schizophrenia, 153
    sickle-cell anemia, 151
    structure, 140
    toxicity, 153–154
Vitamin B-12 (cyanocobalamin), 162, 171, 179–186
    absorption, metabolism, and excretion, 181–182

clinical studies, 182-185
coenzymes, 179, 181, 182
depression, 185
dietary sources, 181
dosage forms, 182
elderly, 184
fatigue, chronic, 185
gastrectomy, 183-184
gastrointestinal abnormalities, 180
genetic diseases, 184-185
interactions, 185-186
pancreatitis, 184
pernicious anemia, 182-183
    cf. folic acid deficiency, 171, 183
psychosis, organic, 185
small intestine bacterial overgrowth, 184
structure, 180
tobacco amblyopia, 182
toxicity, 185
vegan diet, 180
and vitamin C, 233
Vitamin B-12a (hydroxocobalamin), 179, 182, 184, 186
Vitamin B-12b (aquacobalamin), 179
Vitamin B-12c (nitrocobalamin), 179
Vitamin C (L-ascorbic acid), 6, 18, 61, 78, 165, 253, 264, 201-233
absorption, metabolism, and excretion, 204-206
alcoholism, 218, 221-222
allergy, 224
antioxidant functions, 202-203, 205
    hydroxyl radical, 202, 203, 220
asthma, 224
bedsores, 222-223
burns, 223, 224
cancer, 224-227
    cervical dysplasia, 226-227
    nitrosamine synthesis inhibition, 224-225, 232
    polyposis coli, familial, 226
    uterine dysplasia, 226-227
cataracts, 221
cholesterol
    HDL, 212, 214, 219
    hypercholesterolemia, 212-214
    LDL, 212
clinical studies, 207-227

common cold, 208-209
Crohn's disease, 223
deficiency, symptoms, 204
diabetes, 213, 216-217
dietary sources, 203-204
dosage forms, 206-207
elderly, 218-221
    free radical theory of aging, 219-220
fibrinolytic activity, 211-212
frostbite, 224
functions, 202-203
glaucoma, 110
herpes labialis, 209
high-dose
    rebound scurvy following, 229
    and vitamin B-12, 186
histamine, 208-209, 224
hypertension, 215-216
hypertriglyceridemia, 215
infections, immune function in, 210-211, 220
infertility, male, 224
interactions, 231-233
ischemic heart disease, 211
periodontal disease, 222
postmenopausal bone loss, 42
pregnancy, 217-218
renal disease, hyperoxalemia risk, 215
and selenium, 279
sickle cell anemia, 224
smokers, 206
structure, 202
surgery, 223
toxicity, 227-231
trauma, 223
urine acidification, 223-224
Vitamin D, 31-52
absorption, metabolism, and excretion, 34-36
    sunlight, 32, 35, 46, 51
    ultraviolet light, 34-35, 46
aging and bone loss, 39-43
    osteoporosis, postmenopausal, 40-43
alcoholism, 46
breast cancer, 49
breast-fed infants, 38-39
clinical studies, 37-49
cochlear deafness, 47

deficiency, symptoms, 34
dietary sources, 34
dosage forms, 36
hemolytic disorders, 48
hydroxylation, 32–34, 36, 46, 48
hyperphosphatemia, 44
hypertension, 48
hypocalcemia, 37–38, 43, 47
hypoparathyroidism, 46
hypophosphatemia, 38
immunoregulatory functions, 33–34
interactions, 23, 51–52
leukemia, myeloid, 48–49
liver cf. skin synthesis, 34
malabsorption, 45–46
melanoma, 49
osteomalacia, 34, 40, 43, 47
psoriasis, 48
renal osteodystrophy, 43–45
rickets, 34
    genetic forms, 46–47
    nutritional, 39
    of prematurity, 37–38
structural formulas, 32
toxicity, 49–51
    pregnancy, 51
    renal function, 49
Vitamin D-dependent rickets, 46, 47
Vitamin E (alpha-tocopherol), 59–82, 165–166, 202, 203, 220, 232
    absorption, metabolism, and excretion, 61–63
    alcoholics, 62, 77
    antiinflammatory action, osteoarthritis, 74
    autoimmune disorders, 76–77
    burn patients, 80
    cancer, 78–80
    cardiovascular disease, 72–73
    cystic mastitis, 68–69
    dietary sources, 60
    dosage forms, 63–64
    elderly patients, 76
    exercise, 78
    fish oil enhances requirement, 60
    G6PD deficiency, 73–74
    interactions, 81–82
        with vitamin A, 6, 11, 17, 18, 21, 23
    intravenous injection dangerous in infants, 64–66, 81
    intraventricular hemorrhage, infants, 64, 65, 67, 81
    leg cramps, restless legs, 77
    lipid peroxidation, 75–77
    lupus erythematosus, 76–77
    malabsorption, 70–71
    necrotizing enterocolitis, 64, 66, 81
    oxygen administration, 64–65, 67, 75
    oxygen radical scavenger, 60–61, 65, 75, 77
    Parkinson's disease, 61
    phototherapy of hyperbilirubinemia, 67
    phrynoderma, 77
    porphyria, 76–77
    pregnancy, 68
    premenstrual syndrome, 69–70
    prostaglandins, 69–70, 73, 76
    renal failure, 77–78
    respiratory distress syndrome, 67
    retinopathy of prematurity, 64–66
    and selenium, 270, 275, 276, 279
    shock lung syndrome, 74
    structure, 60
    surgery, 75–76
    toxicity, 80–81
Vitamin K, 93–101
    absorption, metabolism, and excretion, 95–96
    acute care, TPN, 98, 100
    clinical studies, 96–99
    deficiency with hypothrombinemia of infancy, 95–99, 101
        breast-fed infant, 97, 98
    dosage forms, 96
    elderly patients with vitamin C deficiency, 98–99
    genetic dependency, 99
    interactions, 100–101
    malabsorption with hypothrombinemia, 99
    plasma clotting factors, 93–95
    structural formulas, 94
    toxicity, 96, 98, 100
        anaphylaxis, 96, 100
        intravenous, 100
Vitamin K-1, 93, 94

DATE DUE

FEB 2 9 2000			
DEC 0 9 1999			

GAYLORD — PRINTED IN U.S.A.